DATE DUE

THE ROAD SINCE

STRUCTURE

THE ROAD SINCE
STRUCTURE

*Philosophical
Essays, 1970–1993,
with an
Autobiographical
Interview*

Thomas S. Kuhn

EDITED BY JAMES CONANT AND JOHN HAUGELAND

*The University of Chicago Press
Chicago and London*

The University of Chicago Press, Chicago 60637
The University of Chicago Press, Ltd., London
© 2000 by The University of Chicago
All rights reserved. Published 2000
Paperback edition 2002
Printed in the United States of America

11 10 09 08 07 06 05 04 03 02   2 3 4 5
ISBN: 0-226-45798-2 (cloth)
ISBN: 0-226-45799-0 (paperback)

Library of Congress Cataloging-in-Publication Data

Kuhn, Thomas S.
    The road since structure : philosophical essays, 1970–1993, with an
autobiographical interview / Thomas S. Kuhn ; edited by James Conant and
John Haugeland.
        p. cm.
    Includes bibliographical references.
    ISBN 0-226-45798-2 (cloth : alk. paper)
    1. Science—Philosophy.   2. Science—History.   3. Kuhn, Thomas S.—
Interviews.   4. Kuhn, Thomas S.—Bibliography.   I. Conant, James.   II.
Haugeland, John, 1945– .   III. Title.

    Q175 .K94 2000
    501—dc21

                                                    00-131583

Contents

Foreword

JEHANE R. KUHN

Tom's preface to an earlier selection of his published papers, *The Essential Tension*, published in 1977, was cast as the narrative of a journey of inquiry—towards, and then on from, *The Structure of Scientific Revolutions*, published fifteen years earlier. Some autobiographical framing was called for, he explained, since his published papers did not tell the story of a journey that had found its way from physics to historiography and philosophy. The preface to that volume closed by focusing on the philosophical/metahistorical issues that "currently . . . concern me the most, and I hope before long to have more to say about them." In the introduction to this new volume, the editors place each paper in relation to those continuing issues, again pointing forward: this time to the work-in-progress which they are preparing for publication. It will represent not the goal of Tom's journey, but the stage at which he left it.

The title of this book again invokes the metaphor of a journey, and its closing section, which records an extended interview at the University of Athens, amounts to another, longer, more personal narrative. I am delighted that the interviewers, and the editorial board of the journal *Neusis*, in which it first appeared, have agreed to its republication here. I was present during the interview and admired the knowledge, perceptiveness, and sympathetic candor of the three colleagues, who were also our hosts in Athens. Tom was exceptionally at ease with these three friends and talked freely on the assumption that he would review the transcript; but time ran out, and that task fell to me, in consultation with the other participants. I know that Tom would have intervened viii

in the transcript substantially—not so much from discretion, which was not high among his virtues, as from courtesy. In his talk as it appears here, there are some expressions of feeling and of judgment, which I'm fairly sure he would have moderated or perhaps omitted; I did not think it my place—or anyone else's—to moderate or omit them on his behalf. Many of the grammatical inconsistencies and unfinished phrases of informal talk have for that reason been left unsmoothed, as a reminder of the interview's unauthorized status. I am grateful to colleagues and friends, in particular Karl Hufbauer, who have caught local errors of chronology or helped to decipher names.

The circumstances in which Jim Conant and John Haugeland accepted the task of editing this volume are told in their introduction; I have only to add that Tom's wholehearted confidence is the best testimonial they could have. I am warmly grateful to them, and no less warmly to Susan Abrams, for her friendship and, inseparably, her professional judgment, during this project as in the past. Sarah, Liza, and Nathaniel Kuhn are sustaining participants in my role as their father's literary executor.

Editors' Introduction

JAMES CONANT AND JOHN HAUGELAND

Shifts happen.

In *The Structure of Scientific Revolutions,* as nearly everyone knows, Thomas Kuhn argued that the history of science is not gradual and cumulative, but rather punctuated by a series of more or less radical "paradigm shifts." What is less well known is that Kuhn's own understanding of how best to characterize these episodes itself underwent a number of significant shifts. The essays collected in this volume represent several of his later attempts to rethink and extend his own "revolutionary" hypotheses.

We discussed the contents of this volume with Kuhn at some length shortly before he died. Although he declined to specify them in full detail, he had a quite definite idea of what he wanted the volume to be. In making this clear to us, he made several explicit stipulations, reviewed with us the pros and cons in several other cases, and then provided four general guidelines for us to follow. For those readers interested in how the final choices were made, we will begin by briefly summarizing these guidelines.

The first three guidelines that we were given flow from Kuhn's vision of this volume as a sequel to, and as modeled upon, his earlier collection, *The Essential Tension,* which appeared in 1977. In that collection, Kuhn restricted himself to substantial essays that he regarded as developing philosophically significant themes (albeit generally in the context of historical or historiographical considerations), as opposed to those that mainly explore particular historical case studies. Accordingly, our first three guidelines were: include only essays that are expressly philosophical in their concerns; include only those philosophical essays written in Kuhn's last two decades[1]; include only substantial essays, as opposed to brief reviews or addresses.

The fourth guideline concerns material that Kuhn regarded as essentially preparatory to—in effect, early drafts of—the book he had been working on for some years. Since it is also part of our charge to edit and publish that work, making use, where appropriate, of this material, we were instructed not to include any of it here. Covered under this restriction are three important lecture series: "The Natures of Conceptual Change" (Perspectives in the Philosophy of Science, University of Notre Dame, 1980), "Scientific Development and Lexical Change" (the Thalheimer lectures, Johns Hopkins University, 1984), and "The Presence of Past Science" (the Shearman lectures, University College, London, 1987). Although typescripts of these lectures have circulated here and there in samizdat form, and have occasionally been cited and discussed in publications by others,[2] Kuhn did not want any of them published in their present form.

———◦———

Speaking very broadly, the essays reprinted here can be seen to address four main topics. First, Kuhn reiterates and defends his view, going all the way back to *The Structure of Scientific Revolutions* (hereafter cited as *Structure*), that science is a cognitive empirical investigation of nature that exhibits a unique sort of progress, despite the fact that this progress cannot be further explicated as "approximating closer and closer to reality." Rather, progress takes the form of ever-improving technical puzzle-solving ability, operating under strict—though always tradition-bound—standards of success or failure. This pattern of progress, in its

1. Kuhn made it clear that those essays with expressly philosophical concerns that he chose to omit from *The Essential Tension* were omitted because he had become dissatisfied with them, and that he did not want them collected in this volume, either. In particular, he was adamant that his 1963 essay "The Function of Dogma in Scientific Research" should not be included here, even though it has been widely read and cited.

2. Perhaps the most notable of these is Ian Hacking's essay "Working in a New World: The Taxonomic Solution" (in *World Changes: Thomas Kuhn and the Nature of Science*, ed. Paul Horwich [Cambridge, MA: Bradford/MIT Press, 1993]), in which he expounds and attempts to refine the central argument of the Shearman lectures.

fullest realization exclusive to science, is prerequisite to the extraordinarily esoteric (and often expensive) investigations that are characteristic of scientific research, and thus to the astonishingly precise and detailed knowledge that it makes possible.

Second, Kuhn develops further the theme, which again goes back to *Structure*, that science is fundamentally a social undertaking. This shows up especially in times of trouble, with the potential for more or less radical change. It is only because individuals working in a common research tradition are able to arrive at differing judgments concerning the degree of seriousness of the various difficulties they collectively face that some of them will be moved individually to explore alternative (often—as Kuhn likes to emphasize—seemingly nonsensical) possibilities, while others will attempt doggedly to resolve the problems within the current framework.

The fact that the latter are in the majority when such difficulties first arise is essential to the fertility of scientific practices. For, *usually*, the problems can be resolved, and eventually are. In the absence of the requisite persistence to find those solutions, scientists would not be able to home in, as they do, on those rarer but crucial cases in which efforts to introduce radical conceptual revision are fully repaid. On the other hand, of course, if no one were ever to develop possible alternatives, major reconceptions could never emerge, even in those cases in which they genuinely become necessary. Thus, a *social* scientific tradition is able to "distribute the conceptual risks" in a way that would be impossible for any single individual, and yet is prerequisite to the long-term viability of science.

Third, Kuhn spells out and emphasizes the analogy, barely hinted at in the closing pages of *Structure*, between scientific progress and evolutionary biological development. In elaborating this theme, he plays down his original picture, which had periods of normal science within a single area of research punctuated by occasional cataclysmic revolutions, and introduces in its place a new picture, which has periods of development within a coherent tradition divided occasionally by periods of "speciation" into two distinct traditions with somewhat different areas of research. To be sure, the possibility remains that one of the resulting traditions may eventually stagnate and die out, in which case we have, in effect, the older structure of revolution and replacement. But at least as often in the history of science, both successors, neither quite like their common ancestor, flourish as new scientific "specialties." In science, speciation is specialization.

Finally, and most important, Kuhn spent his last decades defending, clarifying, and substantially developing the idea of incommensurability. This theme too was already conspicuous in *Structure*, but not very well articulated. It is the feature of the book that was most widely criticized in the philosophical literature; and Kuhn came to be dissatisfied with his original presentation. Commensurability and incommensurability, as presented in Kuhn's later work, are terms that denote a relation obtaining between *linguistic* structures. There are basically two new points underlying this linguistic reformulation of the notion of incommensurability.

First, Kuhn carefully explicates the difference between distinct but commensurable languages (or portions of languages) and incommensurable ones. Between pairs of the former, translation is perfectly possible: whatever can be said in the one can be said in the other (though it may be considerable work to figure out how). Between *incommensurable* languages, however, strict translation is not possible (even though, on a case-by-case basis, various paraphrases *may* suffice for adequate communication).

The idea of incommensurability, as it was elaborated in *Structure*, was widely criticized on the grounds that it made it unintelligible how scientists working under different paradigms were able to communicate with one another (let alone adjudicate and resolve their disagreements) across a revolutionary divide. A related criticism concerned the putative explanations of past scientific paradigms furnished within the pages of *Structure* itself: didn't the work undermine its own doctrine of incommensurability by offering illuminating explanations (in contemporary English) of how alien scientific terms were used?

Kuhn here responds to these objections by pointing out the difference between language translation and language learning. Just because a foreign language is not translatable into whatever language one already speaks does not mean that one cannot learn it. That is, there is no reason that a single person cannot speak and understand two languages that he or she cannot translate between. Kuhn calls the process of figuring out such an alien language (say, from historical texts) *interpretation*, and also—to emphasize its distinctness from so-called "radical" interpretation (à la Davidson)—*hermeneutics*. His own explanations of the terminology from, say, Aristotelian "physics" or phlogiston "chemistry" are exercises in hermeneutic interpretation and, at the same time, aids to the reader in learning a language incommensurable with his or her own.

Kuhn's second main point about incommensurability is a new and fairly detailed account of how and why it occurs in two sorts of scientific context. Technical scientific terminology, he explains, always occurs in families of essentially interrelated terms; and he discusses two varieties of such families. In the first variety, the terms are kind terms—roughly, sortals—which Kuhn calls "taxonomic categories." These are always arrayed in a strict hierarchy, which is to say that they are subject to what he calls "the no-overlap principle": no two such categories or kinds can have any instances in common unless one of them entirely and necessarily subsumes the other.

Any taxonomy adequate to the purposes of scientific description and explanation is constructed on the basis of an implicit no-overlap principle. The meanings of the relevant kind terms specifying such taxonomic categories, Kuhn argues, are partially constituted by this implicit presupposition: the meanings of the terms depend on their respective subsumption and mutual exclusion relations (plus, of course, the learnable skills of recognizing members). Such a structure—which Kuhn calls a "lexicon"—has, in itself, considerable empirical content, because there are always multiple ways of recognizing (multiple "criteria" for) membership in any given category. Distinct taxonomic structures (ones with different subsumption and exclusion relations) are inevitably incommensurable, because those very differences result in terms with fundamentally disparate meanings.

The other variety of terminological family (also called a lexicon) involves those terms whose meanings are determined in part—but crucially—by scientific laws relating them. The clearest examples are the quantitative variables that occur in laws expressed as equations—for instance, weight, force, and mass in Newtonian dynamics. Though this sort of case is not as well worked out in the extant Kuhnian texts, Kuhn believed that here, as well, the meanings of the relevant fundamental terms are partially constituted through their occurrence in claims—in this case, scientific laws—that categorically exclude certain possibilities; hence any changes in the understandings or formulations of the relevant laws must result, according to Kuhn, in fundamental differences in the understandings (hence, meanings) of the corresponding terms, and thus incommensurability.

———◦———

This volume is divided into three parts: two groups of essays, each arranged chronologically, and an interview. Part 1 includes five free-standing essays presenting various of Kuhn's views as they developed

from the early 1980s through the early 1990s. Two of these essays include brief replies to comments that were made when the essays were
first presented. Although, of course, such replies can be fully appreciated
only in the context of those comments themselves, Kuhn is careful in
each case to summarize the specific points he is replying to, and the
resulting remarks add useful clarity to the main paper. Part 2 includes
six essays, varying widely in length, each of which consists mainly of
Kuhn's response to the work of one or more other philosophers—often,
though not always, itself developments or criticisms of Kuhn's own
prior work. Finally, in part 3, we have included a lengthy and candid
interview with Kuhn, conducted in Athens in 1995 by Aristides Baltas,
Kostas Gavroglu, and Vassiliki Kindi.

Part 1: Reconceiving Scientific Revolutions

Essay 1, "What are Scientific Revolutions?" (~1981), consists primarily
of a philosophical analysis of three historical scientific sea-changes (concerning the theories of motion, the voltaic cell, and black-body radiation) as illustrations of Kuhn's then nascent account of taxonomic structures.

Essay 2, "Commensurability, Comparability, Communicability"
(1982), is an elaboration and defense of the importance of incommensurability with regard to the two principal charges that (1) it is impossible, because intelligibility at all entails translatability, hence commensurability; and (2), if it were possible, it would imply that major scientific
changes cannot be responsive to evidence, and therefore must be fundamentally irrational. Versions of these charges made by Donald Davidson, Philip Kitcher, and Hilary Putnam receive particular attention.

Essay 3, "Possible Worlds in History of Science" (1989), develops
the idea—dramatically propounded but not well explained in *Structure*—that incommensurable scientific languages (now called lexicons)
give access to different sets of possible worlds. In his discussion,
Kuhn clearly distances himself from possible-worlds semantics and
from the causal theory of reference (along with the associated forms
of "realism").

Essay 4, "The Road since Structure" (1990), is announced as a brief
sketch of the book Kuhn had (in 1990) been working on for just over
a decade (the book he never finished). Though at the highest level the
topic of the book is realism and truth, what it mostly will discuss is

incommensurability—with particular emphasis on why it is not a threat to scientific rationality and its basis in evidence. Thus, in part, the book is conceived as a repudiation of what Kuhn regarded as certain excesses in the so-called "strong program" in the philosophy (or sociology) of science. At the conclusion of the essay (and, in more detail, in the Shearman lectures), he describes his position as "post-Darwinian Kantianism," because it presupposes something like an ineffable but permanent and fixed *"Ding an sich."* Kuhn had earlier rejected the notion of a *Ding an sich* (see essay 8), and he again later repudiated (in conversations with us) both that notion and the reasons he had put forward for it.

Essay 5, "The Trouble with the Historical Philosophy of Science" (1992), considers both traditional philosophy of science and the now fashionable "strong program" in the sociology of science, and what is wrong with each. Kuhn suggests that "the trouble" with the latter may be that it *retains* a traditional conception of knowledge, while noticing that science does not live up to that conception. The reconceptualization that is required—and which gets rationality and evidence back into the picture—is to focus not on rational evaluation of beliefs, but rather on rational evaluation of *changes* in beliefs.

Part 2: Comments and Replies

Essay 6, "Reflections on My Critics" (1970), is the oldest essay in the collection, and the only one that antedates the compilation of *The Essential Tension.* We discussed its inclusion explicitly with Kuhn, who was pulled both ways. And, in trying to decide whether to include it, so were we. On the one hand, it violates the third "guideline" mentioned above, and, moreover, it consists primarily of corrections of various misreadings of *Structure*—corrections that, in a perfect world, ought not to be necessary. On the other hand, many of those misunderstandings persist, and so their correction *is* still needed—something this essay achieves with unique clarity, thoroughness, and vigor. In the end, Kuhn left the decision up to us. We have decided to reprint it because of its still relevant special merits, and because the volume in which it originally appeared—*Criticism and the Growth of Knowledge*—has for some time been out of print.

Essay 7, "Theory Change as Structure Change: Comments on the Sneed Formalism" (1976), is a tentative but mostly very favorable discussion of Joseph Sneed's model-theoretic formalism for the semantics of scientific theories, along with Wolfgang Stegmüller's uses and elaborations of it. Although the essay will be of particular interest to readers already familiar with the Sneed-Stegmüller approach, Kuhn's remarks are nontechnical and of more general interest as well. He is especially gratified by the way in which, according to this approach, the central terms of a theory acquire a significant part of their determinate content from multiple exemplary *applications*. It is important that there be *several* such applications, because they mutually constrain one another (via the theory), thereby avoiding a kind of circularity. It is important that the applications be *exemplary* because this emphasizes the role of learnable skills, which can then be extended to new cases. Kuhn's only expressed reservation about the approach—albeit a serious one—is that it leaves no obvious place for the essential phenomenon of theoretical incommensurability.

Essay 8, "Metaphor in Science" (1979), is a response to a presentation by Richard Boyd about the analogies he sees between scientific terminology and ordinary-language metaphors. Though they agree on several important points, Kuhn demurs from the specific *way* in which Boyd extends the view to include the causal theory of reference, especially with regard to natural-kind terms. In his conclusion, Kuhn describes himself as, like Boyd, an "unregenerate realist," but he thinks this doesn't mean the same thing in their two cases. In particular, he rejects Boyd's own metaphor of scientific theories (coming closer and closer to) "carving nature at the joints." He likens this idea of nature's "joints" to Kant's *Ding an sich*, an aspect of Kantianism he here rejects.

Essay 9, "Rationality and Theory Choice" (1983), is Kuhn's contribution to a symposium on the philosophy of Carl G. Hempel. In it, he responds to a question that Hempel had put to him on several occasions: does he (Kuhn) recognize the difference between *explaining* theory-choice behavior and *justifying* it? Granted that choices of theories are *in fact* based on their puzzle-solving ability (including accuracy, scope, and so on), this does not get any philosophical bite as a *justification* unless and until those criteria themselves are justified as somehow nonarbitrary. Kuhn replies that they are nonarbitrary ("necessary") in the relevant way because they belong together in an empirically contentful taxonomy of disciplines; reliance on *just such* criteria (plural) is what distinguishes *scientific* investigation from other professional pursuits (fine arts, law, engineering, and so on)—hence is, in effect, definitive of 'science' as a genuine kind term.

Essay 10, "The Natural and the Human Sciences" (1989), mainly discusses Charles Taylor's influential essay "Interpretation and the Sciences of Man," which Kuhn much admires. While he is inclined to agree with Taylor that the natural and human sciences are different, he probably doesn't agree about what that difference is. After arguing that the natural sciences, too, have a "hermeneutic base," he acknowledges that, unlike the present human sciences, they are not, in themselves, hermeneutic. But he questions whether this reflects an essential difference or, rather, simply indicates that most of the human sciences have not yet reached the developmental stage that he used to associate with acquisition of a paradigm.

Essay 11, "Afterwords" (1993), like essay 6, is the final chapter in a volume of essays largely devoted to discussions of Kuhn's own work (*World Changes: Thomas Kuhn and the Nature of Science*, edited by Paul Horwich). Unlike its somewhat feisty predecessor, however, this essay is primarily an appreciative and constructive engagement with essays that are themselves primarily constructive. The main themes are taxonomic structures, incommensurability, the social character of scientific research, and truth cum rationality cum realism. The discussion of these themes is presented here in the form of a brief sketch of some of the central ideas of Kuhn's long promised but never finished new book— on which he continued to work until he no longer could.

Part 3: A Discussion with Thomas S. Kuhn

"A Discussion with Thomas S. Kuhn" (1997) is a candid intellectual autobiography in the form of an interview, conducted by Aristides Baltas, Kostas Gavroglu, and Vassiliki Kindi in Athens in the fall of 1995. It is reprinted, lightly edited, in its entirety.

The volume ends with a complete bibliography of Kuhn's published work.

Reconceiving Scientific Revolutions

Chapter One

What Are Scientific Revolutions?

"What Are Scientific Revolutions?" was first published in The Probabilistic Revolution, volume I: Ideas in History, *edited by Loren₃ Kruger, Lorraine J. Daston, and Michael Heidelberger (Cambridge, MA: MIT Press, 1987). The three examples that constitute its bulk were developed in this form for the first of three lectures delivered under the title "The Natures of Conceptual Change" at the University of Notre Dame in late November 1980, as part of the series "Perspectives in the Philosophy of Science." In very nearly its present form, but under the title "From Revolutions to Salient Features," the paper was read to the third annual conference of the Cognitive Science Society in August 1981.*

IT IS NOW ALMOST TWENTY YEARS since I first distinguished what I took to be two types of scientific development, normal and revolutionary.[1] Most successful scientific research results in change of the first sort, and its nature is well captured by a standard image: normal science is what produces the bricks that scientific research is forever adding to the growing stockpile of scientific knowledge. That cumulative conception of scientific development is familiar, and it has guided the elaboration of a considerable methodological literature. Both it and its methodological by-products apply to a great deal of significant scientific work. But scientific development also displays a noncumulative mode, and the episodes that exhibit it provide unique clues to a central aspect of scientific knowledge. Returning to a long-standing concern, I shall therefore

---

1. T. S. Kuhn, *The Structure of Scientific Revolutions* (Chicago: University of Chicago Press, 1962).

here attempt to isolate several such clues, first by describing three examples of revolutionary change and then by briefly discussing three characteristics which they all share. Doubtless revolutionary changes share other characteristics as well, but these three provide a sufficient basis for the more theoretical analyses on which I am currently engaged, and on which I shall be drawing somewhat cryptically when concluding this paper.

Before turning to a first extended example, let me try—for those not previously familiar with my vocabulary—to suggest what it is an example of. Revolutionary change is defined in part by its difference from normal change, and normal change is, as already indicated, the sort that results in growth, accretion, cumulative addition to what was known before. Scientific laws, for example, are usually products of this normal process: Boyle's law will illustrate what is involved. Its discoverers had previously possessed the concepts of gas pressure and volume as well as the instruments required to determine their magnitudes. The discovery that, for a given gas sample, the product of pressure and volume was a constant at constant temperature simply added to the knowledge of the way these antecedently understood[2] variables behave. The overwhelming majority of scientific advance is of this normal cumulative sort, but I shall not multiply examples.

Revolutionary changes are different and far more problematic. They involve discoveries that cannot be accommodated within the concepts in use before they were made. In order to make or to assimilate such

---

2. The phrase 'antecedently understood' was introduced by C. G. Hempel, who shows that it will serve many of the same purposes as 'observational' in discussions involving the distinction between observational and theoretical terms (cf., particularly, his *Aspects of Scientific Explanation* [New York: Free Press, 1965], pp. 208 ff.). I borrow the phrase because the notion of an antecedently understood term is intrinsically developmental or historical, and its use within logical empiricism points to important areas of overlap between that traditional approach to philosophy of science and the more recent historical approach. In particular, the often elegant apparatus developed by logical empiricists for discussions of concept formation and of the definition of theoretical terms can be transferred as a whole to the historical approach and used to analyze the formation of new concepts and the definition of new terms, both of which usually take place in intimate association with the introduction of a new theory. A more systematic way of preserving an important part of the observational/theoretical distinction by embedding it in a developmental approach has been developed by Joseph D. Sneed (*The Logical Structure of Mathematical Physics* [Dordrecht: Reidel, 1971], pp. 1–64, 249–307). Wolfgang Stegmüller has clarified and extended Sneed's approach by positing a hierarchy of theoretical terms, each level introduced within a particular historical theory (*The Structure and Dynamics of Theories* [New York: Springer, 1976], pp. 40–67, 196–231). The resulting picture of linguistic strata shows intriguing parallels to the one discussed by Michel Foucault in *The Archeology of Knowledge*, trans. A. M. Sheridan Smith (New York: Pantheon, 1972).

a discovery one must alter the way one thinks about and describes some range of natural phenomena. The discovery (in cases like these "invention" may be a better word) of Newton's second law of motion is of this sort. The concepts of force and mass deployed in that law differed from those in use before the law was introduced, and the law itself was essential to their definition. A second, fuller, but more simplistic example is provided by the transition from Ptolemaic to Copernican astronomy. Before it occurred, the sun and moon were planets, the earth was not. After it, the earth was a planet, like Mars and Jupiter; the sun was a star; and the moon was a new sort of body, a satellite. Changes of that sort were not simply corrections of individual mistakes embedded in the Ptolemaic system. Like the transition to Newton's laws of motion, they involved not only changes in laws of nature but also changes in the criteria by which some terms in those laws attached to nature. These criteria, furthermore, were in part dependent upon the theory with which they were introduced.

When referential changes of this sort accompany change of law or theory, scientific development cannot be quite cumulative. One cannot get from the old to the new simply by an addition to what was already known. Nor can one quite describe the new in the vocabulary of the old or vice versa. Consider the compound sentence, "In the Ptolemaic system planets revolve about the earth; in the Copernican they revolve about the sun." Strictly construed, that sentence is incoherent. The first occurrence of the term 'planet' is Ptolemaic, the second Copernican, and the two attach to nature differently. For no univocal reading of the term 'planet' is the compound sentence true.

No example so schematic can more than hint at what is involved in revolutionary change. I therefore turn at once to some fuller examples, beginning with the one that, a generation ago, introduced me to revolutionary change, the transition from Aristotelian to Newtonian physics. Only a small part of it, centering on problems of motion and mechanics, can be considered here, and even about it I shall be schematic. In addition, my account will invert historical order and describe, not what Aristotelian natural philosophers required to reach Newtonian concepts, but what I, raised a Newtonian, required to reach those of Aristotelian natural philosophy. The route I traveled backward with the aid of written texts was, I shall simply assert, nearly enough the same one that earlier scientists had traveled forward with no text but nature to guide them.

I first read some of Aristotle's physical writings in the summer of 1947, at which time I was a graduate student of physics trying to prepare a case study on the development of mechanics for a course in science for nonscientists. Not surprisingly, I approached Aristotle's texts with the Newtonian mechanics I had previously read clearly in mind. The question I hoped to answer was how much mechanics Aristotle had known, how much he had left for people like Galileo and Newton to discover. Given that formulation, I rapidly discovered that Aristotle had known almost no mechanics at all. Everything was left for his successors, mostly those of the sixteenth and seventeenth centuries. That conclusion was standard, and it might in principle have been right. But I found it bothersome because, as I was reading him, Aristotle appeared not only ignorant of mechanics, but a dreadfully bad physical scientist as well. About motion, in particular, his writings seemed to me full of egregious errors, both of logic and of observation.

These conclusions were unlikely. Aristotle, after all, had been the much admired codifier of ancient logic. For almost two millennia after his death, his work played the same role in logic that Euclid's played in geometry. In addition, Aristotle had often proved an extraordinarily acute naturalistic observer. In biology, especially, his descriptive writings provided models that were central in the sixteenth and seventeenth centuries to the emergence of the modern biological tradition. How could his characteristic talents have deserted him so systematically when he turned to the study of motion and mechanics? Equally, if his talents had so deserted him, why had his writings in physics been taken so seriously for so many centuries after his death? Those questions troubled me. I could easily believe that Aristotle had stumbled, but not that, on entering physics, he had totally collapsed. Might not the fault be mine rather than Aristotle's, I asked myself. Perhaps his words had not always meant to him and his contemporaries quite what they meant to me and mine.

Feeling that way, I continued to puzzle over the text, and my suspicions ultimately proved well-founded. I was sitting at my desk with the text of Aristotle's *Physics* open in front of me and with a four-colored pencil in my hand. Looking up, I gazed abstractedly out the window of my room—the visual image is one I still retain. Suddenly the fragments in my head sorted themselves out in a new way, and fell into place together. My jaw dropped, for all at once Aristotle seemed a very good physicist indeed, but of a sort I'd never dreamed possible. Now I could understand why he had said what he'd said, and what his authority had been. Statements that had previously seemed egregious mistakes, now seemed at worst near misses within a powerful and generally successful tradition. That sort of experience—the pieces suddenly sorting themselves out and coming together in a new way—is the first general characteristic of revolutionary change that I shall be singling out after further consideration of examples. Though scientific revolutions leave much piecemeal mopping up to do, the central change cannot be experienced piecemeal, one step at a time. Instead, it involves some relatively sudden and unstructured transformation in which some part of the flux of experience sorts itself out differently and displays patterns that were not visible before.

To make all this more concrete let me now illustrate some of what was involved in my discovery of a way of reading Aristotelian physics, one that made the texts make sense. A first illustration will be familiar to many. When the term 'motion' occurs in Aristotelian physics, it refers to change in general, not just to the change of position of a physical body. Change of position, the exclusive subject of mechanics for Galileo and Newton, is one of a number of subcategories of motion for Aristotle. Others include growth (the transformation of an acorn to an oak), alterations of intensity (the heating of an iron bar), and a number of more general qualitative changes (the transition from sickness to health). As a result, though Aristotle recognizes that the various subcategories are not alike in *all* respects, the basic characteristics relevant to the recognition and analysis of motion must apply to changes of all sorts. In some sense that is not merely metaphorical; all varieties of change are seen as like each other, as constituting a single natural family.[3]

A second aspect of Aristotle's physics—harder to recognize and even more important—is the centrality of qualities to its conceptual structure. By that I do not mean simply that it aims to explain quality and change of quality, for other sorts of physics have done that. Rather I have in mind that Aristotelian physics inverts the ontological hierarchy of matter and quality that has been standard since the middle of the seventeenth century. In Newtonian physics a body is constituted of particles of matter, and its qualities are a consequence of the way those particles are arranged, move, and interact. In Aristotle's physics, on the other hand, matter is very nearly dispensable. It is a neutral substrate,

---

3. For all of this see Aristotle's *Physics*, book V, chapters 1–2 (224a21–226b16). Note that Aristotle does have a concept of change that is broader than that of motion. Motion is change of substance, change from something to something (225a1). But change also includes coming to be and passing away, i.e., change from nothing to something and from something to nothing (225a34–225b9), and these are not motions.

present wherever a body could be—which means wherever there's space or place. A particular body, a substance, exists in whatever place this neutral substrate, a sort of sponge, is sufficiently impregnated with qualities like heat, wetness, color, and so on to give it individual identity. Change occurs by changing qualities, not matter, by removing some qualities from some given matter and replacing them with others. There are even some implicit conservation laws that the qualities must apparently obey.[4]

Aristotle's physics displays other similarly general aspects, some of great importance. But I shall work toward the points that concern me from these two, picking up one other well-known one in passing. What I want now to begin to suggest is that, as one recognizes these and other aspects of Aristotle's viewpoint, they begin to fit together, to lend each other mutual support, and thus to make a sort of sense collectively that they individually lack. In my original experience of breaking into Aristotle's text, the new pieces I have been describing and the sense of their coherent fit actually emerged together.

Begin from the notion of a qualitative physics that has just been sketched. When one analyzes a particular object by specifying the qualities that have been imposed on omnipresent neutral matter, one of the qualities that must be specified is the object's position, or, in Aristotle's terminology, its place. Position is thus, like wetness or hotness, a quality of the object, one that changes as the object moves or is moved. Local motion (motion *tout court* in Newton's sense) is therefore change-of-quality or change-of-state for Aristotle, rather than being itself a state as it is for Newton. But it is precisely seeing motion as change-of-quality that permits its assimilation to all other sorts of change—acorn to oak or sickness to health, for examples. That assimilation is the aspect of Aristotle's physics from which I began, and I could equally well have traveled the route in the other direction. The conception of motion-as-change and the conception of a qualitative physics prove deeply interdependent, almost equivalent notions, and that is a first example of the fitting or the locking together of parts.

If that much is clear, however, then another aspect of Aristotle's physics—one that regularly seems ridiculous in isolation—begins to make sense as well. Most changes of quality, especially in the organic realm, are asymmetric, at least when left to themselves. An acorn natu-

4. Compare Aristotle's *Physics*, book I, and especially his *On Generation and Corruption*, book II, chapters 1–4.

rally develops into an oak, not vice versa. A sick man often grows healthy by himself, but an external agent is needed, or believed to be needed, to make him sick. One set of qualities, one end point of change, represents a body's natural state, the one that it realizes voluntarily and thereafter rests. The same asymmetry should be characteristic of local motion, change of position, and indeed it is. The quality that a stone or other heavy body strives to realize is position at the center of the universe; the natural position of fire is at the periphery. That is why stones fall toward the center until blocked by an obstacle and why fire flies to the heavens. They are realizing their natural properties just as the acorn does through its growth. Another initially strange part of Aristotelian doctrine begins to fall into place.

One could continue for some time in this manner, locking individual bits of Aristotelian physics into place in the whole. But I shall instead conclude this first example with a last illustration, Aristotle's doctrine about the vacuum or void. It displays with particular clarity the way in which a number of theses that appear arbitrary in isolation lend each other mutual authority and support. Aristotle states that a void is impossible: his underlying position is that the notion itself is incoherent. By now it should be apparent how that might be so. If position is a quality, and if qualities cannot exist separate from matter, then there must be matter wherever there's position, wherever body might be. But that is to say that there must be matter everywhere in space: the void, space without matter, acquires the status of, say, a square circle.[5]

That argument has force, but its premise seems arbitrary. Aristotle need not, one supposes, have conceived position as a quality. Perhaps, but we have already noted that that conception underlies his view of motion as change-of-state, and other aspects of his physics depend on it as well. If there could be a void, then the Aristotelian universe or cosmos could not be finite. It is just because matter and space are co-extensive that space can end where matter ends, at the outermost sphere

---

5. There is an ingredient missing from my sketch of this argument: Aristotle's doctrine of place, developed in the *Physics*, book IV, just before his discussion of the vacuum. Place, for Aristotle, is always the place of body or, more precisely, the interior surface of the containing or surrounding body (212a2–7). Turning to his next topic, Aristotle says, "Since the void (if there is any) must be conceived as place in which there might be body but is not, it is clear that, so conceived, the void cannot exist at all, either as inseparable or separable" (214a16–20). (I quote from the Loeb Classical Library translation by Philip H. Wicksteed and Francis M. Cornford, a version that, on this difficult aspect of the *Physics*, seems to me clearer than most, both in text and commentary.) That it is not merely a mistake to substitute 'position' for 'place' in a sketch of the argument is indicated by the last part of the next paragraph of my text.

beyond which there is nothing at all, neither space nor matter. That doctrine, too, may seem dispensable. But expanding the stellar sphere to infinity would make problems for astronomy, since that sphere's rotations carry the stars about the earth. Another, more central, difficulty arises earlier. In an infinite universe there is no center—any point is as much the center as any other—and there is thus no natural position at which stones and other heavy bodies realize their natural quality. Or, to put the point in another way, one that Aristotle actually uses, in a void a body could not be aware of the location of its natural place. It is just by being in contact with all positions in the universe through a chain of intervening matter that a body is able to find its way to the place where its natural qualities are fully realized. The presence of matter is what provides space with structure.[6] Thus, both Aristotle's theory of natural local motion and ancient geocentric astronomy are threatened by an attack on Aristotle's doctrine of the void. There is no way to "correct" Aristotle's views about the void without reconstructing much of the rest of his physics.

Those remarks, though both simplified and incomplete, should sufficiently illustrate the way in which Aristotelian physics cuts up and describes the phenomenal world. Also, and more important, they should indicate how the pieces of that description lock together to form an integral whole, one that had to be broken and reformed on the road to Newtonian mechanics. Rather than extend them further, I shall therefore proceed at once to a second example, returning to the beginning of the nineteenth century for the purpose. The year 1800 is notable, among other things, for Volta's discovery of the electric battery. That discovery was announced in a letter to Sir Joseph Banks, President of the Royal Society.[7] It was intended for publication and was accompanied by the illustration reproduced here as figure 1. For a modern audience there is something odd about it, though the oddity is seldom noticed, even by historians. Looking at any one of the so-called "piles" (of coins) in the lower two-thirds of the diagram, one sees, reading upward from the bottom right, a piece of zinc, Z, then a piece of silver, A, then a

---

6. For this and closely related arguments see Aristotle, *Physics*, book IV, chapter 8 (especially 214b27–215a24).

7. Alessandro Volta, "On the Electricity Excited by the mere Contact of Conducting Substances of Different Kinds," *Philosophical Transactions*, 90 (1800): 403–31. On this subject, see T. M. Brown, "The Electric Current in Early Nineteenth-Century French Physics," *Historical Studies in the Physical Sciences* 1 (1969): 61–103.

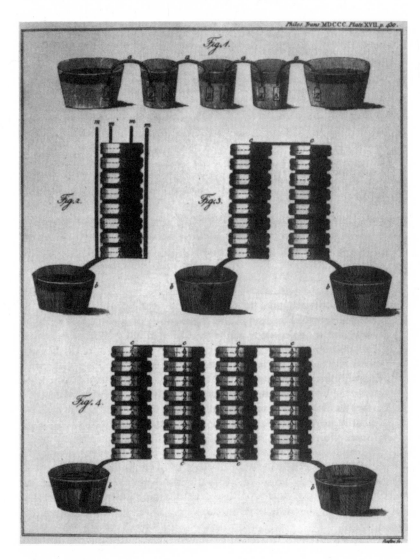

FIGURE I

piece of wet blotting paper, then a second piece of zinc, and so on. The cycle zinc, silver, wet blotting paper is repeated an integral number of times, eight in Volta's original illustration. Now suppose that, instead of having all this spelled out, you had been asked simply to look at the diagram, then to put it aside and reproduce it from memory. Almost certainly, those of you who know even the most elementary physics would have drawn zinc (or silver), followed by wet blotting paper, followed by silver (or zinc). In a battery, as we all know, the liquid belongs between the two different metals.

If one recognizes this difficulty and puzzles over it with the aid of Volta's texts, one is likely to realize suddenly that for Volta and his followers, the unit cell consists of the two pieces of metal in contact. The source of power is the metallic interface, the bimetallic junction that Volta had previously found to be the source of an electrical tension, what we would call a voltage. The role of the liquid then is simply to connect one unit cell to the next without generating a contact potential, which would neutralize the initial effect. Pursuing Volta's text still further, one realizes that he is assimilating his new discovery to electrostatics. The bimetallic junction is a condenser or Leyden jar, but one that charges itself. The pile of coins is, then, a linked assemblage or "battery" of charged Leyden jars, and that is where, by specialization from the group to its members, the term 'battery' comes from in its application to electricity. For confirmation, look at the top part of Volta's diagram, which illustrates an arrangement he called "the crown of cups." This time the resemblance to diagrams in elementary modern textbooks is striking, but there is again an oddity. Why do the cups at the two ends of the diagram contain only one piece of metal? Why does Volta include two half-cells? The answer is the same as before. For Volta the cups are not cells but simply containers for the liquids that connect cells. The cells themselves are the bimetallic horseshoe strips. The apparently unoccupied positions in the outermost cups are what we would think of as binding posts. In Volta's diagram there are no half-cells.

As in the previous example, the consequences of this way of looking at the battery are widespread. For example, as shown in figure 2, the transition from Volta's viewpoint to the modern one reverses the direction of current flow. A modern cell diagram (figure 2, bottom) can be derived from Volta's (top left) by a process like turning the latter inside out (top right). In that process what was previously current flow internal to the cell becomes the external current and vice versa. In the Voltaic diagram the external current flow is from black metal to white, so that the black is positive. In the modern diagram both the direction of flow and the polarity are reversed. Far more important conceptually is the change in the current source effected by the transition. For Volta the metallic interface was the essential element of the cell and necessarily the source of the current the cell produced. When the cell was turned

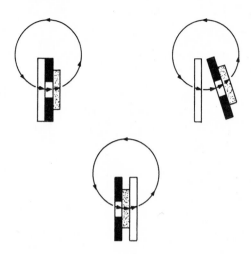

FIGURE 2

inside out, the liquid and its two interfaces with the metals provided its essentials, and the source of the current became the chemical effects at these interfaces. When both viewpoints were briefly in the field at once, the first was known as the contact theory, the second as the chemical theory of the battery.

Those are only the most obvious consequences of the electrostatic view of the battery, and some of the others were even more immediately important. For example, Volta's viewpoint suppressed the conceptual role of the external circuit. What we would think of as an external circuit is simply a discharge path like the short circuit to ground that discharges a Leyden jar. As a result, early battery diagrams do not show an external circuit unless some special effect, like electrolysis or heating a wire, is occurring there, and then, very often the battery is not shown. Not until the 1840s do modern cell diagrams begin to appear regularly in books on electricity. When they do, either the external circuit or explicit points for its attachment appears with them.[8] Examples are shown in figures 3 and 4.

Finally, the electrostatic view of the battery leads to a concept of

8. The illustrations are from A. de la Rive, *Traité d'électricité théorique et appliquée*, vol. 2 (Paris: J. B. Bailière, 1856), pp. 600, 656. Structurally similar but schematic diagrams appear in Faraday's experimental researches from the early 1830s. My choice of the 1840s as the period when such diagrams became standard results from a casual survey of electricity texts lying ready to hand. A more systematic study would, in any case, have had to distinguish between British, French, and German responses to the chemical theory of the battery.

FIGURE 3

FIGURE 4

electrical resistance very different from the one now standard. There is an electrostatic concept of resistance, or there was in this period. For an insulating material of given cross section, resistance was measured by the shortest length the material could have without breaking down or leaking—ceasing to insulate—when subjected to a given voltage. For a conducting material of given cross section, it was measured by the shortest length the material could have without melting when connected across a given voltage. It is possible to measure resistance conceived in this way, but the results are not compatible with Ohm's law. To get those results one must conceive the battery and circuit on a more hydrodynamic model. Resistance must become like the frictional resistance to the flow of water in pipes. The assimilation of Ohm's law required a noncumulative change of that sort, and that is part of what made his law so difficult for many people to accept. It has for some time provided a standard example of an important discovery that was initially rejected or ignored.

At this point I end my second example and proceed at once to a third, this one both more modern and more technical than its predecessors.

Substantively, it is controversial, involving a new version, not yet everywhere accepted, of the origins of the quantum theory.[9] Its subject is Max Planck's work on the so-called black-body problem, and its structure may usefully be anticipated as follows. Planck first solved the black-body problem in 1900 using a classical method developed by the Austrian physicist Ludwig Boltzmann. Six years later a small but crucial error was found in his derivation, and one of its central elements had to be reconceived. When that was done Planck's solution did work, but it then also broke radically with tradition. Ultimately that break spread through and caused the reconstruction of a good deal of physics.

Begin with Boltzmann, who had considered the behavior of a gas, conceived as a collection of many tiny molecules, moving rapidly about within a container, and colliding frequently both with each other and with the container's walls. From previous works of others, Boltzmann knew the average velocity of the molecules (more precisely, the average of the square of their velocity). But, many of the molecules were, of course, moving much more slowly than the average, others much faster. Boltzmann wanted to know what proportion of them were moving at, say, $\frac{1}{2}$ the average velocity, what proportion at $\frac{1}{3}$ the average, and so on. Neither that question nor the answer he found to it was new. But Boltzmann reached the answer by a new route, from probability theory, and that route was fundamental for Planck, since whose work it has been standard.

Only one aspect of Boltzmann's method is of present concern. He considered the total kinetic energy E of the molecules. Then, to permit the introduction of probability theory, he mentally subdivided that energy into little cells or elements of size ε, as in figure 5. Next, he imagined distributing the molecules at random among those cells, drawing numbered slips from an urn to specify the assignment of each molecule and then excluding all distributions with total energy different from E. For example, if the first molecule were assigned to the last cell (energy E), then the only acceptable distribution would be the one that assigned all other molecules to the first cell (energy o). Clearly, that particular distribution is a most improbable one. It is far more likely that most molecules will have appreciable energy, and by probability theory one can discover the most probable distribution of all. Boltzmann showed

9. For the full version with supporting evidence see my *Black-Body Theory and the Quantum Discontinuity, 1894–1912* (Oxford and New York: Clarendon and Oxford University Presses, 1978).

FIGURE 5

how to do so, and his result was the same as the one he and others had previously gotten by more problematic means.

That way of solving the problem was invented in 1877, and twenty-three years later, at the end of 1900, Max Planck applied it to an apparently rather different problem, black-body radiation. Physically the problem is to explain the way in which the color of a heated body changes with temperature. Think, for example, of the radiation from an iron bar, which, as the temperature increases, first gives off heat (infrared radiation), then glows dull red, and then gradually becomes a brilliant white. To analyze that situation Planck imagined a container or cavity filled with radiation, that is, with light, heat, radio waves, and so on. In addition, he supposed that the cavity contained a lot of what he called "resonators" (think of them as tiny electrical tuning forks, each sensitive to radiation at one frequency, not at others). These resonators absorb energy from the radiation, and Planck's question was: How does the energy picked up by each resonator depend on its frequency? What is the frequency distribution of the energy over the resonators?

Conceived in that way, Planck's problem was very close to Boltzmann's, and Planck applied Boltzmann's probabilistic techniques to it. Roughly speaking, he used probability theory to find the proportion of resonators that fell in each of the various cells, just as Boltzmann had found the proportion of molecules. His answer fit experimental results better than any other then or since known, but there turned out to be one unexpected difference between his problem and Boltzmann's. For Boltzmann's, the cell size ε could have many different values without changing the result. Though permissible values were bounded, could not be too large or too small, an infinity of satisfactory values was available in between. Planck's problem proved different: other aspects of physics determined ε, the cell size. It could have only a single value given by the famous formula ε = $h\nu$, where ν is the resonator frequency and h is the universal constant subsequently known by Planck's name. Planck was, of course, puzzled about the reason for the restriction on cell size, though he had a strong hunch about it, one he attempted to develop. But, excepting that residual puzzle, he had solved his problem, and his approach remained very close to Boltzmann's. In particular—the presently crucial point—in both solutions the division of the total energy E into cells of size ε was a mental division made for statistical purposes. The molecules and resonators could lie anywhere along the line and were governed by all the standard laws of classical physics.

The rest of this story is very quickly told. The work just described was done at the end of 1900. Six years later, in the middle of 1906, two other physicists argued that Planck's result could not be gained in Planck's way. One small but absolutely crucial alteration of the argument was required. The resonators could not be permitted to lie anywhere on the continuous energy line but only at the divisions between cells. A resonator might, that is, have energy o, ε, 2ε, 3ε, . . . , and so on, but not ($\frac{1}{3}$)ε, ($\frac{4}{5}$)ε, etc. When a resonator changed energy it did not do so continuously but by discontinuous jumps of size ε or a multiple of ε.

After those alterations, Planck's argument was both radically different and very much the same. Mathematically it was virtually unchanged, with the result that it has been standard for years to read Planck's 1900 paper as presenting the subsequent modern argument. But physically, the entities to which the derivation refers are very different. In particular, the element ε has gone from a mental division of the total energy to a separable physical energy atom, of which each resonator may have o, 1, 2, 3, or some other number. Figure 6 tries to capture that change

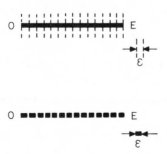

FIGURE 6

in a way that suggests its resemblance to the inside-out battery of my last example. Once again the transformation is subtle, difficult to see. But also once again, the change is consequential. Already the resonator has been transformed from a familiar sort of entity governed by standard classical laws to a strange creature the very existence of which is incompatible with traditional ways of doing physics. As most of you know, changes of the same sort continued for another twenty years as similar nonclassical phenomena were found in other parts of the field.

Those later changes, I shall not attempt to follow, but instead conclude this example, my last, by pointing to one other sort of change that occurred near its start. In discussing the earlier examples, I pointed out that revolutions were accompanied by changes in the way in which terms like 'motion' or 'cell' attached to nature. In this example there was actually a change in the words themselves, one that highlights those features of the physical situation that the revolution had made prominent. When Planck around 1909 was at last persuaded that discontinuity had come to stay he switched to a vocabulary that has been standard since. Previously he had ordinarily referred to the cell size ε as the energy "element." Now, in 1909, he began regularly to speak instead of the energy "quantum"; for 'quantum', as used in German physics, was a separable element, an atomlike entity that could exist by itself. While ε had been merely the size of a mental subdivision, it had not been a quantum but an element. Also in 1909, Planck abandoned the acoustic analogy. The entities he had introduced as "resonators" now became "oscillators," the latter a neutral term that refers to any entity that simply vibrates regularly back and forth. By contrast, 'resonator' refers in the first instance to an acoustic entity or, by extension, to a vibrator that responds gradually to stimulation, swelling and diminishing with the applied stimulus. For one who believed that energy changes discontinuously, 'resonator' was not an appropriate term, and Planck gave it up in and after 1909.

That vocabulary change concludes my third example. Rather than give others, I shall conclude this discussion by asking what characteristics of revolutionary change are displayed by the examples at hand. Answers will fall under three headings, and I shall be relatively brief about each. The extended discussion they require, I am not quite ready to provide.

A first set of shared characteristics was mentioned near the start of this paper. Revolutionary changes are somehow holistic. They cannot, that is, be made piecemeal, one step at a time, and they thus contrast with normal or cumulative changes like, for example, the discovery of Boyle's law. In normal change, one simply revises or adds a single generalization, all others remaining the same. In revolutionary change one must either live with incoherence or else revise a number of inter-related generalizations together. If these same changes were introduced one at a time, there would be no intermediate resting place. Only the initial and final sets of generalizations provide a coherent account of nature. Even in my last example, the most nearly cumulative of the three, one cannot simply change the description of the energy element ε. One must also change one's notion of what it is to be a resonator, for resonators, in any normal sense of the term, cannot behave as these do. Simultaneously, to permit the new behavior, one must change, or try to, laws of mechanics and of electromagnetic theory. Again, in the second example, one cannot simply change one's mind about the order of elements in a battery cell. The direction of the current, the role of the external circuit, the concept of electrical resistance, and so on, must also be changed. Or still again, in the case of Aristotelian physics, one cannot simply discover that a vacuum is possible or that motion is a state, not a change-of-state. An integrated picture of several aspects of nature has to be changed at the same time.

A second characteristic of these examples is closely related. It is the one I have in the past described as meaning change and which I have here been describing, somewhat more specifically, as change in the way words and phrases attach to nature, change in the way their referents are determined. Even that version is, however, somewhat too general. As recent studies of reference have emphasized, anything one knows about the referents of a term may be of use in attaching that term to nature. A newly discovered property of electricity, of radiation, or of the effects of force on motion may thereafter be called upon (usually with others) to determine the presence of electricity, radiation, or force and thus to pick out the referents of the corresponding term. Such dis-coveries need not be and usually are not revolutionary. Normal science, too, alters the way in which terms attach to nature. What character-izes revolutions is not, therefore, simply change in the way referents are determined, but change of a still more restricted sort.

How best to characterize that restricted sort of change is among the problems that currently occupy me, and I have no full solution. But roughly speaking, the distinctive character of revolutionary change in language is that it alters not only the criteria by which terms attach to nature but also, massively, the set of objects or situations to which those terms attach. What had been paradigmatic examples of motion for Aristotle—acorn to oak or sickness to health—were not motions at all for Newton. In the transition, a natural family ceased to be natural; its members were redistributed among preexisting sets; and only one of them continued to bear the old name. Or again, what had been the unit cell of Volta's battery was no longer the referent of any term forty years after his invention was made. Though Volta's successors still dealt with metals, liquids, and the flow of charge, the units of their analyses were different and differently interrelated.

What characterizes revolutions is, thus, change in several of the taxonomic categories prerequisite to scientific descriptions and generalizations. That change, furthermore, is an adjustment not only of criteria relevant to categorization, but also of the way in which given objects and situations are distributed among preexisting categories. Since such redistribution always involves more than one category and since those categories are interdefined, this sort of alteration is necessarily holistic. That holism, furthermore, is rooted in the nature of language, for the criteria relevant to categorization are ipso facto the criteria that attach the names of those categories to the world. Language is a coinage with two faces, one looking outward to the world, the other inward to the world's reflection in the referential structure of the language.

Look now at the last of the three characteristics shared by my three examples. It has been the most difficult of the three for me to see, but now seems the most obvious and probably the most consequential. Even more than the others, it should repay further exploration. All of my examples have involved a central change of model, metaphor, or analogy—a change in one's sense of what is similar to what, and of what is different. Sometimes, as in the Aristotle example, the similarity is internal to the subject matter. Thus, for Aristotelians, motion was a special case of change, so that the falling stone was *like* the growing oak, or *like* the person recovering from illness. That is the pattern of similarities that constitutes these phenomena a natural family, that places them in the same taxonomic category, and that had to be replaced in the development of Newtonian physics. Elsewhere the similarity is external. Thus, Planck's resonators were *like* Boltzmann's molecules, or Volta's battery cells were *like* Leyden jars, and resistance was *like* electrostatic leakage. In these cases, too, the old pattern of similarities had to be discarded and replaced before or during the process of change.

All these cases display interrelated features familiar to students of metaphor. In each case two objects or situations are juxtaposed and said to be the same or similar. (An even slightly more extended discussion would have also to consider examples of dissimilarity, for they, too, are often important in establishing a taxonomy.) Furthermore, whatever their origin—a separate issue with which I am not presently concerned—the primary function of all these juxtapositions is to transmit and maintain a taxonomy. The juxtaposed items are exhibited to a previously uninitiated audience by someone who can already recognize their similarity, and who urges that audience to learn to do the same. If the exhibit succeeds, the new initiates emerge with an acquired list of features salient to the required similarity relation—with a feature-space, that is, within which the previously juxtaposed items are durably clustered together as examples of the same thing and are simultaneously separated from objects or situations with which they might otherwise have been confused. Thus, the education of an Aristotelian associates the flight of an arrow with a falling stone and both with the growth of an oak and the return to health. All are thereafter changes-of-state; their end points and the elapsed time of transition are their salient features. Seen in that way, motion cannot be relative and must be in a category distinct from rest, which is a state. Similarly, on that view, an infinite motion, because it lacks an end point, becomes a contradiction in terms.

The metaphor-like juxtapositions that change at times of scientific revolution are thus central to the process by which scientific and other language is acquired. Only after that acquisition or learning process has passed a certain point can the practice of science even begin. Scientific practice always involves the production and the explanation of generalizations about nature, those activities presuppose a language with some minimal richness, and the acquisition of such a language brings knowledge of nature with it. When the exhibit of examples is part of the process of learning terms like 'motion', 'cell', or 'energy element', what is acquired is knowledge of language and of the world together. On the one hand, the student learns what these terms mean, what features are relevant to attaching them to nature, what things cannot be said of them on pain of self-contradiction, and so on. On the other hand, the student learns what categories of things populate the world, what their salient features are, and something about the behavior that is and is not permitted to them. In much of language learning these two sorts of knowledge—knowledge of words and knowledge of nature—are acquired together, not really two sorts of knowledge at all, but two faces of the single coinage that a language provides.

The reappearance of the double-faced character of scientific language provides an appropriate terminus for this paper. If I am right, the central characteristic of scientific revolutions is that they alter the knowledge of nature that is intrinsic to the language itself and that is thus prior to anything quite describable as description or generalization, scientific or everyday. To make the void or an infinite linear motion part of science required observation reports that could only be formulated by altering the language with which nature was described. Until those changes had occurred, language itself resisted the invention and intro- duction of the sought-after new theories. The same resistance by lan- guage is, I take it, the reason for Planck's switch from 'element' and 'resonator' to 'quantum' and 'oscillator'. Violation or distortion of a previously unproblematic scientific language is the touchstone for revo- lutionary change.

Chapter Two

Commensurability, Comparability, Communicability

"Commensurability, Comparability, Communicability" was the main paper in a symposium at the biennial meetings of the Philosophy of Science Association in 1982, for which Philip Kitcher and Mary Hesse served as commentators; Kuhn's reply to their comments is included here, as a postscript to the essay. The proceedings of the symposium were published in PSA 1982, *volume 2 (East Lansing, MI: The Philosophy of Science Association, 1983).*

———◄○►———

TWENTY YEARS HAVE PASSED since Paul Feyerabend and I first used in print a term we had borrowed from mathematics to describe the relationship between successive scientific theories. 'Incommensurability' was the term; each of us was led to it by problems we had encountered in interpreting scientific texts.[1] My use of the term was broader than

---

Since this paper was first drafted, many people have contributed to its improvement, among them colleagues at MIT and auditors at the PSA meeting and at the Columbia seminar in History and Philosophy of Science where a preliminary version was first tried out. I am grateful to all of them, above all to Ned Block, Paul Horwich, Nathaniel Kuhn, Stephen Stich, and my two official commentators.

1. P. K. Feyerabend, "Explanation, Reduction, and Empiricism," in *Scientific Explanation, Space, and Time,* ed. H. Feigl and G. Maxwell, Minnesota Studies in the Philosophy of Science, vol. 3 (Minneapolis: University of Minnesota Press, 1962), pp. 28–97; T. S. Kuhn, *The Structure of Scientific Revolutions* (Chicago: University of Chicago Press, 1962). I believe that Feyerabend's and my resort to 'incommensurability' was independent, and I have an uncertain memory of Paul's finding it in a draft manuscript of mine and telling me he too had been using it. Passages illustrating our early usages are: Kuhn, *The Structure of Scientific Revolutions,* 2d ed., rev. (Chicago: University of Chicago Press, 1970), pp. 102 f., 112, 128 f., 148–51, all unchanged from the first edition, and Feyerabend, pp. 56–59, 74–76, 81.

his; his claims for the phenomenon were more sweeping than mine; but our overlap at that time was substantial.[2] Each of us was centrally concerned to show that the meanings of scientific terms and concepts— 'force' and 'mass', for example, or 'element' and 'compound'—often changed with the theory in which they were deployed.[3] And each of us claimed that when such changes occurred, it was impossible to define all the terms of one theory in the vocabulary of the other. The latter claim we independently embodied in talk about the incommensurability of scientific theories.

All that was in 1962. Since then problems of meaning variance have been widely discussed, but virtually no one has fully faced the difficulties that led Feyerabend and me to speak of incommensurability. Doubtless, that neglect is due in part to the role played by intuition and metaphor in our initial presentations. I, for example, made much use of the double sense, visual and conceptual, of the verb 'to see', and I repeatedly likened theory changes to gestalt switches. But for whatever reasons, the concept of incommensurability has been widely and often dismissed, most recently in a book published late last year by Hilary Putnam.[4] Putnam redevelops cogently two lines of criticism that had figured widely in earlier philosophical literature. A brief restatement of those criticisms here should prepare the way for some extended comments.

Most or all discussions of incommensurability have depended upon the literally correct but regularly overinterpreted assumption that, if two theories are incommensurable, they must be stated in mutually untranslatable languages. If that is so, a first line of criticism runs, if there is no way in which the two can be stated in a single language, then they cannot be compared, and no arguments from evidence can be relevant to the choice between them. Talk about differences and comparisons presupposes that some ground is shared, and that is what proponents

2. Both Feyerabend and I wrote of the impossibility of defining the terms of one theory on the basis of the terms of the other. But he restricted incommensurability to language; I spoke also of differences in "methods, problem-field, and standards of solution " (*Structure*, 2d ed., p. 103), something I would no longer do except to the considerable extent that the latter differences are necessary consequences of the language-learning process. Feyerabend (p. 59) on the other hand, wrote that "it is possible neither to define the primitive terms of T' on the basis of the primitive terms of T nor to establish correct empirical relations involving both these terms." I made no use of a notion of primitive terms and restricted incommensurability to a few specific terms.

3. This point had been previously emphasized in N. R. Hanson, *Patterns of Discovery* (Cambridge: Cambridge University Press, 1958).

4. H. Putnam, *Reason, Truth, and History* (Cambridge: Cambridge University Press, 1981), pp. 113–24.

of incommensurability, who often do talk of comparisons, have seemed to deny. At these points their talk is necessarily incoherent.[5] A second line of criticism cuts at least as deep. People like Kuhn, it is said, tell us that it is impossible to translate old theories into a modern language. But they then proceed to do exactly that, reconstructing Aristotle's or Newton's or Lavoisier's or Maxwell's theory without departing from the language they and we speak every day. What can they mean, under these circumstances, when they speak about incommensurability?[6]

My concerns in this paper arise primarily from the second of these lines of argument, but the two are not independent, and I shall need also to speak of the first. With it I begin, attempting first to set aside some widespread misunderstanding of at least my own point of view. Even with misunderstanding eliminated, however, a damaging residue of the first line of criticism will remain. To it I shall return only at the end of this paper.

Local Incommensurability

Remember briefly where the term 'incommensurability' came from. The hypotenuse of an isosceles right triangle is incommensurable with its side or the circumference of a circle with its radius in the sense that there is no unit of length contained without residue an integral number of times in each member of the pair. There is thus no common measure. But lack of a common measure does not make comparison impossible. On the contrary, incommensurable magnitudes can be compared to any required degree of approximation. Demonstrating that this could be done and how to do it were among the splendid achievements of Greek mathematics. But that achievement was possible only because, from the start, most geometric techniques applied without change to both of the items between which comparison was sought.

Applied to the conceptual vocabulary deployed in and around a sci-

5. For this line of criticism, see D. Davidson, "The Very Idea of a Conceptual Scheme," in *Proceedings and Addresses of the American Philosophical Association* 47 (1974): 5–20; D. Shapere, "Meaning and Scientific Change," in *Mind and Cosmos: Essays in Contemporary Science and Philosophy*, University of Pittsburgh Series in the Philosophy of Science, vol. 3, ed. R. G. Colodny (Pittsburgh: University of Pittsburgh Press, 1966), pp. 41–85; I. Scheffler, *Science and Subjectivity* (Indianapolis: Bobbs-Merrill, 1967), pp. 81–83.
6. For this line of criticism, see Davidson, "The Very Idea," pp. 17–20; P. Kitcher, "Theories, Theorists, and Theoretical Change," *Philosophical Review* 87 (1978): 519–47; Putnam, *Reason, Truth, and History*.

entific theory, the term 'incommensurability' functions metaphorically. The phrase 'no common measure' becomes 'no common language'. The claim that two theories are incommensurable is then the claim that there is no language, neutral or otherwise, into which both theories, conceived as sets of sentences, can be translated without residue or loss. No more in its metaphorical than its literal form does incommensurability imply incomparability, and for much the same reason. Most of the terms common to the two theories function the same way in both; their meanings, whatever those may be, are preserved; their translation is simply homophonic. Only for a small subgroup of (usually interdefined) terms and for sentences containing them do problems of translatability arise. The claim that two theories are incommensurable is more modest than many of its critics have supposed.

I shall call this modest version of incommensurability 'local incommensurability'. Insofar as incommensurability was a claim about language, about meaning change, its local form is my original version. If it can be consistently maintained, then the first line of criticism directed at incommensurability must fail. The terms that preserve their meanings across a theory change provide a sufficient basis for the discussion of differences and for comparisons relevant to theory choice.[7] They even provide, as we shall see, a basis from which the meanings of incommensurable terms can be explored.

It is not clear, however, that incommensurability can be restricted to a local region. In the present state of the theory of meaning, the distinction between terms that change meaning and those that preserve it is at best difficult to explicate or apply. Meanings are a historical product, and they inevitably change over time with changes in the demands on the terms that bear them. It is simply implausible that some terms should change meaning when transferred to a new theory without infecting the terms transferred with them. Far from supplying a solution, the phrase 'meaning invariance' may supply only a new home for the problems presented by the concept of incommensurability. This difficulty is real, no product of misunderstanding. I shall be returning to it at the end of this paper, and it will then appear that 'meaning' is not the rubric under which incommensurability is best discussed. But no more suitable alternative is presently at hand. In search of one, I now turn to

---

7. Note that these terms are not theory-independent but are simply used in the same way within the two theories at issue. It follows that testing is a process that compares two theories, not one that can evaluate theories one at a time.

the second main line of criticism directed regularly at incommensurability. It survives the return to the original local version of that notion.

Translation versus Interpretation

If any non-vacuous terms of an older theory elude translation into the language of its successor, how can historians and other analysts succeed so well in reconstructing or interpreting that older theory, including the use and function of those very terms? Historians claim to be able to produce successful interpretations. So, in a closely related enterprise, do anthropologists. I shall here simply premise that their claims are justified, that there are no limits of principle on the extent to which those criteria can be fulfilled. Whether or not correct, as I think they are, these assumptions are, in any case, fundamental to the arguments directed at incommensurability by such critics as Davidson, Kitcher, and Putnam.[8] All three sketch the technique of interpretation, all describe its outcome as a translation or a translation schema, and all conclude that its success is incompatible with even local incommensurability. As I now try to show what is the matter with their argument, I come to the central concerns of this paper.

The argument or argument sketch I have just supplied depends critically upon the equation of interpretation with translation. That equation is traceable at least to Quine's *Word and Object*.[9] I believe it is wrong and that the mistake is important. My claim is that interpretation, a process about which I shall be having more to say, is not the same as translation, at least not as translation has been conceived in much recent philosophy. The confusion is easy because actual translation often or perhaps always involves at least a small interpretive component. But in that case actual translation must be seen to involve two distinguishable processes. Recent analytic philosophy has concentrated exclusively on one and conflated the other with it. To avoid confusion I shall here follow recent usage and apply 'translation' to the first of these processes, 'interpretation' to the second. But so long as the existence of two processes is recognized, nothing in my argument depends upon preserving the term 'translation' for the first.

8. Davidson, "The Very Idea," p. 19; Kitcher, "Theories, Theorists, and Theoretical Change," pp. 519–29; Putnam, *Reason, Truth, and History*, pp. 116 f.
9. W. V. O. Quine, *Word and Object* (Cambridge, MA: Technology Press of the Massachusetts Institute of Technology, 1960).

For present purposes, then, translation is something done by a person who knows two languages. Confronted with a text, written or oral, in one of these languages, the translator systematically substitutes words or strings of words in the other language for words or strings of words in the text in such a way as to produce an equivalent text in the other language. What it is to be an "equivalent text" can, for the moment, remain unspecified. Sameness of meaning and sameness of reference are both obvious desiderata, but I do not yet invoke them. Let us simply say that the translated text tells more or less the same story, presents more or less the same ideas, or describes more or less the same situation as the text of which it is a translation.

Two features of translation thus conceived require special emphasis. First, the language into which the translation is cast existed before the translation was begun. The fact of translation has not, that is, changed the meanings of words or phrases. It may, of course, have increased the number of known referents of a given term, but it has not altered the way in which those referents, new and old, are determined. A second feature is closely related. The translation consists exclusively of words and phrases that replace (not necessarily one-for-one) words and phrases in the original. Glosses and translators' prefaces are not part of the translation, and a perfect translation would have no need for them. If they are nonetheless required, we shall need to ask why. Doubtless, these features of translation seem idealizations, and they surely are. But the idealization is not mine. Among other sources, both derive directly from the nature and function of a Quinean translation manual.

Turn now to interpretation. It is an enterprise practiced by historians and anthropologists, among others. Unlike the translator, the interpreter may initially command only a single language. At the start, the text on which he or she works consists in whole or in part of unintelligible noises or inscriptions. Quine's "radical translator" is in fact an interpreter, and 'gavagai' exemplifies the unintelligible material he starts from. Observing behavior and the circumstances surrounding the production of the text, and assuming throughout that good sense can be made of apparently linguistic behavior, the interpreter seeks that sense, strives to invent hypotheses, like 'Gavagai' means "Lo, a rabbit," which make utterance or inscription intelligible. If the interpreter succeeds, what he or she has in the first instance done is learn a new language, perhaps the language in which 'gavagai' is a term, or perhaps an earlier version of the interpreter's own language, one in which still current terms like 'force' and 'mass' or 'element' and 'compound' functioned differently. Whether that language can be translated into the one with which the interpreter began is an open question. Acquiring a new language is not the same as translating from it into one's own. Success with the first does not imply success with the second.

It is with respect to just those problems that Quine's examples are consistently misleading, for they conflate interpretation and translation. To *interpret* the utterance 'Gavagai', Quine's imagined anthropologist need not come from a speech community that knows of rabbits and possesses a word which refers to them. Rather than finding a term that corresponds to 'gavagai', the interpreter/anthropologist could acquire the native's term much as, at an earlier stage, some terms of his or her own language were acquired.[10] The anthropologist or interpreter, that is, can and often does learn to recognize the creatures that evoke 'Gavagai' from natives. Rather than translate, the interpreter can simply learn the animal and use the natives' term for it.

The availability of that alternative does not, of course, preclude translation. The interpreter may not, for reasons previously explained, merely introduce the term 'gavagai' into his or her own language, say, English. That would be to alter English, and the result would not be translation. But the interpreter can attempt to describe in English the referents of the term 'gavagai'—they are furry, long-eared, bushy-tailed, and the like. If the description is successful, if it fits all and only creatures that elicit utterances involving 'gavagai', then 'furry, long-eared, bushy-tailed . . . creature' is the sought-after translation, and 'gavagai' can thereafter be introduced into English as an abbreviation for it.[11] Under these circumstances, no issue of incommensurability arises.

But these circumstances need not obtain. There need be no English description coreferential with the native term 'gavagai'. In learning to recognize gavagais, the interpreter may have learned to recognize distinguishing features unknown to English speakers and for which English

10. Quine (*Word and Object*, pp. 47, 70 f.) notes that his radical translator might choose the "costly" way and "learn the language direct as an infant might." But he takes this process to be simply an alternate route to the same end as those reached by his standard means, that end being a translation manual.

11. Some would object that a string like "furry, long-eared, bushy-tailed . . . creature" is too long and complex to count as a translation of a single term in another language. But I incline to the view that any term which can be introduced by a string can be internalized so that, with practice, its referents can be recognized directly. In any case, I am concerned with a stronger version of untranslatability, one in which not even long strings are available.

supplies no descriptive terminology. Perhaps, that is, the natives structure the animal world differently from the way English speakers do, using different discriminations in doing so. Under those circumstances, 'gavagai' remains an irreducibly native term, not translatable into English. Though English speakers may learn to use the term, they speak the native language when they do so. Those are the circumstances for which I would reserve the term 'incommensurability'.

Reference Determination versus Translation

My claim, then, has been that circumstances of this sort are regularly encountered, if not always recognized, by historians of science attempting to understand out-of-date scientific texts. The phlogiston theory has provided one of my standard examples, and Philip Kitcher has used it as the basis for a penetrating critique of the whole notion of incommensurability. What is currently at issue will be considerably clarified if I first exhibit the kernel of that critique and then indicate the point at which I think it goes astray.

Kitcher argues, successfully I think, that the language of twentieth-century chemistry can be used to identify referents of the terms and expressions of eighteenth-century chemistry, at least to the extent that those terms and expressions actually refer. Reading a text by, say, Priestley and thinking of the experiments he describes in modern terms, one can see that 'dephlogisticated air' sometimes refers to oxygen itself, sometimes to an oxygen-enriched atmosphere. 'Phlogisticated air' is regularly air from which oxygen has been removed. The expression 'α is richer in phlogiston than β' is coreferential with 'α has a greater affinity for oxygen than β'. In some contexts—for example, in the expression 'phlogiston is emitted during combustion'—the term 'phlogiston' does not refer at all, but there are other contexts in which it refers to hydrogen.[12]

I have no doubt that historians dealing with old scientific texts can and must use modern language to identify referents of out-of-date terms. Like the native's pointing to gavagais, these reference determinations often provide the concrete examples from which historians may hope to learn what the problematic expressions in their texts mean. In addition, the introduction of modern terminology makes it possible to

12. Kitcher, "Theories, Theorists, and Theoretical Change," pp. 531–36.

explain why and in what areas older theories were successful.[13] Kitcher, however, describes this process of reference determination as translation, and he suggests that its availability should bring talk of incommensurability to a close. In both these respects he seems to me mistaken.

Think for a moment of what a text translated by Kitcher's techniques would look like. How, for example, would nonreferring occurrences of 'phlogiston' be rendered? One possibility—suggested both by Kitcher's silence on the subject and by his concern to preserve truth values, which are in these places problematic—would be to leave the corresponding spaces blank. To leave blanks is, however, to fail as a translator. If only referring expressions possess translations, then no work of fiction could be translated at all, and for present purposes, old scientific texts must be treated with at least the courtesy normally extended to works of fiction. They report what scientists of the past believed, independent of its truth value, and that is what a translation must communicate.

Alternatively, Kitcher might use the same context-dependent strategy he developed for referring terms like 'dephlogisticated air'. 'Phlogiston' would then sometimes be rendered as 'substance released from burning bodies', sometimes as 'metallizing principle', and sometimes by still other locutions. This strategy, however, also leads to disaster, not only with terms like 'phlogiston' but with referring expressions as well. Use of a single word, 'phlogiston', together with compounds like 'phlogisticated air' derived from it, is one of the ways by which the original text communicated the beliefs of its author. Substituting unrelated or differently related expressions for those related, sometimes identical terms of the original must at least suppress those beliefs, leaving the

---

13. Kitcher supposes that his translation techniques permit him to specify which statements of the older theory were true, which false. Thus statements about the substance released on combustion were false but statements about the effect of dephlogisticated air on vital activities were true because in those statements 'dephlogisticated air' referred to oxygen. I think, however, that Kitcher is only using modern theory to explain why some statements made by practitioners of the older theory were confirmed by experience, others not. The ability to explain such successes and failures is basic to the historian of science's interpretation of texts. (If an interpretation attributes to the author of a text repeated assertions which easily available observations would have infirmed, then the interpretation is almost certainly wrong, and the historian must go to work again. For an example of what may then be required, see my "A Function for Thought Experiments," in *Mélanges Alexandre Koyré*, vol. 2, *L'aventure de la science*, ed. I. B. Cohen and R. Taton [Paris: Hermann, 1964], pp. 307–34; reprinted in *The Essential Tension: Selected Studies in Scientific Tradition and Change* [Chicago: University of Chicago Press, 1977], pp. 240–65.) But neither interpretation nor Kitcher's translation techniques allows individual sentences containing terms from the older theory to be declared true or false. Theories are, I believe, structures that must be evaluated as wholes.

text that results incoherent. Examining a Kitcher translation, one would repeatedly be at a loss to understand why those sentences were juxtaposed in a single text.[14]

To see more clearly what is involved in dealing with out-of-date texts, consider the following epitome of some central aspects of the phlogiston theory. For the sake of clarity and brevity, I have constructed it myself, but it could, style aside, have been drawn from an eighteenth-century chemical manual.

All physical bodies are composed of chemical elements and principles, the latter endowing the former with special properties. Among the elements are the earths and airs, and among the principles is phlogiston. One set of earths, for example, carbon and sulphur, are, in their normal state, especially rich in phlogiston and leave an acid residue when deprived of it. Another set, the calxes or ores, are normally poor in phlogiston, and they become lustrous, ductile, and good heat conductors—thus metallic—when impregnated with it. Transfer of phlogiston to air occurs during combustion and such related processes as respiration and calcination. Air of which the phlogistic content has been thus increased (phlogisticated air) has reduced elasticity and reduced ability to support life. Air from which part of the normal phlogistic component has been removed (dephlogisticated air) supports life especially energetically.

The manual continues from here, but this excerpt will serve for the whole.

My constructed epitome consists of sentences from phlogistic chemistry. Most of the words in those sentences appear in both eighteenth-century and twentieth-century chemical texts, and they function in the same way in both. A few other terms in such texts, most notably 'phlogistication', 'dephlogistication', and their relatives, can be replaced by phrases in which only the term 'phlogiston' is foreign to modern chemistry. But after all such replacements are completed, a small group of terms remains for which the modern chemical vocabulary offers no equivalent. Some have vanished from the language of chemistry entirely, 'phlogiston' being the presently most obvious example. Others, like the term 'principle', have lost all purely chemical significance. (The imperative "purify your reactants" is a chemical principle in a sense

---

14. Kitcher, of course, does explain these juxtapositions by referring to the beliefs of the author of the text and to modern theory. But the passages in which he does this are glosses, not parts of his translation at all.

very different from that in which phlogiston was one.) Still other terms, 'element' for example, remain central to the chemical vocabulary, and they inherit some functions from their older homonyms. But terms like 'principle', previously learned with them, have disappeared from modern texts, and with them has gone the previously constitutive generalization that qualities like color and elasticity provide direct evidence concerning chemical composition. As a result, the referents of these surviving terms, as well as the criteria for identifying them, are now drastically and systematically altered. In both respects, the term 'element' in eighteenth-century chemistry functioned as much like the modern phrase 'state of aggregation' as like the modern term 'element'.

Whether or not these terms from eighteenth-century chemistry refer—terms like 'phlogiston', 'principle', and 'element'—they are not eliminable from any text that purports to be a translation of a phlogistic original. At the very least they must serve as placeholders for the interrelated sets of properties which permit the identification of the putative referents of these interrelated terms. To be coherent, a text that deploys the phlogiston theory must represent the stuff given off in combustion as a chemical principle, the same one that renders the air unfit to breathe and that also, when abstracted from an appropriate material, leaves an acid residue. But if these terms are not eliminable, they seem also not to be replaceable individually by some set of modern words or phrases. And if that is the case—a point to be considered at once—then the constructed passage in which those terms appeared above cannot be a translation, at least not in the sense of that term standard in recent philosophy.

The Historian as Interpreter and Language Teacher

Can it, however, be correct to assert that eighteenth-century chemical terms like 'phlogiston' are untranslatable? I have, after all, already described in modern language a number of ways in which the older term 'phlogiston' refers. Phlogiston is, for example, given off in combustion; it reduces the elasticity and life-supporting properties of air; and so on. It appears that modern-language phrases like these might be compounded to produce a modern-language translation of 'phlogiston'. But they cannot. Among the phrases which describe how the referents of the term 'phlogiston' are picked out are a number that include other untranslatable terms like 'principle' and 'element'. Together with 'phlogiston', they constitute an interrelated or interdefined set that must be acquired together, as a whole, before any of them can be used, applied to natural phenomena.[15] Only after they have been thus acquired can one recognize eighteenth-century chemistry for what it was, a discipline that differed from its twentieth-century successor not simply in what it had to say about individual substances and processes, but in the way it structured and parceled out a large part of the chemical world.

A more restricted example will clarify my point. In learning Newtonian mechanics, the terms 'mass' and 'force' must be acquired together, and Newton's second law must play a role in their acquisition. One cannot, that is, learn 'mass' and 'force' independently and then empirically discover that force equals mass times acceleration. Nor can one first learn 'mass' (or 'force') and then use it to define 'force' (or 'mass') with the aid of the second law. Instead, all three must be learned together, parts of a whole new (but not a wholly new) way of doing mechanics. That point is unfortunately obscured by standard formalizations. In formalizing mechanics one may select either 'mass' or 'force' as primitive and then introduce the other as a defined term. But that formalization supplies no information about how either the primitive or the defined terms attach to nature, how the forces and masses are picked out in actual physical situations. Though 'force', say, may be a primitive in some particular formalization of mechanics, one cannot learn to recognize forces without simultaneously learning to pick out masses and without recourse to the second law. That is why Newtonian 'force' and 'mass' are not translatable into the language of a physical theory (Aristotelian or Einsteinian, for example) in which Newton's version of the second law does not apply. To learn any one of these three ways of doing mechanics, the interrelated terms in some local part of the web of language must be learned or relearned together and then laid down on nature whole. They cannot simply be rendered individually by translation.

How, then, can a historian who teaches or writes about the phlogiston theory communicate his results at all? What is it that occurs when the historian presents to readers a group of sentences like those about phlogiston in the epitome above? The answer to that question varies with the

---

15. Perhaps only 'element' and 'principle' have to be learned together. Once they have been learned, but only then, 'phlogiston' could be introduced as a principle that behaved in certain specified ways.

audience, and I begin with the one presently most relevant. It consists of people without any sort of previous exposure to the phlogiston theory. To them the historian is describing the world in which the phlogistic chemist of the eighteenth century believed. Simultaneously, he or she is teaching the language which eighteenth-century chemists used in describing, explaining, and exploring that world. Most of the words in that older language are identical both in form and function with words in the language of the historian and the historian's audience. But others are new and must be learned or relearned. These are the untranslatable terms for which the historian or some predecessor has had to discover or invent meanings in order to render intelligible the texts on which he works. Interpretation is the process by which the use of those terms is discovered, and it has been much discussed recently under the rubric hermeneutics.[16] Once it has been completed and the words acquired, the historian uses them in his own work and teaches them to others. The question of translation simply does not arise.

All this applies, I suggest, when passages like the one emphasized above are presented to an audience that knows nothing of the phlogiston theory. For that audience these passages are glosses on phlogistic texts, intended to teach them the language in which such texts are written and the way they are to be read. But such texts are also encountered by people who have already learned to read them, people for whom they are simply one more example of an already familiar type. These are the people to whom such texts will seem merely translations, or perhaps merely texts, for they have forgotten that they had to learn a special language before they could read them. The mistake is an easy one. The language they learned largely overlapped the native language they had learned before. But it differed from their native language in part by enrichment—e.g., the introduction of terms like 'phlogiston'—and in part by the introduction of systematically transformed uses of terms like 'principle' and 'element'. In their unrevised native language, these texts could not have been rendered.

16. To the sense of 'hermeneutic' I have in mind (there are others) the most useful introduction is C. Taylor, "Interpretation and the Sciences of Man," *Review of Metaphysics* 25 (1971): 3–51; reprinted in *Understanding and Social Inquiry*, ed. F. A. Dallmayr and T. A. McCarthy (Notre Dame, IN: University of Notre Dame Press, 1977), pp. 101–31. Taylor, however, takes for granted that the descriptive language of the natural sciences (and the behavioral language of the social sciences) is fixed and neutral. A useful corrective from within the hermeneutic tradition is provided by Karl-Otto Apel in "The A Priori of Communication and the Foundation of the Humanities," *Man and World* 5 (1972): 3–37, reprinted in Dallmayr and McCarthy, *Understanding and Social Inquiry*, pp. 292–315.

Though the point requires far more discussion than can be attempted here, much of what I have been saying is neatly captured by the form of Ramsey sentences. The existentially quantified variables with which such sentences begin can be seen as what I previously called "placeholders" for terms requiring interpretation, e.g., 'phlogiston', 'principle', and 'element'. Together with its logical consequences, the Ramsey sentence itself is then a compendium of the clues available to an interpreter, clues that, in practice, he or she would have to discover through extended exploration of texts. That, I think, is the proper way to understand the plausibility of the technique introduced by David Lewis for defining theoretical terms through Ramsey sentences.[17] Like contextual definitions, which they closely resemble, and like ostensive definitions as well, Lewis's Ramsey definitions schematize an important (perhaps essential) mode of language learning. But the sense of 'definition' involved is in all three cases metaphorical, or at least extended. None of these three sorts of "definitions" will support substitution: Ramsey sentences cannot be used for translation.

With this last point, of course, Lewis disagrees. This is not the place to respond to the details of his case, many of them technical, but two lines of criticism may at least be indicated. Lewis's Ramsey definitions determine reference only on the assumption that the corresponding Ramsey sentence is uniquely realizable. It is questionable whether that assumption ever holds and unlikely that it holds regularly. When and if it does, furthermore, the definitions it makes possible are uninformative. If there is one and only one referential realization of a given Ramsey sentence, a person may of course hope simply by trial and error to hit upon it. But having hit upon the referent of a Ramsey-defined term at one point in a text would be of no help in finding the referent of that term on its next occurrence. The force of Lewis's argument depends therefore on his further claim that Ramsey definitions determine not only reference but also sense, and this part of his case encounters difficulties closely related to but even more severe than the ones just outlined.

Even if Ramsey definitions escaped these difficulties, another major set would remain. I have previously pointed out that the laws of a scientific theory, unlike the axioms of a mathematical system, are only law sketches in that their symbolic formalizations depend upon the problem to which they are applied.[18] That point has since been considerably extended by Joseph Sneed and Wolfgang Stegmüller, who consider Ramsey sentences and show that their standard sentential formulation varies from one range of applications to the next.[19] Most occurrences of new or problematic terms in a science text are, however, within applications, and the corresponding Ramsey sentences are simply not a rich enough source of clues to block a multitude of trivial interpretations. To permit reasonable interpretation of a text studded with Ramsey definitions, readers would have first to collect a variety of different ranges of application. And having done so they would still have to do what the historian/interpreter attempts in the same situation. They would, that is, have to invent and test hypotheses about the sense of the terms introduced by Ramsey definitions.

The Quinean Translation Manual

Most of the difficulties I have been considering derive more or less directly from a tradition which holds that translation can be construed in purely referential terms. I have been insisting that it cannot, and my arguments have at least implied that something from the realm of meanings, intensionalities, concepts must be invoked as well. To make those points I have considered an example from history of science, an example of the sort that brought me to the problem of incommensurability and thence to translation in the first place. The same sorts of points can, however, be made directly from recent discussions of referential semantics and related discussions of translation. Here I shall consider the single example to which I alluded at the start: Quine's conception of a translation manual. Such a manual—the end product of the efforts of a radical translator—consists of parallel lists of words and phrases, one in the translator's own language, the other in the language of the tribe he is investigating. Each item on each list is linked to one or often to several items on the other, each link specifying a word or phrase in

18. *Structure*, 2d ed., pp. 188 f.
19. J. D. Sneed, *The Logical Structure of Mathematical Physics* (Dordrecht, Boston: D. Reidel), 1971; W. Stegmüller, *Probleme und Resultate der Wissenschaftstheorie und analytischen Philosophie*, vol. 2, *Theorie und Erfahrung*, part 2, *Theorienstrukturen und Theoriendynamik* (Berlin: Springer-Verlag, 1973); reprinted as *The Structure and Dynamics of Theories*, trans. W. Wohlhueter (New York: Springer-Verlag, 1976).

one language that can, the translator supposes, be substituted in appropriate contexts for the linked word or phrase in the other. Where the linkages are one-many, the manual includes specifications of the contexts in which each of the various links is to be preferred.[20]

The network of difficulties I want to isolate concerns the last of these components of the manual, the context specifiers. Consider the French word *'pompe'*. In some contexts (typically those involving ceremonies), its English equivalent is 'pomp'; in other contexts (typically hydraulic), its equivalent is 'pump'. Both equivalents are precise. *'Pompe'* thus provides a typical example of ambiguity like the standard English example, 'bank': sometimes a riverside and sometimes a financial institution.

Now contrast the case of *'pompe'* with that of French words like *'esprit'* or *'doux'/'douce'*. *'Esprit'* can be replaced, depending on context, by such English terms as 'spirit', 'aptitude', 'mind', 'intelligence', 'judgment', 'wit', or 'attitude'. The latter, an adjective, can be applied, inter alia, to honey ('sweet'), to wool ('soft'), to underseasoned soup ('bland'), to a memory ('tender'), or to a slope or a wind ('gentle'). These are not cases of ambiguity, but of conceptual disparity between French and English. *Esprit* and *doux/douce* are unitary concepts for French speakers, and English speakers as a group possess no equivalents. As a result, though the various translations offered above preserve truth value in appropriate contexts, none of them is in any context intensionally precise. *'Esprit'* and *'doux'/'douce'* are thus examples of terms that can be translated only in part and by compromise. The translator's choice of a particular English word or phrase for one of them is ipso facto the choice of some aspects of the intension of the French term at the expense of others. Simultaneously it introduces intensional associations characteristic of English but foreign to the work being translated.[21] Quine's analysis of translation suffers badly, I think, from its inability to distinguish cases of this sort from straightforward ambiguity, from the case of terms like *'pompe'*.

20. Quine, *Word and Object*, pp. 27, 68–82.

21. Glosses which describe how the French view the psychic (or the sensory) world can be of great help with this problem, and French language textbooks usually include material on such cultural matters. But glosses describing the culture are not parts of the translation itself. Long English paraphrases for French terms provide no substitute, partly because of their clumsiness but mostly because terms like *esprit* or *doux/douce* are items in a vocabulary certain parts of which must be learned together. The argument is the same as the one given previously for 'element' and 'principle' or 'force' and 'mass'.

The difficulty is identical with the one encountered by Kitcher's translation of 'phlogiston'. By now its source must be obvious: a theory of translation based on an extensional semantics and therefore restricted to truth-value preservation or some equivalent as a criterion of adequacy. Like 'phlogiston', 'element', and so on, both *'doux'*/*'douce'* and *'esprit'* belong to clusters of interrelated terms, a number of which must be learned together and which, when learned, give a structure to some portion of the world of experience different from the one familiar to contemporary English speakers. Such words illustrate incommensurability between natural languages. In the case of *'doux'*/*'douce'* the cluster includes, for example, *'mou'*/*'molle'*, a word closer than *'doux'*/*'douce'* to English 'soft', but which applies also to warm damp weather. Or, in the cluster with *'esprit'*, consider *'disposition'*. The latter overlaps *'esprit'* in the area of attitudes and aptitudes, but also applies to state of health or to the arrangement of words within a phrase. These intensionalities are what a perfect translation would preserve, and that is why there can be no perfect translations. But approximating the unobtainable ideal remains a constraint on actual translations, and, if the constraint were taken into account, arguments for the indeterminacy of translation would require a form very different from that now current.

By treating the one-many linkages in his translation manuals as cases of ambiguity, Quine discards the intensional constraints on adequate translation. Simultaneously, he discards the primary clue to the discovery of how the words and phrases in other languages refer. Though one-many linkages are sometimes caused by ambiguity, they far more often provide evidence of which objects and situations are similar and which are different for speakers of the other language; they show, that is, how the other language structures the world. Their function is thus very much the same as that played by multiple observations in learning a first language. Just as the child learning 'dog' must be shown many different dogs and probably some cats as well, so the English speaker learning *'doux'*/*'douce'* must observe it in many contexts and also take note of contexts where French employs *'mou'*/*'molle'* instead. These are the ways, or some of them, by which one learns the techniques for attaching words and phrases to nature, first those of one's own language and then, perhaps, the different ones embedded in other languages. By giving them up, Quine eliminates the very possibility of interpretation, and interpretation is, as I argued at the start, what his radical translator must do before translation can begin. Is it then a wonder that Quine discovers previously unanticipated difficulties about "translation"?

The Invariants of Translation

I turn finally to a problem that has been held at arm's length since the beginning of this paper: What is it that translation must preserve? Not merely reference, I have argued, for reference-preserving translations may be incoherent, impossible to understand while the terms they employ are taken in their usual sense. That description of the difficulty suggests an obvious solution: translations must preserve not only reference but also sense or intension. Under the rubric 'meaning invariance', that is the position I have taken in the past and that I adopted *faute de mieux* in the introduction to this paper. It is by no means merely wrong, but it is not quite right either, an equivocation symptomatic, I believe, of a deep duality in the concept of meaning. In another context it will be essential to confront that duality directly. Here I shall skirt it by avoiding talk of 'meaning' entirely. Instead I shall discuss, though as yet in quite general, quasi-metaphorical terms, how members of a language community pick out the referents of the terms they employ.

Consider the following thought experiment which some of you will have encountered previously as a joke. A mother first tells her daughter the story of Adam and Eve, then shows the child a picture of the pair in the Garden of Eden. The child looks, frowns in puzzlement, and says, "Mother, tell me which is which. I would know if they had their clothes on." Even in so condensed a format, this story underscores two obvious characteristics of language. In matching terms with their referents, one may legitimately make use of anything one knows or believes about those referents. Two people may, moreover, speak the same language and nevertheless use different criteria in picking out the referents of its terms. An observer aware of their differences would simply conclude that the two differed in what they knew about the objects under discussion. That different people use different criteria in identifying the referents of shared terms may, I think, safely be taken for granted. I shall posit, in addition, the now widely shared thesis that none of the criteria used in reference determination are merely conventional, associated simply by definition with the terms they help to characterize.[22]

How can it be, though, that people whose criteria are different so

---

22. Two points must be underscored. First, I am not equating meaning with a set of criteria. Second, 'criteria' is to be understood in a very broad sense, one that embraces whatever techniques, not all of them necessarily conscious, people do use in pinning words to the world. In particular, as used here, 'criteria' can certainly include similarity to paradigmatic examples (but

regularly pick out the same referents for their terms? A first answer is straightforward. Their language is adapted to the social and natural world in which they live, and that world does not present the sorts of objects and situations which would, by exploiting their criterial differences, lead them to make different identifications. That answer, in turn, raises a further and more difficult question: What is it that determines the adequacy of the sets of criteria a speaker employs when applying language to the world which that language describes? What must speakers with disparate reference-determining criteria share in order that they be speakers of the same language, members of the same language community?[23]

Members of the same language community are members of a common culture, and each may therefore expect to be presented with the same range of objects and situations. If they are to co-refer, each must associate each individual term with a set of criteria sufficient to distinguish its referents from other sorts of objects or situations which the community's world actually presents, though not from still other objects that are merely imaginable. The ability to identify correctly the members of one set often therefore requires a knowledge of contrast sets as well. Some years ago, for example, I suggested that learning to identify geese may also require knowing such creatures as ducks and swans.[24] The cluster of criteria adequate to the identification of geese depends, I indicated, not only on the characteristics shared by actual geese, but also on the characteristics of certain other creatures in the world inhabited by geese and those who talk about them. Few referring terms or expressions are learned in isolation either from the world or from each other.

This very partial model of the way speakers match language with the world is intended to reintroduce two closely related themes that have emerged repeatedly in this paper. The first, of course, is the essential role of sets of terms that must be learned together by those raised

---

then the relevant similarity relation must be known) or recourse to experts (but then speakers must know how to find the relevant experts).

23. I have found no brief way to discuss this topic without seeming to imply that criteria are somehow logically and psychologically prior to the objects and situations for which they are criterial. But, in fact, I think both must be learned and that they are often learned together. For example, the presence of masses and forces is criterial for what I might call the 'Newtonian-mechanical-situation', one to which Newton's second law applies. But one can learn to recognize mass and force only within the Newtonian-mechanical-situation, and vice versa.

24. T. S. Kuhn, "Second Thoughts on Paradigms," in *The Structure of Scientific Theories*, ed. F. Suppe (Urbana: University of Illinois Press, 1974), pp. 459–82; reprinted in *The Essential Tension*, pp. 293–319.

inside a culture, scientific or other, and which foreigners encountering that culture must consider together during interpretation. That is the holistic element which entered this paper at the start, with local incommensurability, and the basis for it should now be clear. If different speakers using different criteria succeed in picking out the same referents for the same terms, contrast sets must have played a role in determining the criteria each associates with individual terms. At least they must when, as is usual, those criteria do not themselves constitute necessary and sufficient conditions for reference. Under these circumstances, some sort of local holism must be an essential feature of language.

These remarks may also provide a basis for my second recurrent theme, the reiterated assertion that different languages impose different structures on the world. Imagine, for a moment, that for each individual a referring term is a node in a lexical network from which radiate labels for the criteria that he or she uses in identifying the referents of the nodal term. Those criteria will tie some terms together and distance them from others, thus building a multidimensional structure within the lexicon. That structure mirrors aspects of the structure of the world which the lexicon can be used to describe, and it simultaneously limits the phenomena that can be described with the lexicon's aid. If anomalous phenomena nevertheless arise, their description (perhaps even their recognition) will require altering some part of the language, changing the previously constitutive linkages between terms.

Note, now, that homologous structures, structures mirroring the same world, may be fashioned using different sets of criterial linkages. What such homologous structures preserve, bare of criterial labels, is the taxonomic categories of the world and the similarity/difference relationships between them. Though I here verge on metaphor, my direction should be clear. What members of a language community share is homology of lexical structure. Their criteria need not be the same, for those they can learn from each other as needed. But their taxonomic structures must match, for where structure is different, the world is different, language is private, and communication ceases until one party acquires the language of the other.

By now it must be clear where, in my view, the invariants of translation are to be sought. Unlike two members of the same language community, speakers of mutually translatable languages need not share terms: 'Rad' is not 'wheel'. But the referring expressions of one language must be matchable to coreferential expressions in the other, and the lexical structures employed by speakers of the languages must be the same, not only within each language but also from one language to the other. Taxonomy must, in short, be preserved to provide both shared categories and shared relationships between them. Where it is not, translation is impossible, an outcome precisely illustrated by Kitcher's valiant attempt to fit the phlogiston theory to the taxonomy of modern chemistry.

Translation is, of course, only the first resort of those who seek comprehension. Communication can be established in its absence. But where translation is not feasible, the very different processes of interpretation and language acquisition are required. These processes are not arcane. Historians, anthropologists, and perhaps small children engage in them every day. But they are not well understood, and their comprehension is likely to require the attention of a wider philosophical circle than the one currently engaged with them. Upon that expansion of attention depends an understanding, not only of translation and its limitations, but also of conceptual change. It is no accident that the synchronic analysis of Quine's *Word and Object* is introduced by the diachronic epigraph of Neurath's boat.

POSTSCRIPT: RESPONSE TO COMMENTARIES

I am grateful to my commentators for their patience with my delays, for the thoughtfulness of their criticism, and for the proposal that I supply a written reply. With much that they have to say I fully agree, but not with all. Part of our residual disagreement rests on misunderstanding, and with that part I begin.

Kitcher suggests I believe his "procedure of interpretation," his "interpretive strategy," breaks down when confronted with incommensurable parts of an older scientific vocabulary.[25] I take it that by "interpretive strategy" he means his procedure for identifying in modern language the referents of older terms. But I do not mean to have implied that that strategy need ever break down. On the contrary, I have suggested that it is an essential tool of the historian/interpreter. If it anywhere necessarily breaks down, which I doubt, then in that place interpretation is impossible.

25. P. Kitcher, "Implications of Incommensurability," *PSA 1982: Proceedings of the 1982 Biennial Meeting of the Philosophy of Science Association*, vol. 2, ed. P. D. Asquith and T. Nickles (East Lansing, MI: Philosophy of Science Association, 1983), pp. 692–93.

Kitcher may read the preceding sentence as a tautology, for he appears to regard his reference-determining procedure as itself interpretation, rather than merely a prerequisite to it. Mary Hesse sees what is missing when she says that for interpretation, "We have not only to *say* that phlogiston sometimes referred to hydrogen and sometimes to absorption of oxygen, but we have to convey the whole ontology of phlogiston in order to make plausible why it was taken to be a single natural kind."[26] The processes to which she refers are independent, and the older literature of the history of science provides countless examples of the ease with which one may complete the first without taking even a step toward the second. The result is an essential ingredient of Whig history.

So far I have been dealing only with misunderstanding. As I now continue, a more substantive sort of disagreement may begin to emerge. (In this area no clear line separates misunderstanding from substantive disagreement.) Kitcher supposes that interpretation makes possible "full communication across the revolutionary divide" and that the process by which it does so is "extending the resources of the home language," for example, by the addition of terms like 'phlogiston' and its relatives (p. 691). About at least the second of these points Kitcher is, I think, seriously mistaken. Though languages are enrichable, they can only be enriched in certain directions. The language of twentieth-century chemistry has, for example, been enriched by adding the names of new elements like berkelium and nobelium. But there is no coherent or interpretable way to add the name of a quality-bearing principle without altering what it is to be an element and a good deal else besides. Such alterations are not simply enrichments; they change rather than add to what was there before; and the language that results from them can no longer directly render all laws of modern chemistry. In particular, those laws involving the term 'element' escape it.

Is "full communication" nevertheless possible between an eighteenth- and twentieth-century chemist, as Kitcher supposes? Perhaps, yes, but only if one of the two learns the other's language, becoming, in that sense, a participant in the other's practice of chemistry. That transformation can be achieved, but the people who then communicate are only in a Pickwickian sense chemists of different centuries. Such communica-

26. M. Hesse, "Comment on Kuhn's 'Commensurability, Comparability, Communicability'," *PSA 1982: Proceedings of the 1982 Biennial Meeting of the Philosophy of Science Association*, vol. 2, ed. P. D. Asquith and T. Nickles (East Lansing, MI: Philosophy of Science Association, 1983), pp. 707–11; italics in original.

tion does permit significant (though not complete) comparison of the effectiveness of the two modes of practice, but that was never for me in question. What was and is at issue is not significant comparability but rather the shaping of cognition by language, a point by no means epistemologically innocuous. My claim has been that key statements of an older science, including some that would ordinarily be considered merely descriptive, cannot be rendered in the language of a later science and vice versa. By the language of a science, I here mean not only the parts of that language in actual use, but also all extensions that can be incorporated in that language without altering components already in place.

What I have in mind may be clarified if I sketch a response to Mary Hesse's call for a new theory of meaning. I share her conviction that traditional meaning theory is bankrupt and that some sort of replacement, not purely extensional, is needed. I suspect also that Hesse and I are close in our guesses about what that replacement will look like. But she is somehow missing the shape of my guess both when she supposes that my brief remarks about homologous taxonomies are not directed toward a theory of meaning and also when she describes my discussion of '*doux*'/'*douce*' and of '*esprit*' as concerned with a kind of "meaning-trope" rather than directly and literally with meaning (p. 709).

Reverting to my earlier metaphor, which is all that present space permits, let me take '*doux*' to be a node in a multidimensional lexical network where its position is specified by its distance from such other nodes as '*mou*', '*sucré*', etc. To know what '*doux*' means is to possess the relevant network together with *some* set of techniques sufficient to attach to the '*doux*' node the same experiences, objects, or situations as are associated with it by other French speakers. So long as it links the right referents to the right nodes, the particular set of techniques employed makes no difference; the meaning of '*doux*' consists simply of its structural relation to other terms of the network. Since '*doux*' is itself reciprocally implicated in the meanings of these other terms, none of them, taken by itself, has an independently specifiable meaning.

Some of the inter-term relations constitutive of meaning, e.g., '*doux*'/'*mou*', are metaphor-like, but they are not metaphor. On the contrary, what has so far been in question is the establishment of literal meanings without which there could be neither metaphor nor other tropes. Tropes function by suggesting alternate lexical structures constructible with the same nodes, and their very possibility depends upon

the existence of a primary network with which the suggested alternate is contrasted or in tension. Though there are tropes, or something
very like them, in science, they have been no part of the subject of my
paper.

Note now that the English term 'sweet' is also a node in a lexical
network where its position is specified by its distance from such other
terms as 'soft' and 'sugary'. But those relative distances are not the same
as those in the network for French, and the English nodes attach to
only some of the same situations and properties as the most nearly corresponding nodes in the network for French. That lack of structural
homology is what makes these portions of the French and English vocabularies incommensurable. Any attempt to remove the incommensurability, say, by inserting a node for 'sweet' in the French network,
would change preexisting distance relations and thus alter, rather than
simply extend, the preexisting structure. I am uncertain about the sympathy with which Hesse will receive these as yet undeveloped aperçus;
but they should at least indicate the extent to which my talk of taxonomies is directed by concern for a theory of meaning.

I turn finally to a problem raised, though in different ways, by both
my commentators. Hesse suggests that my condition that taxonomy be
shared is probably too strong and that "*approximate* sharing" or "*significant* intersection" of taxonomies will probably do "in the particular
situations in which speakers of different languages find themselves"
(p. 708, italics in original). Kitcher thinks incommensurability is too
common to be a criterion of revolutionary change and suspects that I
am, in any case, no longer concerned to distinguish sharply between
normal and revolutionary development in science (p. 697). I see the
force of these positions, for my own view of revolutionary change has
increasingly moderated, as Kitcher supposes. Nevertheless, I think he
and Hesse push the case for continuity of change too far. Let me sketch
a position I mean elsewhere to develop and defend.

The concept of a scientific revolution originated in the discovery that
to understand any part of the science of the past the historian must first
learn the language in which that past was written. Attempts at translation into a later language are bound to fail, and the language-learning
process is therefore interpretive and hermeneutic. Since success in interpretation is generally achieved in large chunks ("breaking into the hermeneutic circle"), the historian's discovery of the past repeatedly involves the sudden recognition of new patterns or gestalts. It follows
that the historian, at least, does experience revolutions. Those theses

were at the heart of my original position, and on them I would still insist.

Whether scientists, moving through time in a direction opposite to the historian's, also experience revolutions is left open by what I have so far said. If they do, their shifts in gestalt will ordinarily be smaller than the historian's, for what the latter experiences as a single revolutionary change will usually have been spread over a number of such changes during the development of the sciences. It is not clear, furthermore, that even those small changes need have had the character of revolutions. Might not the holistic language changes that the historian experiences as revolutionary have taken place originally by a process of gradual linguistic drift?

In principle they might, and in some realms of discourse—political life, for example—they presumably do, but not, I think, ordinarily in the developed sciences. There, holistic changes tend to happen all at once, as in the gestalt switches to which I have likened revolutions before. Part of the evidence for that position remains empirical: reports of "aha" experiences, cases of mutual incomprehension, and so on. But there is also a theoretical argument which may increase understanding of what I take to be involved.

So long as the members of a speech community agree on a number of standard examples (paradigms), the utility of terms like 'democracy', 'justice', or 'equity' is not much threatened by the occurrence also of cases in which community members differ about the applicability of these terms. Words of this sort need not function unequivocally; fuzziness at the borders is expected, and it is the acceptance of fuzziness that permits drift, the gradual warping of the meanings of a set of interrelated terms over time. In the sciences, on the other hand, persistent disagreement as to whether substance x is an element or a compound, whether celestial body y is a planet or a comet, or whether particle z is a proton or a neutron would quickly cast doubt on the integrity of the corresponding concepts. In the sciences borderline cases of this sort are sources of crisis, and drift is correspondingly inhibited. Instead, pressures build up until a new viewpoint, including new uses for parts of language, is introduced. If I were now rewriting *The Structure of Scientific Revolutions*, I would emphasize language change more and the normal/revolutionary distinction less. But I would still discuss the special difficulties the sciences experience with holistic language change, and I would attempt to explain that difficulty as resulting from the sciences' need for special precision in reference determination.

Chapter Three

Possible Worlds in History of Science

"Possible Worlds in History of Science" is the published version of a paper presented at the 65th Nobel Symposium in 1986, for which Arthur I. Miller and Tore Frängsmyr served as commentators; Kuhn's reply to their comments is included here as a postscript to the essay. The proceedings of the symposium were published as Possible Worlds in Humanities, Arts and Sciences, *edited by Sture Allén (Berlin: Walter de Gruyter, 1989).*

———◦——

THE INVITATION TO OPEN this symposium's discussion of possible worlds in history of science has been particularly welcome, for several issues raised by the topic are central to my current research. Their centrality is, however, also a source of problems. In the book on which I am at work, these issues arise only after much prior discussion has led to conclusions I must here present as premises. Limited illustration and evidence for those premises will follow, but only in the later portions of this paper, where they are put to work.

What I am presupposing will be suggested by the following claim: to understand some body of past scientific belief, the historian must acquire a lexicon that here and there differs systematically from the one current in his own day. Only by using that older lexicon can he or she accurately render certain of the statements that are basic to the science

This paper has been considerably revised since the Symposium. For much relevant criticism and advice during that process, I am grateful to Barbara Partee and to my M.I.T. colleagues Ned Block, Sylvain Bromberger, Dick Cartwright, Jim Higginbotham, Judy Thomson, and Paul Horwich.

under scrutiny. Those statements are not accessible by means of a translation that uses the current lexicon, not even if the list of words it contains is expanded by the addition of selected terms from its predecessor.

That claim is elaborated in the first of the four sections of this paper, and its relevance to an ongoing debate in possible-world semantics is briefly suggested in the second. The third section, an extended analysis of some interrelated terms from Newtonian mechanics, illustrates the entanglements of the lexicon with the substantive claims of a scientific theory, entanglements which may make it impossible to change the theory without changing the lexicon as well. Finally, the closing section of the paper examines the way such entanglements restrict the applicability to scientific development of the conception of possible worlds.

———◦———

A historian reading an out-of-date scientific text characteristically encounters passages that make no sense. That is an experience I have had repeatedly whether my subject was an Aristotle, a Newton, a Volta, a Bohr, or a Planck.[1] It has been standard to ignore such passages or to dismiss them as the products of error, ignorance, or superstition, and that response is occasionally appropriate. More often, however, sympathetic contemplation of the troublesome passages suggests a different diagnosis. The apparent textual anomalies are artifacts, products of misreading.

For lack of an alternative, the historian has been understanding words and phrases in the text as he or she would if they had occurred in contemporary discourse. Through much of the text that way of reading proceeds without difficulty; most terms in the historian's vocabulary are still used as they were by the author of the text. But some sets of interrelated terms are not, and it is failure to isolate those terms and to discover how they were used that has permitted the passages in question to seem anomalous. Apparent anomaly is thus ordinarily evidence of the need

1. For Newton, see my "Newton's '31st Query' and the Degradation of Gold, *Isis* 42 (1951): 296–98. For Bohr, see J. L. Heilbron and T. S. Kuhn, "The Genesis of the Bohr Atom," *Historical Studies in the Physical Sciences* 1 (1969): 211–90, where the nonsense passages that gave rise to the project are quoted on p. 271. For an introduction to the other examples mentioned, see my "What Are Scientific Revolutions?" Occasional Paper 18, Center for Cognitive Science (Cambridge, MA: Massachusetts Institute of Technology, 1981); reprinted in *The Probabilistic Revolution*, vol. 1, *Ideas in History*, ed. L. Krüger, L. J. Daston, and M. Heidelberger (Cambridge, MA: MIT Press, 1987), pp. 7–22; also reprinted in this volume as essay 1.

for local adjustment of the lexicon, and it often provides clues to the nature of that adjustment as well.[2] An important clue to problems in reading Aristotle's physics is provided by the discovery that the term translated 'motion' in his text refers not simply to change of position but to all changes characterized by two end points. Similar difficulties in reading Planck's early papers begin to dissolve with the discovery that, for Planck before 1907, 'the energy element $h\nu$' referred, not to a physically indivisible atom of energy (later to be called 'the energy quantum') but to a mental subdivision of the energy continuum, any point on which could be physically occupied.

These examples all turn out to involve more than mere changes in the use of terms, thus illustrating what I had in mind years ago when speaking of the "incommensurability" of successive scientific theories.[3] In its original mathematical use 'incommensurability' meant "no common measure," for example, of the hypotenuse and side of an isosceles right triangle. Applied to a pair of theories in the same historical line, the term meant that there was no common language into which both could be fully translated.[4] Some statements constitutive of the older theory could not be stated in any language adequate to express its successor, and vice versa.

Incommensurability thus equals untranslatability, but what incommensurability bars is not quite the activity of professional translators. Rather, it is a quasi-mechanical activity governed in full by a manual which specifies, as a function of context, which string in one language may, *salva veritate*, be substituted for a given string in the other. Trans-

---

2. Throughout this paper I shall continue to speak of the lexicon, of terms, and of statements. My concern, however, is actually with conceptual or intensional categories more generally, e.g., with those which may be reasonably attributed to animals or to the perceptual system. For the support this extension receives from possible-worlds semantics, see B. H. Partee, "Possible Worlds in Model-Theoretic Semantics: A Linguistic Perspective," in *Possible Worlds in Humanities, Arts and Sciences: Proceedings of Nobel Symposium 65*, ed. Sture Allén, Research in Text Theory, vol. 14 (Berlin: Walter de Gruyter, 1989), pp. 93–123.

3. For a fuller and more nuanced discussion of this point and those that follow, see my "Commensurability, Comparability, Communicability," in *PSA 1982: Proceedings of the 1982 Biennial Meeting of the Philosophy of Science Association*, vol. 2, ed. P. D. Asquith and T. Nickles (East Lansing, MI: Philosophy of Science Association, 1983), pp. 669–88; reprinted in this volume as essay 2.

4. My original discussion described nonlinguistic as well as linguistic forms of incommensurability. That I now take to have been an overextension resulting from my failure to recognize how large a part of the apparently nonlinguistic component was acquired with language during the learning process. The acquisition during language learning of what I once took to be incommensurability with respect to instrumentation is, for example, illustrated by the discussion of the spring balance in the next section of this paper.

lation of that sort is Quinean, and the point at which I aim will be suggested by the remark that most or all of Quine's arguments for the indeterminacy of translation can, with equal force, be directed to an opposite conclusion: instead of there being an infinite number of translations compatible with all normal dispositions to speech behavior, there are often none at all.

With that much, Quine might very nearly agree. His arguments require that a choice be made, but they do not dictate its outcome. In his view, one must either entirely abandon traditional notions of meaning, of intension, or else one must give up the assumption that language is, or could be, universal, that anything expressible in one language, or by using one lexicon, can be expressed also in any other. His own conclusion—that meaning must be abandoned—follows only because he takes universality for granted, and this paper will suggest that there is no sufficient basis for doing so. To possess a lexicon, a structured vocabulary, is to have access to the varied set of worlds which that lexicon can be used to describe. Different lexicons—those of different cultures or different historical periods, for example—give access to different sets of possible worlds, largely but never entirely overlapping. Though a lexicon may be enriched to yield access to worlds previously accessible only with another, the result is peculiar, a point to be elaborated below. In order that the "enriched" lexicon continue to serve some essential functions, the terms added during enrichment must be rigidly segregated and reserved for a special purpose.

What has made the assumption of universal translatability so nearly inescapable is, I believe, its deceptive similarity to a quite different one, in this case an assumption that I share: anything which can be said in one language can, with imagination and effort, be *understood* by a speaker of another. What is prerequisite to such understanding, however, is not translation but language learning. Quine's radical translator is, in fact, a language learner. If he succeeds, which I think no principle bars, he will become bilingual. But that does not ensure that he or anyone else will be able to translate from his newly acquired language to the one with which he was raised. Though learnability could in principle imply translatability, the thesis that it does so needs to be argued. Much philosophical discussion instead takes it for granted. Quine's *Word and Object* provides a notably explicit case in point.[5]

5. W. V. O. Quine, *Word and Object* (Cambridge, MA: Technology Press of the Massachusetts Institute of Technology, 1960), pp. 47, 70 f.

I am suggesting, in short, that the problems of translating a scientific text, whether into a foreign tongue or into a later version of the language in which it was written are far more like those of translating literature than has generally been supposed. In both cases the translator repeatedly encounters sentences that can be rendered in several alternative ways, none of which captures them completely. Difficult decisions must then be made about which aspects of the original it is most important to preserve. Different translators may differ, and the same translator may make different choices in different places even though the terms involved are in neither language ambiguous. Such choices are governed by standards of responsibility, but they are not determined by them. In these matters there is no such thing as being merely right or wrong. The preservation of truth values when translating scientific prose is very nearly as delicate a task as the preservation of resonance and emotional tone in the translation of literature. Neither can be fully achieved; even responsible approximation requires the greatest tact and taste. In the scientific case, these generalizations apply, not only to passages that make explicit use of theory, but also and more significantly to those their authors took to be merely descriptive.

Unlike many people who share my generally structuralist leanings, I am not attempting to erase or even to reduce the gap generally thought to separate literal from figurative use of language. On the contrary, I cannot imagine a theory of figurative use—a theory, for example, of metaphor and other tropes—that did not presuppose a theory of literal meanings. Nor, to turn from theory to practice, can I imagine how words could be employed effectively in tropes like metaphor except within a community whose members had previously assimilated their literal use.[6] My point is simply that the literal and the figurative use of terms are alike in their dependence on preestablished associations between words.

That remark provides entree to a theory of meaning, but only two aspects of that theory are centrally relevant to the arguments which follow, and I must here restrict myself to them. First, knowing what a word means is knowing how to use it for communication with other members of the language community within which it is current. But that ability does not imply that one knows something that attaches to the word by itself, its meaning, say, or its semantic markers. Words do not, with occasional exceptions, have meanings individually, but only

6. See my "Metaphor in Science," in *Metaphor and Thought*, ed. Andrew Ortony (Cambridge: Cambridge University Press, 1979), pp. 409–19; reprinted in this volume as essay 8.

through their associations with other words within a semantic field. If the use of an individual term changes, then the use of the terms associated with it normally changes as well.

The second aspect of my developing view of meaning is both less standard and more consequential. Two people may use a set of interrelated terms in the same way but employ different sets (in principle, totally disjunct sets) of field coordinates in doing so. Examples will be found in the next section of this paper; meanwhile the following metaphor may prove suggestive. The United States can be mapped in many different coordinate systems. Individuals with different maps will specify the location of, say, Chicago by means of a different pair of coordinates. But all will nevertheless locate the same city provided that the maps are scaled to preserve the relative distances between the items mapped. The metric that accompanies each of the various sets of coordinates must, that is, be chosen to preserve the structural geometrical relations within the mapped area.[7]

The premises just sketched have implications for a continuing debate within possible-world semantics, a subject I shall briefly epitomize before relating it to what has already been said. A possible world is often spoken of as a way our world might have been, and that informal description will very nearly serve present purposes.[8] Thus, in our world the earth has only a single natural satellite (the moon), but there are other possible worlds, almost the same as ours except that the earth has two or more satellites or has none at all. (The "almost" allows the adjustment of phenomena like the tides which, the laws of nature remaining the same, would vary with number of satellites.) There are also possible worlds less like ours: some in which there is no earth, others in which there are no planets, and still others in which not even the laws of nature are the same.

What has recently excited a number of philosophers and linguists about the concept of possible worlds is that it offers a route both to a logic of modal statements and to an intensional semantics for logic and

<hr/>

7. Some preliminary indications of what these cryptic remarks intend are supplied in my "Commensurability, Comparability, Communicability."

8. Barbara Partee's essay provides an elegant summary of the objectives and techniques of possible-world semantics as seen by both linguists and philosophers. Readers unfamiliar with the topic are advised to read it first.

for natural languages. Necessarily true statements, for example, are true in all possible worlds; possibly true statements are true in some; and a true counterfactual is a statement true in some worlds but not in that of the person who made it. Given a set of possible worlds over which to quantify, a formal logic of modal statements appears within reach. Quantification over possible worlds can also lead to an intensional semantics, though by a more complex route. Since the meaning or intension of a statement is what picks out the possible worlds in which that statement is true, each statement corresponds to and may be conceived as a function from possible worlds to truth values. Similarly, a property may be conceived as a function from possible worlds to the sets whose members display that property in each world. Other sorts of referring terms may be conceptually reconstructed in related ways.

Even so brief a sketch of possible-world semantics suggests the likely significance of the range of possible worlds over which quantification occurs, and on this issue opinions vary. David Lewis, for example, would quantify over the entire range of worlds that have been or might be conceived; Saul Kripke, at the other extreme, restricts attention to possible worlds that can be stipulated; intermediate positions are available and some have been filled.[9] Partisans of these positions debate a variety of issues, most of which have no present relevance. But the debate's participants all appear to assume, with Quine, that anything can be said in any language. If, as I have premised, that assumption fails, additional considerations become relevant.

Questions about the semantics of modal statements, or about the intension of words and strings constructed from them, are ipso facto questions about statements and words in a specified language. Only the possible worlds stipulatable in that language can be relevant to them. Extending quantification to include worlds accessible only by resort to other languages seems at best functionless, and in some applications it may be a source of error and confusion. One relevant sort of confusion has already been mentioned—that of the historian who tries to present an older science in his or her own language—and the next two sections will explore some others. At least in their application to historical development, the power and utility of possible-world arguments appears to

9. Partee provides a fuller account of these divisions as well as a useful bibliography. A more analytic account is included in R. C. Stalnaker, *Inquiry* (Cambridge, MA: MIT Press, 1984). The debate focuses on the ontological status of possible worlds, i.e., on their reality: differences about the range of quantification appropriate for possible-world theories follow directly.

require their restriction to the worlds accessible with a given lexicon, the worlds that can be stipulated by participants in a given language community or culture.[10]

————◦————

I have so far dealt in general assertions, omitting both illustration and defense. Let me now begin to supply them, acknowledging again that I shall not complete the task in this place. My argument will proceed in two stages. This section examines part of the lexicon of Newtonian mechanics, especially the interrelated terms 'force', 'mass', and 'weight'. It asks, first, what one need and need not know to be a member of the community that uses these terms, and, then, how possession of that knowledge constrains the worlds which members of that community can describe without doing violence to the language. Some of the worlds they cannot describe are, of course, described at a later time, but only after a change of lexicon which bars coherent description of some worlds describable before. That sort of change is the subject of this paper's last section. It focuses upon the so-called causal theory of reference, an application of possible-world concepts said to eliminate the significance of such changes.

The vocabulary in which the phenomena of a field like mechanics are described and explained is a historical product, developed over time, and repeatedly transmitted, in its then current state, from one generation to its successor. In the case of Newtonian mechanics, the required cluster of terms has been stable for some time, and transmission techniques are relatively standard. Examining them will suggest characteristics of what

---

10. Partee emphasizes that possible worlds are not conceivable worlds, points out "that we can conceive of there being possibilities we can't conceive of," and suggests that restricting possible worlds to conceivable worlds may make it impossible to deal with such cases. Talk with her since the symposium has made me realize the need for still another distinction. Not all the worlds accessible or stipulatable with a given lexicon are conceivable: a world containing square circles can be stipulated but not conceived; other examples will recur below. It is only worlds access to which requires restructuring of the lexicon which I mean to exclude when quantifying over possible worlds. Note also that to speak of different lexicons as giving access to different sets of possible worlds is not simply to add one more to the standard kind of accessibility relations discussed at the start of Partee's paper. There is no type of necessity corresponding to lexical accessibility. Excepting statements that stipulate an inconceivable world, no statement framable in a given lexicon is necessarily true or false simply because it can be accessed in that lexicon. More generally, the issue of lexical accessibility seems to arise for all applications of possible-world arguments, thus cutting across the standard set of accessibility relations.

the student acquires in the course of becoming a licensed practitioner of the field.[11]

Before the exposure to Newtonian terminology can usefully begin, other significant portions of the lexicon must be in place. Students must, for example, already have a vocabulary adequate to refer to physical objects and to their locations in space and time. Onto this they must have grafted a mathematical vocabulary rich enough to permit the quantitative description of trajectories and the analyses of velocities and accelerations of bodies moving along them.[12] Also, at least implicitly, they must command a notion of extensive magnitude, a quantity whose value for the whole of a body is the sum of its values for the body's parts. Quantity of matter provides a standard example. These terms can all be acquired without resort to Newtonian theory, and the student must control them before that theory can be learned. The other lexical items required by that theory—most notably 'force', 'mass', and 'weight' in their Newtonian senses—can only be acquired together with the theory itself.

Five aspects of the way in which these Newtonian terms are learned require particular illustration and emphasis. First, as already indicated, learning cannot begin until a considerable antecedent vocabulary is in place. Second, in the process through which the new terms are acquired, definition plays a negligible role. Rather than being defined, these terms are introduced by exposure to examples of their use, examples provided by someone who already belongs to the speech community in which they are current. That exposure often includes actual exhibits, for example, in the student laboratory, of one or more exemplary situations to which the terms in question are applied by someone who already knows how to use them. The exhibits need not be actual, however. The exemplary situations may instead be introduced by a description conducted primarily in terms drawn from the antecedently available vocabulary but in which the terms to be learned also appear here and there. The

11. I discuss the acquisition of a lexicon because it is a source of clues to what the individual's possession of a lexicon entails. Nothing about the end product depends, however, upon the lexicon's being acquired by generation-to-generation transmission. The consequences would be the same if, for example, the lexicon were a genetic endowment or had been implanted by a skilled neurosurgeon. I shall, for example, shortly emphasize that transmitting a lexicon requires repeated recourse to concrete examples. Implanting the same lexicon surgically would, I am suggesting, have involved implanting the memory traces left by exposure to such examples.

12. In practice, the techniques for describing velocities and accelerations along trajectories are usually learned in the same courses that introduce the terms to which I turn next. But the first set can be acquired without the second, whereas the second cannot be acquired without the first.

two processes are for the most part interchangeable, and most students encounter them both, in some mix or other. Both include an indispensable ostensive or stipulative element: terms are taught through the exhibit, direct or by description, of situations to which they apply.[13] The learning that results from such a process is not, however, about words alone but equally about the world in which they function. When I use the phrase 'stipulative descriptions' in what follows, the stipulations I have in mind will be simultaneously and inseparably about both the substance and the vocabulary of science, about both the world and the language.

A third significant aspect of the learning process is that exposure to a single exemplary situation seldom or never supplies enough information to permit the student to use a new term. Several examples of varied sorts are required, often accompanied by examples of apparently similar situations to which the term in question does not apply. The terms to be learned, furthermore, are seldom applied to these situations in isolation but are instead embedded in whole sentences or statements, among which are some usually referred to as laws of nature.

Fourth, among the statements involved in learning one previously unknown term are some that include other new terms as well, terms that must be acquired together with the first. The learning process thus interrelates a set of new terms, giving structure to the lexicon that contains them. Finally, though there is usually considerable overlap between the situations to which individual language learners are exposed (and even more between the accompanying statements), individuals can in principle communicate fully even though they acquired the terms with which they do so along very different routes. To the extent that the process I am describing supplies individuals with anything resembling a definition, it is not a definition that need be shared by other members of the speech community.

For illustrations, consider first the term 'force'. The situations which exemplify a force's presence are of varied sorts. They include, for exam-

---

13. The terms 'ostension' and 'ostensive' seem to have two different uses, which for present purposes need to be distinguished. In one, these terms imply that *nothing but* the exhibit of a word's referent is needed to learn or to define it. In the other, they imply only that *some* exhibit is required during the acquisition process. I shall, of course, be using the second sense of the terms. The propriety of extending them to cases in which description in an antecedent vocabulary replaces an actual exhibit depends on recognizing that description does not supply a string of words equivalent to the statements containing the words to be learned. Rather it enables students to visualize the situation and apply to the visualization the same mental processes (whatever they may be) that would otherwise have been applied to the situation as perceived.

ple, muscular exertion, a stretched string or spring, a body possessed of weight (note the occurrence of another of the terms to be learned), or, finally, certain sorts of motion. The last is particularly important and presents particular difficulties to the student. As Newtonians use 'force', not all motions signify the presence of its referent, and examples which display the distinction between forced and force-free motions are therefore required. Their assimilation, furthermore, demands the suppression of a highly developed pre-Newtonian intuition. For children and Aristotelians the standard example of a forced motion is the hurled projectile. Force-free motion is for them exemplified by the falling stone, the spinning top, or the rotating flywheel. For the Newtonian all of these are cases of forced motion. The only example of a Newtonian force-free motion is motion in a straight line at constant speed, and that can be exhibited directly only in interplanetary space. Teachers nevertheless try. (I still remember the contrived lecture demonstration—a block of ice sliding on a sheet of glass—that helped me undo prior intuitions and acquire the Newtonian concept of 'force'.) But for most students the main path to this key aspect of the use of the term is provided by the string of words known as Newton's first law of motion: 'in the absence of an external force applied to it, a body moves continuously at constant speed in a straight line'. It exhibits, by description, the motions which require no force.[14]

More will need to be said about 'force', but let me first look briefly at its two Newtonian companions, 'weight' and 'mass'. The first refers to a particular sort of force, the one which causes a physical body to press on its supports while at rest or to fall when unsupported. In this still qualitative form the term 'weight' is available prior to Newtonian 'force' and is used during the latter's acquisition. 'Mass' is usually introduced as equivalent to 'quantity of matter', where matter is the substrate underlying physical bodies, the stuff of which quantity is conserved as the qualities of material bodies change. Any feature which, like weight, picks out a physical body is an index also of the presence of matter and of mass. As in the case of 'weight' and unlike the case of 'force', the

---

14. Newton's first law is a logical consequence of his second, and Newton's reason for stating them separately has long been a puzzle. The answer may well lie in pedagogic strategy. If Newton had permitted the second law to subsume the first, his readers would have had to sort out his use of 'force' and of 'mass' together, an intrinsically difficult task further complicated by the fact that the terms had previously been different not only in their individual use but in their interrelation. Separating them to the extent possible displayed the nature of the required changes more clearly.

qualitative features by which one picks out the referents of 'mass' are identical with those of pre-Newtonian usage.

But the Newtonian use of all three terms is quantitative, and the Newtonian form of quantification alters both their individual uses and the interrelationships between them.[15] Only the unit measures may be established by convention; the scales must be chosen so that weight and mass are extensive quantities and so that forces can be added vectorially. (Contrast the case of temperature, in which both unit and scale can be chosen by convention.) Once again, the learning process requires the juxtaposition of statements involving the terms to be learned with situations drawn directly or indirectly from nature.

Begin with the quantification of 'force'. Students acquire the full quantitative concept by learning to measure forces with a spring balance or some other elastic device. Such instruments had appeared nowhere in scientific theory or practice before Newton's time, when they took over the conceptual role previously played by the pan balance. But they have since been central, for reasons that are conceptual rather than pragmatic. The use of a spring balance to exhibit the proper measure of force requires, however, recourse to two statements ordinarily described as laws of nature. One of these is Newton's third law, which states, for example, that the force exerted by a weight on a spring is equal and opposite to the force exerted by the spring on the weight. The other is Hooke's law, which states that the force exerted by a stretched spring is proportional to the spring's displacement. Like Newton's first law, these are first encountered during language learning where they are juxtaposed with examples of situations to which they apply. Such juxtapositions play a double role, simultaneously stipulating how the word 'force' is to be used and how the world populated by forces behaves.

Turn now to the quantification of the terms 'mass' and 'weight'. It illustrates with special clarity a key aspect of the lexical acquisition process, one that has not yet been considered. To this point, my discussion of Newtonian terminology has probably suggested that, once the required antecedent vocabulary is in place, students learn the terms that

---

15. Though my analysis diverges from theirs, many of the considerations that follow (as well as a few of those introduced above) were suggested by contemplation of the techniques developed by J. D. Sneed and Wolfgang Stegmüller for formalizing physical theories, especially by their manner of introducing theoretical terms. Note also that these remarks suggest a route to the solution of a central problem of their approach, how to distinguish the core of a theory from its expansions. For this problem see my paper, "Theory Change as Structure Change: Comments on the Sneed Formalism," *Erkenntnis*, 10 (1976): 179–99; reprinted in this volume as essay 7.

remain by exposure to some single specifiable set of examples of their use. Those particular examples may well have seemed to provide necessary conditions for the acquisition of those terms. In practice, however, cases of that sort are very rare. Usually there are alternate sets of examples that will serve for the acquisition of the same term or terms. And, though it usually makes no difference to which set of these examples an individual has, in fact, been exposed, there are special circumstances in which the differences between sets prove very important.

In the case of 'mass' and 'weight', one of these alternate sets is standard. It is able to supply the missing elements of both vocabulary and theory together, and it probably therefore enters the lexical acquisition process for all students. But logically other examples would have done as well, and for most students some of them also play a role. Begin with the standard route, which first quantifies 'mass' in the guise of what today is called 'inertial mass'. Students are presented with Newton's second law—force equals mass times acceleration—as a description of the way moving bodies actually behave, but the description makes essential use of the still incompletely established term 'mass'. That term and the second law are thus acquired together, and the law can thereafter be used to supply the missing measure: the mass of a body is proportional to its acceleration under the influence of a known force. For purposes of concept acquisition, centripetal force apparatus provides a particularly effective way to make the measurement.

Once mass and the second law have been added to the Newtonian lexicon in this way, the law of gravity can be introduced as an empirical regularity. Newtonian theory is applied to observation of the heavens and the attractions manifest there are compared to those between the earth and bodies resting on it. The mutual attraction between bodies is thus shown to be proportional to the product of their masses, an empirical regularity that can be used to introduce the still missing aspects of the Newtonian term 'weight'. 'Weight' is now seen to denote a relational property, one that depends on the presence of two or more bodies. It can therefore, unlike mass, differ from one location to another, at the surface of the earth and of the moon, for example. That difference is captured only by the spring balance, not by the previously standard pan balance (which yields the same reading at all locations). What the pan balance measures is mass, a quantity that depends only on the body and on the choice of a unit measure.

Because it establishes both the second law and the use of 'mass', the sequence just sketched provides the most direct route to many applications of Newtonian theory.[16] That is why it plays so central a role in introducing the theory's vocabulary. But it is not, as previously indicated, required for that purpose, and, in any case, it rarely functions alone. Let me now consider a second route along which the use of 'mass' and 'weight' can be established. It starts from the same point as the first, by quantifying the notion of force with the aid of a spring balance. Next, 'mass' is introduced in the guise of what is today labeled 'gravitational mass'. A stipulative description of the way the world is provides students with the notion of gravity as a universal force of attraction between pairs of material bodies, its magnitude proportional to the mass of each. With the missing aspects of 'mass' thus supplied, weight can be explained as a relational property, the force resulting from gravitational attraction.

That is a second way to establish the use of the Newtonian terms 'mass' and 'weight'. With them in hand Newton's second law, the still missing component of Newtonian theory, can be introduced as empirical, a consequence simply of observation. For that purpose, centripetal force apparatus is again appropriate, but no longer to measure mass, as it did on the first route, but now rather to determine the relation between applied force and the acceleration of a mass previously measured by gravitational means. The two routes thus differ in what must be stipulated about nature in order to learn Newtonian terms, what can be left instead for empirical discovery. On the first route the second law enters stipulatively, the law of gravitation empirically. On the second, their epistemic status is reversed. In each case one, but only one, of the laws is, so to speak, built into the lexicon. I do not quite want to call such laws analytic, for experience with nature was essential to their initial formulation. Yet they do have something of the necessity that the label 'analytic' implies. Perhaps 'synthetic a priori' comes closer.

There are, of course, still other ways in which the quantitative elements of 'mass' and 'weight' can be acquired. For example, Hooke's law having been introduced together with 'force', the spring balance can be stipulated as the measure of weight, and mass can be measured, again by stipulation, in terms of the vibration period of a weight at the end of a spring. In practice, several of these applications of Newtonian theory usually enter into the process of acquiring Newtonian language,

16. All applications of Newtonian theory depend on understanding 'mass', but for many of them 'weight' is dispensable.

information about the lexicon and information about the world being distributed in an indivisible mix among them. Under those circumstances, one or another of the examples introduced during lexical acquisition can, when occasion requires, be adjusted or replaced in the light of new observations. Other examples will maintain the lexicon stable, keeping in place a set of quasi-necessities equivalent to those initially induced by language learning.

Clearly, however, only a certain number of examples may be altered piecemeal in this way. If too many require adjustment, then it is no longer individual laws or generalizations that are at stake but the very vocabulary in which they are stated. A threat to that vocabulary is, however, a threat also to the theory or laws essential to its acquisition and use. Could Newtonian mechanics withstand revision of the second law, of the third law, of Hooke's law, or the law of gravity? Could it withstand the revision of any two of these, of three, or of all four? These are not questions that individually have yes or no answers. Rather, like Wittgenstein's "Could one play chess without the queen?" they suggest the strains placed on a lexicon by questions that its designer, whether God or cognitive evolution, did not anticipate its being required to answer.[17] What should one have said when confronted by an egg-laying creature that suckles its young? Is it a mammal or is it not? These are the circumstances in which, as Austin put it, *"we don't know what to say. Words literally fail us."*[18] Such circumstances, if they endure for long, call forth a locally different lexicon, one that permits an answer but to a slightly altered question: "Yes, the creature is a mammal" (but to be a mammal is not what it was before). The new lexicon opens new possibilities, ones that could not have been stipulated by the use of the old.

17. Twenty-five years ago the quotation was a standard part of what I now discover was a merely oral tradition. Though clearly "Wittgensteinian," it is not to be found in any of Wittgenstein's published writings. I preserve it here, because of its recurrent role in my own philosophical development and because I've found no published substitute that so clearly bars responding that additional information might permit the question to be answered.

18. J. L. Austin, "Other Minds," in *Philosophical Papers* (Oxford: Clarendon Press, 1961), pp. 44–84. The quoted passage occurs on p. 56, and the italics are Austin's. For examples from literature of situations in which words fail us, see J. B. White, *When Words Lose Their Meaning: Constitutions and Reconstitutions of Language, Character, and Community* (Chicago: University of Chicago Press, 1984). I have compared an example from the sciences with one from developmental psychology in "A Function for Thought Experiments," in *Mélanges Alexander Koyré*, vol. 2, *L'aventure de la science*, ed. I. B. Cohen and R. Taton (Paris: Hermann, 1964), pp. 307–34; reprinted in *The Essential Tension: Selected Studies in Scientific Tradition and Change* (Chicago: University of Chicago Press, 1977), pp. 240–65.

To clarify what I have in mind, let me suppose that there are only two ways in which use of the terms 'mass' and 'weight' can be acquired: one which stipulates the second law and finds the law of gravity empirically; another which stipulates the law of gravity and discovers the second law empirically. Suppose further that the two routes are exclusive; students traverse one or the other so that on each the necessities of the lexicon and the contingencies of experiment are kept separate. Clearly, these two routes are very different, but the differences will not ordinarily interfere with full communication among those who use the terms. All will pick out the same objects and situations as the referents of the terms they share, and all will agree about the laws and other generalizations governing these objects and situations. All are thus fully participants in a single speech community. What individual speakers may differ about is the epistemic status of generalizations that community members share, and such differences are not usually important. Indeed, in *ordinary* scientific discourse, they do not emerge at all. While the world behaves in anticipated ways—the ones for which the lexicon evolved—these differences between individual speakers make little or no difference.

But change of circumstance may make them consequential. Imagine that a discrepancy is discovered between Newtonian theory and observation, for example, celestial observations of the motion of the lunar perigee. Scientists who had learned Newtonian 'mass' and 'weight' along the first of my two lexical-acquisition routes would be free to consider altering the law of gravity as a way to remove the anomaly. On the other hand, they would be bound by language to preserve the second law. But scientists who had acquired 'mass' and 'weight' along my second route would be free to suggest altering the second law but would be bound by language to preserve the law of gravity. A difference in the language-learning route, one which had had no effect while the world behaved as anticipated, would lead to differences of opinion when anomalies were found.

Now suppose that neither the revisions that preserved the second law nor those that preserved the law of gravity proved effective in eliminating anomaly. The next step would be an attempt at revisions which altered both laws together, and those revisions the lexicon will not, in its present form, permit.[19] Such attempts are often successful nonethe-

---

19. At this point I will seem to be reintroducing the previously banished notion of analyticity, and perhaps I am. Using the Newtonian lexicon, the statement "Newton's second law and the law of gravity are both false" is itself false. Furthermore, it is false by virtue of the meaning

less, but they require recourse to such devices as metaphorical extension, devices that alter the meanings of lexical items themselves. After such revision—say, the transition to an Einsteinian vocabulary—one can write down strings of symbols that *look like* revised versions of the second law and the law of gravity. But the resemblance is deceptive because some symbols in the new strings attach to nature differently than do the corresponding symbols in the old, thus distinguishing between situations which, in the antecedently available vocabulary, were the same.[20] They are the symbols for terms whose acquisition involved laws that have changed form with the change of theory: the differences between the old laws and the new are reflected by the terms acquired with them. Each of the resulting lexicons then gives access to its own set of possible worlds, and the two sets are disjoint. Translations involving terms introduced with the altered laws are impossible.

The impossibility of translation does not, of course, bar users of one lexicon from learning the other. And having done so, they can join the two together, enriching their initial lexicon by adding to it sets of terms from the one they just have acquired. For some purposes such enrichment is essential. At the beginning of this paper, for example, I suggested that historians often required an enriched lexicon to understand the past, and I have argued elsewhere that they must transmit that lexicon to their readers.[21] But the sense of enrichment involved is peculiar. Each of the lexicons combined for the historian's purposes embodies knowledge of nature, and the two sorts of knowledge are incompatible, cannot coherently describe the same world. Except under very special circumstances, like those of the historian at work, the price of combining them is incoherence in the description of phenomena to which either one might alone have been applied.[22] Even the historian avoids incoherence only by being sure at all times which lexicon he is using and why. Under these circumstances, one may reasonably ask whether the term

---

of the Newtonian terms 'force' and 'mass'. But it is not—unlike the statement "Some bachelors are married"—false by virtue of the *definitions* of those terms. The meanings of 'force' and 'mass' are not embodiable in definitions but rather in their relation to the world. The necessity to which I here appeal is not so much analytic as synthetic a priori.

20. In fact, for the Newton to Einstein transition, the most significant lexical change is in the antecedent kinematic vocabulary for space and time, and it moves from there upward into the vocabulary of mechanics.

21. "Commensurability, Comparability, Communicability," this volume, pp. 43–47.

22. In describing the expanded lexicon it is essential to use terms like 'incompatible' and 'incoherent' rather than 'contradictory' and 'false'. The two latter terms would apply only if translation were possible.

'enriched' quite applies to the enlarged lexicon formed by combinations of this sort.

A closely related problem—that of the grue emeralds—has recently been much discussed in philosophy. An object is grue if it has been observed to be green before time t or if, alternatively, it is blue. The puzzle is that the same set of observations, if made prior to t, support two incompatible generalizations: "All emeralds are green" and "All emeralds are grue." (Note that a grue emerald, if not examined before time t, can only be blue.) Here, too, the solution depends upon segregating the lexicon containing the normal descriptive color vocabulary, 'blue', 'green', and the like, from the lexicon that contains 'grue', 'bleen', and the names of other occupants of the corresponding spectrum. One set of terms is projectible, supports induction, the other not. One set of terms is available for descriptions of the world, the other is reserved for the special purposes of the philosopher. Difficulties emerge only when the two, embodying incompatible bodies of knowledge of nature, are used in combination, for there is no world to which the enlarged lexicon could apply.[23]

Students of literature have long taken for granted that metaphor and its companion devices (those which alter the interrelations among words) provide entree to new worlds and make translation impossible by doing so. Similar characteristics have been widely attributed to the language of political life and, by some, to the entire range of the human sciences. But the natural sciences, dealing objectively with the real world (as they do), are generally held to be immune. Their truths (and falsities) are thought to transcend the ravages of temporal, cultural, and linguistic change. I am suggesting, of course, that they cannot do so. Neither the descriptive nor the theoretical language of a natural science provides the bedrock such transcendence would require. I shall not in this place even attempt to deal with the philosophical problems consequent upon that point of view. Let me, instead, attempt to increase their urgency.

---

23. For the original paradox see N. Goodman, *Fact, Fiction, and Forecast*, 4th ed. (Cambridge, MA: Harvard University Press, 1983), chapters 3 and 4. Note that the similarity just emphasized is in one very important respect incomplete. Both the Newtonian terms discussed above and the terms in any color vocabulary from an interrelated set. But in the latter case the difference between vocabularies does not affect vocabulary structure, and it is therefore possible to translate between the projectible 'blue'/'green' vocabulary and the unprojectible vocabulary containing 'bleen' and 'grue'.

The threat to realism is the foremost of the problems I have in mind, and it can here stand for the entire set.[24] A lexicon acquired by techniques like those discussed in the preceding section gives members of the community that employs it conceptual access to an infinite set of lexically stipulatable worlds, worlds describable with the community's lexicon. Of these worlds, only a small fraction are compatible with what they know of their own, the actual world: the others are barred by requirements of internal consistency or of conformity with experiment and observation. As time passes, continuing research excludes more and more possible worlds from the subset that could be actual. If all scientific development proceeded in this way, the progress of science would consist in ever closer specification of a single world, the actual or real one.

A reiterated theme of this paper has been, however, that a lexicon which gives access to one set of possible worlds also bars access to others. (Remember the Newtonian lexicon's inability to describe a world in which the second law and the law of gravity were not simultaneously satisfied.) And scientific development turns out to depend not only on weeding out candidates for reality from the current set of possible worlds, but also upon occasional transitions to another set, one made accessible by a lexicon with different structure. Once such a transition has occurred, some statements previously descriptive of possible worlds prove untranslatable in the terminology developed for the subsequent science. These are the statements the historian first encounters as anomalous word strings; one cannot imagine what those who uttered or wrote them were trying to say. Only when a new lexicon has been mastered can they be understood, and that understanding does not provide them with later equivalents. Individually, they are neither compatible nor incompatible with statements embodying the beliefs of a later age, and they are therefore immune to an evaluation conducted with its conceptual categories.

The immunity of such statements is, of course, only to being judged one at a time, labeled individually with truth values or some other index of epistemic status. Another sort of judgment is possible, and in scientific development something very like it repeatedly occurs. Faced with un-

---

24. Contrary to a widespread impression, the sort of position here sketched does not raise problems of relativism, at least not if 'relativism' is used in any standard sense. There are shared and justifiable, though not necessarily permanent, standards that scientific communities use when choosing between theories. On this subject see my papers "Objectivity, Value Judgment, and Theory Choice," in *The Essential Tension*, pp. 320–39, and "Rationality and Theory Choice," *Journal of Philosophy* 80 (1983): 563–70; reprinted in this volume as essay 9.

translatable statements, the historian becomes bilingual, first learning the lexicon required to frame the problematic statements and then, if it seems relevant, comparing the whole older system (a lexicon plus the science developed with it) to the system in current use. Most of the terms used within either system are shared by both, and most of these shared terms occupy the same positions in both lexicons. Comparisons made using those terms alone ordinarily provide a sufficient basis for judgment. But what is then being judged is the relative success of two whole systems in pursuing an almost stable set of scientific goals, a very different matter from the evaluation of individual statements within a given system.

Evaluation of a statement's truth values is, in short, an activity that can be conducted only with a lexicon already in place, and its outcome depends upon that lexicon. If, as standard forms of realism suppose, a statement's being true or false depends simply on whether or not it corresponds to the real world—independent of time, language, and culture—then the world itself must be somehow lexicon-dependent. Whatever form that dependence takes, it poses problems for a realist perspective, problems that I take to be both genuine and urgent. Rather than explore them further here—a task for another paper—I shall close by examining a standard attempt to dismiss them.

What I have been describing as the problem of lexical dependence is often called the problem of meaning variance. To avoid it and related problems from other sources, many philosophers have in recent years emphasized that truth values depend only on reference, and that an adequate theory of reference need not call upon the way in which the referents of individual terms are in fact picked out.[25] The most influential

---

25. Views which, like mine, depend on talking about the way words are actually used, the situations in which they apply, are regularly charged with invoking a "verification theory of meaning," not currently a respectable thing to do. But in my case at least that charge does not hold. Verification theories attribute meanings to individual sentences and through them to the individual terms those sentences contain. Each term has a meaning determined by the way in which sentences containing it are verified. I have been suggesting, however, that with occasional exceptions terms do not individually have meanings at all. More important, the view sketched above insists that people may use the same lexicon, refer to the same items with it, and yet pick out those items in different ways. Reference is a function of the shared structure of the lexicon but not of the varied feature-spaces within which individuals represent that structure. There is, however, a second charge, closely related to verificationism, of which I am guilty. Those who maintain the independence of reference and meaning also maintain that metaphysics is independent of epistemology. No view like mine (in the respects presently at issue there are a number) is compatible with that separation. The separation of metaphysics from epistemology can come only after a position that involves both has been elaborated.

version of such theories is the so-called causal theory of reference developed primarily by Kripke and Putnam. It is rooted firmly in possible-world semantics, and its expositors resort repeatedly to examples drawn from scientific development. A look at it should both reinforce and extend the viewpoint sketched above. For the purpose, I restrict myself primarily to the version developed by Hilary Putnam, for Putnam deals more explicitly than others with problems of scientific development.[26]

According to causal theory, the referents of natural-kind terms like 'gold', 'tiger', 'electricity', 'gene', or 'force' are determined by some original act of baptizing or dubbing samples of the kind in question with the name they will thereafter bear. That act, to which later speakers are linked by history, is the "cause" of the term's referring as it does. Thus, some samples of a naturally occurring yellow, malleable metal were once baptized 'gold' (or some equivalent in another language), and the term has since referred to all samples of the same stuff as the original whether they displayed the same superficial qualities or not. What establishes the reference of a term is, thus, the original sample together with the primitive relationship, sameness-of-kind. If the original samples were not all or mostly of the same kind, then the term in question, for example 'phlogiston', fails to refer. Theories about what makes the samples the same are, on this view, irrelevant to reference, as are the techniques used in identifying further samples. Both may vary over time as well as from individual to individual at a given time. But the original samples and the relation sameness-of-kind are stable. If meanings are the sorts of things that individuals can carry around in their heads, then meaning does not determine reference.

Excluding proper names, I doubt that there is any set of terms for which this theory works precisely, but it comes very close to doing so for terms like 'gold', and the plausibility of the application of causal theory to natural-kind terms depends on the existence of such cases. Terms that behave like 'gold' ordinarily refer to naturally occurring, widely distributed, functionally significant, and easily recognized substances. They occur in the languages of most or all cultures, retain their

---

26. S. Kripke, *Naming and Necessity* (Cambridge, MA: Harvard University Press, 1972), and H. Putnam, "The Meaning of 'Meaning'," in *Mind, Language and Reality*, Philosophical Papers, vol. 2 (Cambridge: Cambridge University Press, 1975). Putnam has, I believe, now abandoned significant components of the theory, moving from it to a view ("internal realism") with significant parallels to my own. But few philosophers have followed him. The views discussed below are very much alive.

original use over time, and refer throughout to the same sorts of samples. There is little problem about translating them, for they occupy closely equivalent positions in all lexicons. 'Gold' is among the closest approximations we have to an item in a neutral, mind-independent observation vocabulary.

When a term is of this sort, modern science can often be used not only to specify the common essence of its referents but actually to single them out. Modern theory, for example, identifies gold as the substance with atomic number 79 and licenses specialists to identify it by the application of such techniques as X-ray spectroscopy. Neither the theory nor the instrument was available seventy-five years ago, but it is nevertheless reasonable to suggest that "The referents of 'gold' are and have always been the same as the referents of 'substance with atomic number 79'." Exceptions to that equation are few, and they result primarily from our increased ability to detect impurities and forgeries. For the causal theorist, therefore, "having atomic number 79" is *the* essential property of gold—the single property such that, if gold in fact does have it, then it has it necessarily. Other properties—yellowness and ductility, for example—are superficial and correspondingly contingent. Kripke suggests that gold might even be blue, its apparent yellowness resulting from an optical illusion.[27] Though individuals may, in fact, use color and other superficial characteristics when picking out samples of gold, that practice tells nothing essential about the referents of the term.

'Gold' presents a relatively special case, however, and what is special about it obscures essential limitations on the conclusions it will support. More representative is Putnam's most developed example, 'water', and the problems which arise with it are still more severe in the case of such other widely discussed terms as 'heat' and 'electricity'.[28] For water the discussion divides in two parts. In the first, which is the more familiar, Putnam imagines a possible world containing Twin Earth, a planet just like our own except that the stuff called 'water' by Twin Earthians is not H_2O but a different liquid with a very long and complicated chemical formula abbreviated XYZ. "Indistinguishable from water at

27. Kripke, *Naming and Necessity*, p. 118.
28. The force of Putnam's discussion depends in part upon an equivocation that needs to be eliminated. As used in everyday life or by the laity, 'water' has through history behaved much like 'gold'. But that is not the case within the community of scientists and philosophers to which Putnam's argument needs to be applied.

normal temperatures and pressures," XYZ is the stuff that on Twin
Earth quenches thirst, rains from the skies, and fills oceans and lakes,
much as water does here. If a spaceship from Earth ever visits Twin
Earth, Putnam writes:

> then, the supposition at first will be that 'water' has the same meaning on
> Earth and Twin Earth. This supposition will be corrected when it is discov-
> ered that 'water' on Twin Earth is XYZ, and the Earthian spaceship will
> report somewhat as follows:
>
> On Twin Earth the word 'water' means XYZ.

As in the case of gold, superficial qualities like quenching thirst or rain-
ing from the skies have no role in determining to what substance the
term 'water' properly refers.

Two aspects of Putnam's fable require special notice. First, the fact
that Twin Earthians call XYZ by the name 'water' (the same symbol
that Earthians use for the stuff that lies in lakes, quenches thirst, etc.)
is an irrelevancy. The difficulties presented by this story will emerge
more clearly if the visitors from Earth use their own language throughout. Second, and presently central, whatever the visitors call the stuff
that lies in Twin Earthian lakes, the report they send home must take
some form like:

> Back to the drawing board! Something is badly wrong with chemical theory.

The terms 'XYZ' and 'H_2O' are drawn from modern chemical theory,
and that theory is incompatible with the existence of a substance with
properties very nearly the same as water but described by an elaborate
chemical formula. Such a substance would, among other things, be too
heavy to evaporate at normal terrestrial temperatures. Its discovery
would present the same problems as the simultaneous violation of Newton's second law and the law of gravity described in the last section.
It would, that is, demonstrate the presence of fundamental errors in the
chemical theory which gives meanings to compound names like 'H_2O'
and the unabbreviated form of 'XYZ'. Within the lexicon of modern
chemistry, a world containing both our earth and Putnam's Twin Earth
is lexically possible, but the composite statement that describes it is
necessarily false. Only with a differently structured lexicon, one shaped
to describe a very different sort of world, could one, without contradic-

tion, describe the behavior of XYZ at all, and in that lexicon 'H₂O' might no longer refer to what we call 'water'.

So much for the first part of Putnam's argument. In the second he applies it more concretely to the referential history of 'water', suggesting that "we roll the time back to 1750" and continuing:

> At that time chemistry was not developed on either Earth or Twin Earth. The typical Earthian speaker of English did not know water consisted of hydrogen and oxygen, and the typical Twin Earthian speaker of English did not know 'water' consisted of XYZ. . . . Yet the extension of the term 'water' was just as much H2O on Earth in 1750 as in 1950; and the extension of the term 'water' was just as much XYZ on Twin Earth in 1750 as in 1950.

In journeys through time as in those through space, Putnam suggests, it is chemical formula, not superficial qualities, that determines whether a given substance is water.

For present purposes attention can be restricted to Earthian history, and on Earth Putnam's argument for 'water' is the same as it was for 'gold'. The extension of 'water' is determined by the original sample together with the relation sameness-of-kind. That sample dates from before 1750 and the nature of its members has been stable. So has the relation sameness-of-kind, though *explanations* of what it is for two bodies to be of the same kind have varied widely. What matters, however, is not explanations but what gets picked out, and identifying samples of H₂O is, according to causal theory, the best means yet found to pick out samples of the same kind as the original set. Give or take a few discrepancies at the margins, discrepancies due to refinement of technique or perhaps to change of interest, 'H₂O' refers to the same samples that 'water' referred to in either 1750 or 1950. Apparently causal theory has rendered the referents of 'water' immune to changes in the concept of water, the theory of water, and the way samples of water are picked out. The parallel between causal theory's treatment of 'gold' and of 'water' seems complete.

But in the case of water difficulties arise. 'H₂O' picks out samples not only of water but also of ice and steam. H₂O can exist in all three states of aggregation—solid, liquid, and gaseous—and it is therefore not the same as water, at least not as picked out by the term 'water' in 1750. The difference in items referred to is, furthermore, by no means marginal, like that due to impurities for example. Whole categories of substance are involved, and their involvement is by no means accidental. In 1750 the primary differences between chemical species were the states of aggregation or modeled upon them. Water, in particular, was an elementary body of which liquidity was an essential property. For some chemists the term 'water' referred to the generic liquid, and it had done so for many more only a few generations before. Not until the 1780s, in an episode long known as "The Chemical Revolution," was the taxonomy of chemistry transformed so that a chemical species might exist in all three states of aggregation. Thereafter, the distinction between solids, liquids, and gases became physical, not chemical. The discovery that *liquid* water was a compound of two *gaseous* substances, hydrogen and oxygen, was an integral part of that larger transformation and could not have been made without it.

This is not to suggest that modern science is incapable of picking out the stuff that people in 1750 (and most people still) label 'water'. That term refers to *liquid* H_2O. It should be described not simply as H_2O but as close-packed H_2O particles in rapid relative motion. Marginal differences again aside, samples answering that compound description are the ones picked out in 1750 and before by the term 'water'. But this modern description leads to a new network of difficulties, difficulties that may ultimately threaten the concept of natural kinds and that meanwhile must bar the automatic application of causal theory to them.

Causal theory was initially developed with notable success for application to proper names. Its transfer from them to natural-kind terms was facilitated—perhaps made possible—by the fact that natural kinds, like single individual creatures, are denoted by short and apparently arbitrary names, names coextensive with those of the corresponding kind's single essential property. Our examples have been 'gold' paired with 'having atomic number 79' and 'water' paired with 'being H_2O'. The latter member of each pair names a property, of course, as the name coupled with it does not. But so long as only a single essential property is required by each natural kind, that difference is inconsequential. When two non-coextensive names are required, however—'H_2O' and 'liquidity' in the case of water—then each name, if used alone, picks out a larger class than the pair does when conjoined, and the fact that they name properties becomes central. For if two properties are required, why not three or four? Are we not back to the standard set of problems that causal theory was intended to resolve: which properties are essential, which accidental; which properties belong to a kind by definition, which are only contingent? Has the transition to a developed scientific vocabulary really helped at all?

I think it has not. The lexicon required to label attributes like being-H_2O or being-close-packed-particles-in-rapid-relative-motion is rich and systematic. No one can use any of the terms that it contains without being able to use a great many. And given that vocabulary, the problems of choosing essential properties arise again, except that the properties involved can no longer be dismissed as superficial. Is deuterium hydrogen, for example, and is heavy water really water? And what may one say about a sample of close-packed particles of H_2O in rapid relative motion at the critical point, under the conditions of temperature and pressure, that is, at which the liquid, solid, and gaseous states are indistinguishable? Is it really water? The use of theoretical rather than superficial properties offers great advantages, of course. There are fewer of the former, the relations between them are more systematic, and they permit both richer and more precise discriminations. But they come no closer to being essential or necessary properties than the superficial ones they appear to supplant. The problems of meaning and meaning variance are still in place.

The inverse argument proves even more significant. The so-called superficial properties are no less necessary than their apparently essential successors. To say that water is liquid H_2O is to locate it within an elaborate lexical and theoretical system. Given that system, as one must be in order to use the label, one can in principle predict the superficial properties of water (just as one could those of XYZ), compute its boiling and freezing points, the optical wavelengths which it will transmit, and so on.[29] If water is liquid H_2O, then these properties are necessary to it. If they were not realized in practice, that would be a reason to doubt that water really was H_2O.

This last argument applies also to the case of gold, in which causal theory apparently succeeded. 'Atomic number' is a term from the lexicon of atomic-molecular theory. Like 'force' and 'mass', it must be learned together with other terms deployed in that theory, and the theory itself must play a role in the acquisition process. When the process is complete, one can replace the label 'gold' with 'atomic number 79',

29. Laypeople can, of course, say that water is H_2O without controlling the fuller lexicon or the theory which it supports. But their ability to communicate by doing so depends upon the presence of experts in their society. The laity must be able to identify the experts and say something of the nature of the relevant expertise. And the experts must, in turn, command the lexicon, the theory, and the computations.

but one can then also replace the label 'hydrogen' with 'atomic number 1', 'oxygen' with 'atomic number 8', and so on to a total well over a hundred. And one can do something more important as well. Invoking such other theoretical properties as electronic charge and mass, one can in principle, and to a considerable extent in fact, predict the superficial qualities—density, color, ductility, conductivity, and so on—that samples of the corresponding substance will possess at normal temperatures. Those properties are no more accidental than having-atomic-number-79. That color is a superficial property does not make it a contingent one. Furthermore, in a comparison of superficial and theoretical qualities, the former have a double priority. If the theory that posits the relevant theoretical properties could not predict these superficial qualities, or some of them, there would be no reason to take it seriously. If gold were blue for a normal observer under normal conditions of illumination, its atomic number would not be 79. In addition, superficial properties are the ones called upon in those difficult cases of discrimination characteristically raised by new theories. Is deuterium really hydrogen, for example? Are viruses alive?[30]

What remains special about 'gold' is simply that, unlike 'water', only one of the underlying properties recognized by modern science—having atomic number 79—need be called upon to pick out members of the sample to which the term has continued through history to refer.[31] 'Gold' is not the only term that possesses or closely approximates this characteristic. So do many of the basic-level referring terms used in

30. At issue, of course, is where to draw the boundary lines that delimit the referents of 'water', 'living thing', and so on, a problem which arises from and seems to threaten the notion of natural kinds. That notion is closely modeled on the concept of a biological species, and discussions of causal theory repeatedly invoke the relation between a particular gene-type and a corresponding species (often tigers) to illustrate the relation said to hold between a natural kind and its essence, between H_2O and water, for example, or between atomic-number-79 and gold. But even individuals who are unproblematically members of the same species have differently constituted sets of genes. Which sets are compatible with membership in that species is a subject of continuing debate, both in principle and practice, and the subject of the argument is always which superficial *properties* (e.g., the ability to interbreed) the members of the species must share.

31. Even for gold this generalization is not *altogether* correct. As mentioned above, scientific progress does result in marginal adjustments of the original samples of gold by virtue of "our increased ability to detect impurities." But what it is for gold to be pure is determined in part by theory. If gold is the substance with atomic number 79, then even a single atom with a different atomic number constitutes an impurity. But if gold is, as it was in antiquity, a metal that ripens naturally in the earth, changing gradually from lead through iron and silver to gold in the process; then there is no single form of matter that is gold *tout court*. When the ancients applied the term 'gold' to samples from which we might withhold it, they were not always simply mistaken.

everyday speech, including the everyday use of the term 'water'. But not all everyday terms are of this sort. 'Planet' and 'star' now categorize the world of celestial objects differently from the way they did before Copernicus, and the differences are not well described by phrases like "marginal adjustment" or "zeroing in." Similar transitions have characterized the historical development of virtually all the referring terms of the sciences, including the most elementary: 'force', 'species', 'heat', 'element', 'temperature', and so on.

Over the course of history, these and other scientific terms have participated, sometimes repeatedly, in the sorts of changes epitomized incompletely above by the change in the chemist's use of 'water' between 1750 and 1950. Such lexical transformations systematically split up and then regroup in new ways the members of sets to which terms in the lexicon refer. Usually the terms themselves remain the same through such transitions, though sometimes with strategic additions and deletions. So do many of the items referred to by those terms, which is why the terms endure. But the changes in membership of the sets of items to which these enduring terms refer are often massive, and they affect not just the referents of an individual term but of an interrelated set of terms between which the preexisting population is redistributed. Items previously regarded as quite unlike are grouped together after the transformation, while previously exemplary members of some single category are later divided between systematically different ones.

It is lexical change of this sort that results in the apparent textual anomalies with which this paper began. Encountered by a historian in a text of the past, they strenuously resist elimination by any translation or paraphrase that uses the historian's own lexicon, the one he or she initially brought to the text. The phenomena described in those anomalous passages are stipulated neither as present nor as absent in any of the possible worlds to which that lexicon gives access, and the historian cannot therefore quite understand what the author of the text can be trying to say. Those phenomena belong to another set of possible worlds, one in which many of the same phenomena occur that occur in the historian's own, but in which things also occur that the historian, until reeducated, cannot imagine. Under such circumstances the only recourse is reeducation: the recovery of the older lexicon, its assimilation, and the exploration of the set of worlds to which it gives access. Causal theory provides no bridge across the divide, for the transworld voyages it envisages are limited to worlds in a single lexically possible set. And in the absence of the bridge that causal theory has sought to provide, there is no basis for talk of science's gradual elimination of all worlds excepting the single real one. That way of talking, neatly illustrated by the discussion of gold but not by that of water, has provided causal theory's version of what the tradition described as successively closer approximations to the truth, cutting the world closer to its joints, or just zeroing in.

Such descriptions of scientific development can no longer be sustained. I know of only one other strategy available for their defense, and it seems to me self-defeating, an artifice born of desperation. In the case of 'water' that strategy would be implemented as follows: until sometime after 1750 chemists were misled by superficial properties into believing that water is a natural kind, but it is not; what they called 'water' did not exist, any more than did phlogiston; both were chimerical, and the terms used to refer to them did not, in fact, refer at all.[32] But that cannot be right. Putatively non-referring terms like 'water' can neither be isolated nor replaced by more primitive terms of indubitable referential status. If 'water' failed to refer, then so did other chemical terms like 'element', 'principle', 'earth', 'compound', and many others. Nor was referential failure restricted to chemistry. Terms like 'heat', 'motion', 'weight', and 'force' were equally empty; statements in which they appeared were about nothing. On this showing, the history of science is the history of developing vacuity, and from vacuity one cannot zero in. Some other explanation of the achievements of science is needed.

POSTSCRIPT: SPEAKER'S REPLY

I am most grateful to Professors Frängsmyr and Miller for their comments on my paper. Occasional misunderstandings aside (we are, for example, using the phrase "possible worlds" quite differently), I agree fully with what they have to say. Scientific development has numerous aspects besides the ones on which my paper focuses; their remarks complement mine by cogently illustrating other topics I might have discussed. On only one point is further response required.

Early in his commentary, Professor Miller writes, "I see the principal

---

32. I take this to be the sort of response Putnam would have provided when the paper I have been discussing was written.

problem in Kuhn's analysis to be his emphasis on discontinuous change from one theory (or world) to another." There is, however, no talk of discontinuous change in my paper, much less any emphasis on it. The contrast throughout is between the lexicons used at two widely separated times: nothing is said about the nature of the intervening process by which a transition between them is made. The point is worth elaborating: my past work has often invoked discontinuity, and my present paper points the way toward significant reformulation.

In recent years I have increasingly recognized that my conception of the process by which scientists move forward has been too closely modeled on my experience with the process by which historians move into the past.[33] For the historian, the period of wrestling with nonsense passages in out-of-date texts is ordinarily marked by episodes in which the sudden recovery of a long-forgotten way to use some still-familiar terms brings new understanding and coherence. In the sciences, similar "aha experiences" mark the periods of frustration and puzzlement that ordinarily precede fundamental innovation and that often precede the understanding of innovation as well. The testimony of scientists to such experiences, together with my own experience as a historian, was the basis for my repeated reference to gestalt switches, conversion experiences, and the like. In many of the places in which such phrases appeared, their use was literal or very nearly so, and in those places I would use them again, though perhaps with more care for rhetorical overtones.

In other places, however, a special characteristic of scientific development led me to use such terms metaphorically, often without quite recognizing the difference in use. The sciences are unique among creative disciplines in the extent to which they cut themselves off from their past, substituting for it a systematic reconstruction. Few scientists read past scientific works; science libraries ordinarily displace the books and journals in which such work is recorded; scientific life knows no institutional equivalent for the art museum. Another symptom is presently more central. When reconceptualization occurs in a scientific field, displaced concepts rapidly vanish from professional view. Later practitioners reconstruct their predecessors' work in the conceptual vocabulary they use themselves, a vocabulary incapable of representing what those predecessors actually did. Such reconstruction is a precondition

---

33. For a particularly clear example of that modeling, see my "What are Scientific Revolutions?"

for the cumulative image of scientific development familiar from science textbooks, but it badly misrepresents the past.[34] No wonder that the historian, breaking through to that past, experiences the breakthrough as a gestalt switch. And, since what the historian breaks through to is not simply the concepts deployed by a single scientist but those of a once active community, it is natural to speak of the community itself as having undergone a gestalt switch when it displaced its previous conceptual vocabulary for a new one. The temptation to use 'gestalt switch' and related phrases in that way is particularly strong, both because the interval in which the conceptual vocabulary shifted is usually short and because, during that interval, a number of individual scientists did experience gestalt switches.

The transfer of terms like 'gestalt switch' from individuals to groups is, however, clearly metaphorical, and in this case the metaphor proves damaging. Insofar as the historian's gestalt switch provides the model, the magnitude of the conceptual transpositions characteristic of scientific development is exaggerated. Historians, working backward, regularly experience as a single conceptual shift a transposition for which the developmental process required a series of stages. More important, treating groups or communities as though they were individuals-writ-large misrepresents the process of conceptual change. Communities do not have experiences, much less gestalt switches. As the conceptual vocabulary of a community changes, its members may undergo gestalt switches, but only some of them do and not all at the same time. Of those who do not, some cease to be members of the community; others acquire the new vocabulary in less dramatic ways. Meanwhile, communication goes on, however imperfectly, metaphor serving as a partial bridge across the divide between an old literal usage and a new one. To speak, as I repeatedly have, of a community's undergoing a gestalt switch is to compress an extended process of change into an instant, leaving no room for the microprocesses by which the change is achieved.

Recognition of these difficulties opens two directions for further de-

---

34. On this subject see "The Invisibility of Revolutions" in my *Structure of Scientific Revolutions*, 2d ed., rev. (Chicago: University of Chicago Press, 1970), pp. 136–43; "Comment" [on the Relations of Science and Art], *Comparative Studies in Society and History* 11 (1969): 403–12; reprinted as "Comment on the Relations of Science and Art," in *The Essential Tension*, pp. 340–51; and "Revisiting Planck," *Historical Studies in the Physical Sciences* 14 (1984): 231–52; reprinted as a new afterword in *Black-Body Theory and the Quantum Discontinuity 1894–1912* (1978; reprint, Chicago: University of Chicago Press, 1987), pp. 349–70, esp. part 4.

velopment. The first is the one for which Professor Miller calls and which his comments illustrate: study of the microprocesses which occur within a community during periods of conceptual change. Excepting in its repeated references to metaphor, my paper has nothing to say about them, but its formulation, unlike that of my older work, is designed to leave room for their exploration.[35] The second, which may prove even more important, is a systematic attempt to separate the concepts appropriate to the description of groups from those appropriate to the description of individuals. That attempt is currently among my central concerns, and one of its products plays a central, though mostly implicit, role in my paper.[36] People may, I there insist, "use the same lexicon, refer to the same items with it, and yet pick out those items in different ways. Reference is a function of the shared structure of the lexicon but not of the varied feature-spaces within which individuals represent that structure" (n. 25). A number of the classical problems of meaning, I am suggesting, may be seen as a product of the failure to distinguish between the lexicon as a shared property constitutive of community, on the one hand, and the lexicon as something carried by each individual member of the community, on the other.

---

35. The contrast is, of course, only with my older metahistorical work. As historian I have often dealt with the detail of the transition process. See, especially, my *Black-Body Theory*.
36. Others are indicated in my "Scientific Knowledge as Historical Product," to appear in *Synthèse*. [Editors' Note: This essay never appeared.]

Chapter Four

The Road since *Structure*

"The Road since Structure" was Kuhn's presidential address to the biennial meetings of the Philosophy of Science Association in October 1990. It was published in PSA 1990, *volume 2 (East Lansing, MI: Philosophy of Science Association, 1991).*

———◦———

ON THIS OCCASION, and in this place, I feel that I ought, and am probably expected, to look back at the things which have happened to the philosophy of science since I first began to take an interest in it over half a century ago. But I am both too much an outsider and too much a protagonist to undertake that assignment. Rather than attempt to situate the present state of philosophy of science with respect to its past—a subject on which I've little authority—I shall try to situate my present state in philosophy of science with respect to its own past—a subject on which, however imperfect, I'm probably the best authority there is.

As a number of you know, I'm at work on a book, and what I mean to attempt here is an exceedingly brief and dogmatic sketch of its main themes. I think of my project as a return, now under way for a decade, to the philosophical problems left over from the *Structure of Scientific Revolutions*. But it might better be described more generally, as a study of the problems raised by the transition to what's sometimes called the historical and sometimes (at least by Clark Glymour, speaking to me) just the "soft" philosophy of science. That's a transition for which I get far more credit, and also more blame, than I have coming to me. I was, if you will, present at the creation, and it wasn't very crowded.

But others were present too: Paul Feyerabend and Russ Hanson, in particular, as well as Mary Hesse, Michael Polanyi, Stephen Toulmin, and a few more besides. Whatever a *Zeitgeist* is, we provided a striking illustration of its role in intellectual affairs.

Returning to my projected book, you will not be surprised to hear that the main targets at which it aims are such issues as rationality, relativism, and, most particularly, realism and truth. But they're not primarily what the book is about, what occupies most space in it. That role is taken instead by incommensurability. No other aspect of *Structure* has concerned me so deeply in the thirty years since the book was written, and I emerge from those years feeling more strongly than ever that incommensurability has to be an essential component of any historical, developmental, or evolutionary view of scientific knowledge. Properly understood—something I've by no means always managed myself—incommensurability is far from being the threat to rational evaluation of truth claims that it has frequently seemed. Rather, it's what is needed, within a developmental perspective, to restore some badly needed bite to the whole notion of cognitive evaluation. It is needed, that is, to defend notions like truth and knowledge from, for example, the excesses of postmodernist movements like the strong program. Clearly, I can't hope to make all that out here: it's a project for a book. But I shall try, however sketchily, to describe the main elements of the position the book develops. I begin by saying something about what I now take incommensurability to be, and then attempt to sketch its relationship to questions of relativism, truth, and realism. In the book, the issue of rationality will figure, too, but there is no space here even to sketch its role.

Incommensurability is a notion that for me emerged from attempts to understand apparently nonsensical passages encountered in old scientific texts. Ordinarily they had been taken as evidence of the author's confused or mistaken beliefs. My experiences led me to suggest, instead, that those passages were being misread: the appearance of nonsense could be removed by recovering older meanings for some of the terms involved, meanings different from those subsequently current. During the years since, I've often spoken metaphorically of the process by which later meanings had been produced from earlier ones as a process of language change. And, more recently, I've spoken also of the historian's recovery of older meanings as a process of language learning rather like that undergone by the fictional anthropologist whom Quine misdescribes as a radical translator.[1] The ability to learn a language does not, I've emphasized, guarantee the ability to translate into or out of it.

By now, however, the language metaphor seems to me far too inclusive. To the extent that I'm concerned with language and with meanings at all—an issue to which I'll shortly return—it is with the meanings of a restricted class of terms. Roughly speaking, they are taxonomic terms or kind terms, a widespread category that includes natural kinds, artifactual kinds, social kinds, and probably others. In English the class is coextensive, or nearly so, with the terms that by themselves or within appropriate phrases can take the indefinite article. These are primarily the count nouns together with the mass nouns, words which combine with count nouns in phrases that take the indefinite article. Some terms require still further tests hinging, for example, on permissible suffixes.

Terms of this sort have two essential properties. First, as already indicated, they are marked or labeled as kind terms by virtue of lexical characteristics like taking the indefinite article. Being a kind term is thus part of what the word means, part of what one must have in the head to use the word properly. Second—a limitation I sometimes refer to as the no-overlap principle—no two kind terms, no two terms with the kind label, may overlap in their referents unless they are related as species to genus. There are no dogs that are also cats, no gold rings that are also silver rings, and so on: that's what makes dogs, cats, silver, and gold each a kind. Therefore, if the members of a language community encounter a dog that's also a cat (or, more realistically, a creature like the duck-billed platypus), they cannot just enrich the set of category terms but must instead redesign a part of the taxonomy. *Pace* the causal theorists of reference, 'water' did not always refer to H_2O.[2]

Notice now that a lexical taxonomy of some sort must be in place before description of the world can begin. Shared taxonomic categories, at least in an area under discussion, are prerequisite to unproblematic

1. T. S. Kuhn, "Commensurability, Comparability, Communicability," in *PSA 1982: Proceedings of the 1982 Biennial Meeting of the Philosophy of Science Association*, vol. 2, ed. P. D. Asquith and T. Nickles (East Lansing, MI: Philosophy of Science Association, 1983), pp. 669–88; reprinted in this volume as essay 2.

2. T. S. Kuhn, "What Are Scientific Revolutions?" Occasional Paper 18, Center for Cognitive Science (Cambridge, MA: Massachusetts Institute of Technology, 1981); reprinted in *The Probabilistic Revolution*, vol. 1, *Ideas in History*, ed. L. Krüger, L. J. Daston, and M. Heidelberger (Cambridge, MA: MIT Press, 1987), pp. 7–22; also reprinted in this volume as essay 1; T. S. Kuhn, "Dubbing and Redubbing: The Vulnerability of Rigid Designation," in *Scientific Theories*, ed. C. W. Savage, Minnesota Studies in the Philosophy of Science, vol. 14 (Minneapolis: University of Minnesota Press, 1990), pp. 309–14.

communication, including the communication required for the evaluation of truth claims. If different speech communities have taxonomies
that differ in some local area, then members of one of them can (and
occasionally will) make statements that, though fully meaningful within
that speech community, cannot in principle be articulated by members
of the other. To bridge the gap between communities would require
adding to one lexicon a kind term that overlaps, shares a referent, with
one that is already in place. It is that situation which the no-overlap
principle precludes.

Incommensurability thus becomes a sort of untranslatability, localized to one or another area in which two lexical taxonomies differ. The
differences which produce it are not any old differences, but ones that
violate either the no-overlap condition, the kind-label condition, or else
a restriction on hierarchical relations that I cannot spell out here. Violations of those sorts do not bar intercommunity understanding. Members
of one community can acquire the taxonomy employed by members of
another, as the historian does in learning to understand old texts. But
the process which permits understanding produces bilinguals, not translators, and bilingualism has a cost, which will be particularly important
to what follows. The bilingual must always remember within which
community discourse is occurring. The use of one taxonomy to make
statements to someone who uses the other places communication at risk.

Let me formulate these points in one more way, and then make a
last remark about them. Given a lexical taxonomy, or what I'll mostly
now call simply a lexicon, there are all sorts of different statements that
can be made, and all sorts of theories that can be developed. Standard
techniques will lead to some of these being accepted as true, others
rejected as false. But there are also statements which could be made,
theories which could be developed, within some other taxonomy but
which cannot be made with this one, and vice versa. The first volume
of Lyons's *Semantics* contains a wonderfully simple example, which
some of you will know: the impossibility of translating the English statement "the cat sat on the mat," into French, because of the incommensurability between the French and English taxonomies for floor coverings.[3]
In each particular case for which the English statement is true, one can
find a coreferential French statement, some using 'tapis', others 'paillas-
son', still others 'carpette', and so on. But there is no single French
statement which refers to all and only the situations in which the English

statement is true. In that sense, the English statement cannot be made in French. In a similar vein, I've elsewhere pointed out[4] that the content of the Copernican statement "planets travel around the sun," cannot be expressed in a statement that invokes the celestial taxonomy of the Ptolemaic statement "planets travel around the earth." The difference between the two statements is not simply one of fact. The term 'planet' appears as a kind term in both, and the two kinds overlap in membership without either's containing all the celestial bodies contained in the other. All of which is to say that there are episodes in scientific development which involve fundamental change in some taxonomic categories and which therefore confront later observers with problems like those the ethnologist encounters when trying to break into another culture.

A final remark will close this sketch of my current views on incommensurability. I have described those views as concerned with words and with *lexical* taxonomy, and I shall continue in that mode: the sorts of knowledge I deal with come in explicit verbal or related symbolic forms. But it may clarify what I have in mind to suggest that I might more appropriately speak of concepts than of words. What I have been calling a lexical taxonomy might, that is, better be called a conceptual scheme, where the "very notion" of a conceptual scheme is not that of a set of beliefs but of a particular operating mode of a mental module prerequisite to having beliefs, a mode that at once supplies and bounds the set of beliefs it is possible to conceive. Some such taxonomic module I take to be prelinguistic and possessed by animals. Presumably it evolved originally for the sensory, most obviously for the visual, system. In the book I shall give reasons for supposing that it developed from a still more fundamental mechanism which enables individual living organisms to reidentify other substances by tracing their spatio-temporal trajectories.

I shall be coming back to incommensurability, but let me for now set it aside in order to sketch the developmental framework within which it functions. Since I must again move quickly and often cryptically, I begin by anticipating the direction in which I am headed. Basically, I shall be trying to sketch the form which I think any viable evolutionary epistemology has to take. I shall, that is, be returning to the evolutionary analogy introduced in the very last pages of the first edition of *Structure*, attempting both to clarify it and to push it further. During the thirty years since I first made that evolutionary move, theories of the evolution

4. Kuhn, "What Are Scientific Revolutions?" p. 8 (this volume, p. 15).

both of species and of knowledge have, of course, been transformed in ways I am only beginning to discover. I still have much to learn, but to date the fit seems extremely good.

I start from points familiar to many of you. When I first got involved, a generation ago, with the enterprise now often called historical philosophy of science, I and most of my coworkers thought history functioned as a source of empirical evidence. That evidence we found in historical case studies, which forced us to pay close attention to science as it really was. Now I think we overemphasized the empirical aspect of our enterprise (an evolutionary epistemology need not be a naturalized one). What has for me emerged as essential is not so much the details of historical cases as the perspective or the ideology that attention to historical cases brings with it. The historian, that is, always picks up a process already under way, its beginnings lost in earlier time. Beliefs are already in place; they provide the basis for the ongoing research whose results will in some cases change them; research in their absence is unimaginable, though there has nevertheless been a long tradition of imagining it. For the historian, in short, no Archimedean platform is available for the pursuit of science other than the historically situated one already in place. If you approach science as a historian must, little observation of its actual practice is required to reach conclusions of this sort.

Such conclusions have by now been pretty generally accepted: I scarcely know a foundationalist any more. But for me, this way of abandoning foundationalism has a further consequence which, though widely discussed, is by no means widely or fully accepted. The discussions I have in mind usually proceed under the rubric of the rationality or relativity of truth claims, but these labels misdirect attention. Though both rationality and relativism are somehow implicated, what is fundamentally at stake is rather the correspondence theory of truth, the notion that the goal, when evaluating scientific laws or theories, is to determine whether or not they correspond to an external, mind-independent world. It is that notion, whether in an absolute or probabilistic form, that I'm persuaded must vanish together with foundationalism. What replaces it will still require a strong conception of truth, but not, except in the most trivial sense, correspondence truth.

Let me at least suggest what the argument involves. On the developmental view, scientific knowledge claims are necessarily evaluated from a moving, historically situated, Archimedean platform. What requires evaluation cannot be an individual proposition embodying a knowledge claim in isolation: embracing a new knowledge claim typically requires adjustment of other beliefs as well. Nor is it the entire body of knowledge claims that would result if that proposition were accepted. Rather, what's to be evaluated is the desirability of a particular change-of-belief, a change which would alter the existing body of knowledge claims so as to incorporate, with minimum disruption, the new claim as well. Judgments of this sort are necessarily comparative: which of two bodies of knowledge—the original or the proposed alternative—is *better* for doing whatever it is that scientists do. And that is the case whether what scientists do is solve puzzles (my view), improve empirical adequacy (Bas van Frassen's),[5] or increase the dominance of the ruling elite (in parody, the strong program's). I do, of course, have my own preference among these alternatives, and it makes a difference.[6] But no choice between them is relevant to what's presently at stake.

In comparative judgments of the kind just sketched, shared beliefs are left in place: they serve as the given for purposes of the current evaluation; they provide a replacement for the traditional Archimedean platform. The fact that they may—indeed probably will—later be at risk in some other evaluation is simply irrelevant. Nothing about the rationality of the outcome of the current evaluation depends upon their, in fact, being true or false. They are simply in place, part of the historical situation within which this evaluation is made. But if the actual truth value of the shared presumptions required for the evaluation is irrelevant, then the question of the truth or falsity of the changes made or rejected on the basis of that evaluation cannot arise either. A number of classic problems in philosophy of science—most obviously Duhemian holism—turn out on this view to be due not to the nature of scientific knowledge but to a misperception of what justification of belief is all about. Justification does not aim at a goal external to the historical situation but simply, in that situation, at improving the tools available for the job at hand.

To this point I have been trying to firm up and extend the parallel between scientific and biological development suggested at the end of the first edition of *Structure:* scientific development must be seen as a process driven from behind, not pulled from ahead—as evolution from, rather than evolution toward. In making that suggestion, as elsewhere

5. B. van Fraassen, *The Scientific Image* (Oxford: Clarendon, 1980).
6. T. S. Kuhn, "Rationality and Theory Choice," *Journal of Philosophy* 80 (1983): 563–70; reprinted in this volume as essay 9.

in the book, the parallel I had in mind was diachronic, involving the relation between older and more recent scientific beliefs about the same or overlapping ranges of natural phenomena. Now I want to suggest a second, less widely perceived parallel between Darwinian evolution and the evolution of knowledge, one that cuts a synchronic slice across the sciences rather than a diachronic slice containing one of them. Though I have in the past occasionally spoken of the incommensurability between the theories of contemporary scientific specialties, I've only in the last few years begun to see its significance to the parallels between biological evolution and scientific development. Those parallels have also been persuasively emphasized recently in a splendid article by Mario Biagioli.[7] To both of us they seem extremely important, though we emphasize them for somewhat different reasons.

To indicate what is involved I must revert briefly to my old distinction between normal and revolutionary development. In *Structure* it was the distinction between those developments that simply add to knowledge, and those which require giving up part of what's been believed before. In the new book it will emerge as the distinction between developments which do and developments which do not require local taxonomic change. (The alteration permits a significantly more nuanced description of what goes on during revolutionary change than I've been able to provide before.) During this second sort of change, something else occurs that in *Structure* got mentioned only in passing. After a revolution there are usually (perhaps always) more cognitive specialties or fields of knowledge than there were before. Either a new branch has split off from the parent trunk, as scientific specialties have repeatedly split off in the past from philosophy and from medicine. Or else a new specialty has been born at an area of apparent overlap between two preexisting specialties, as occurred, for example, in the cases of physical chemistry and molecular biology. At the time of its occurrence this second sort of split is often hailed as a reunification of the sciences, as was the case in the episodes just mentioned. As time goes on, however, one notices that the new shoot seldom or never gets assimilated to either of its parents. Instead, it becomes one more separate specialty, gradually acquiring its own new specialists' journals, a new professional society, and often also new university chairs, laboratories, and even departments. Over time a diagram of the evolution of scientific fields, special-

7. M. Biagioli, "The Anthropology of Incommensurability," *Studies in History and Philosophy of Science* 21 (1990): 183–209.

ties, and subspecialties comes to look strikingly like a layman's diagram for a biological evolutionary tree. Each of these fields has a distinct lexicon, though the differences are local, occurring only here and there. There is no lingua franca capable of expressing, in its entirety, the content of them all or even of any pair.

With much reluctance I have increasingly come to feel that this process of specialization, with its consequent limitation on communication and community, is inescapable, a consequence of first principles. Specialization and the narrowing of the range of expertise now look to me like the necessary price of increasingly powerful cognitive tools. What's involved is the same sort of development of special tools for special functions that's apparent also in technological practice. And, if that is the case, then a couple of additional parallels between biological evolution and the evolution of knowledge come to seem especially consequential. First, revolutions, which produce new divisions between fields in scientific development, are much like episodes of speciation in biological evolution. The biological parallel to revolutionary change is not mutation, as I thought for many years, but speciation. And the problems presented by speciation (e.g., the difficulty in identifying an episode of speciation until some time after it has occurred, and the impossibility, even then, of dating the time of its occurrence) are very similar to those presented by revolutionary change and by the emergence and individuation of new scientific specialties.

The second parallel between biological and scientific development, to which I return again in the concluding section, concerns the unit which undergoes speciation (not to be confused with a unit of selection). In the biological case, it is a reproductively isolated population, a unit whose members collectively embody the gene pool, which ensures both the population's self-perpetuation and its continuing isolation. In the scientific case, the unit is a community of intercommunicating specialists, a unit whose members share a lexicon that provides the basis for both the conduct and the evaluation of their research and which simultaneously, by barring full communication with those outside the group, maintains their isolation from practitioners of other specialties.

To anyone who values the unity of knowledge, this aspect of specialization—lexical or taxonomic divergence, with consequent limitations on communication—is a condition to be deplored. But such unity may be in principle an unattainable goal, and its energetic pursuit might well place the growth of knowledge at risk. Lexical diversity and the principled limit it imposes on communication may be the isolating mechanism required for the development of knowledge. Very likely it is the specialization consequent on lexical diversity that permits the sciences, viewed collectively, to solve the puzzles posed by a wider range of natural phenomena than a lexically homogeneous science could achieve.

Though I greet the thought with mixed feelings, I am increasingly persuaded that the limited range of possible partners for fruitful intercourse is the essential precondition for what is known as progress in both biological development and the development of knowledge. When I suggested earlier that incommensurability, properly understood, could reveal the source of the cognitive bite and authority of the sciences, its role as an isolating mechanism was prerequisite to the topic I had principally in mind, the one to which I now turn.

This reference to 'intercourse', for which I shall henceforth substitute the term 'discourse', brings me back to problems concerning truth, and thus to the locus of the newly restored bite. I said earlier that we must learn to get along without anything at all like a correspondence theory of truth. But something like a redundancy theory of truth is badly needed to replace it, something that will introduce minimal laws of logic (in particular, the law of non-contradiction) and make adhering to them a precondition for the rationality of evaluations.[8] On this view, as I wish to employ it, the essential function of the concept of truth is to require choice between acceptance and rejection of a statement or a theory in the face of evidence shared by all. Let me try briefly to sketch what I have in mind.

Ian Hacking, in an attempt to denature the apparent relativism associated with incommensurability, spoke of the way in which new "styles" introduce into science new candidates for true/false.[9] Since that time, I've been gradually realizing (the reformulation is still in process) that some of my own central points are far better made without speaking of statements as themselves being true or as being false. Instead, the evaluation of a putatively scientific statement should be conceived as comprising two seldom-separated parts. First, determine the status of the statement: is it a candidate for true/false? To that question, as you'll shortly see, the answer is lexicon-dependent. And second, supposing a positive answer to the first, is the statement rationally assertable? To that question, given a lexicon, the answer is properly found by something like the normal rules of evidence.

8. P. Horwich, *Truth* (Oxford: Blackwell, 1990).
9. I. Hacking, "Language, Truth, and Reason," in *Rationality and Relativism*, ed. M. Hollis and S. Lukes (Cambridge, MA: MIT Press, 1982), pp. 49–66.

In this reformulation, to declare a statement a candidate for true/ false is to accept it as a counter in a language game whose rules forbid asserting both a statement and its contrary. A person who breaks that rule declares him- or herself outside the game. If one nevertheless tries to continue play, then discourse breaks down; the integrity of the language community is threatened. Similar, though more problematic, rules apply, not simply to contrary statements, but more generally to logically incompatible ones. There are, of course, language games without the rule of non-contradiction and its relatives: poetry and mystical discourse, for example. And there are also, even within the declarative-statement game, recognized ways of bracketing the rule, permitting and even exploiting the use of contradiction. Metaphor and other tropes are the most obvious examples; more central for present purposes are the historian's restatements of past beliefs. (Though the originals were candidates for true/false, the historian's later restatements—made by a bilingual speaking the language of one culture to the members of another—are not.) But in the sciences and in many more ordinary community activities, such bracketing devices are parasitic on normal discourse. And these activities—the ones that presuppose normal adherence to the rules of the true/false game—are an essential ingredient of the glue that binds communities together. In one form or another, the rules of the true/false game are thus universals for all human communities. But the result of applying those rules varies from one speech community to the next. In discussion between members of communities with differently structured lexicons, assertability and evidence play the same role for both only in areas (there are always a great many) where the two lexicons are congruent.

Where the lexicons of the parties to discourse differ, a given string of words will sometimes make different statements for each. A statement may be a candidate for truth/falsity with one lexicon without having that status in the others. And even when it does, the two statements will not be the same: though identically phrased, strong evidence for one need not be evidence for the other. Communication breakdowns are then inevitable, and it is to avoid them that the bilingual is forced to remember at all times which lexicon is in play, which community the discourse is occurring within.

These breakdowns in communication do, of course, occur: they're a significant characteristic of the episodes *Structure* referred to as 'crises'. I take them to be the crucial symptoms of the speciation-like process through which new disciplines emerge, each with it own lexicon, and each with its own area of knowledge. It is by these divisions, I've been suggesting, that knowledge grows. And it's the need to maintain discourse, to keep the game of declarative statements going, that forces these divisions and the fragmentation of knowledge that results.

I close with some brief and tentative remarks about what emerges from this position as the relationship between the lexicon—the shared taxonomy of a speech community—and the world the members of that community jointly inhabit. Clearly it cannot be the one Putnam has called metaphysical realism.[10] Insofar as the structure of the world can be experienced and the experience communicated, it is constrained by the structure of the lexicon of the community which inhabits it. Doubtless some aspects of that lexical structure are biologically determined, the products of a shared phylogeny. But, at least among advanced creatures (and not just those linguistically endowed), significant aspects are determined also by education, by the process of socialization, that is, which initiates neophytes into the community of their parents and peers. Creatures with the same biological endowment may experience the world through lexicons that are here and there very differently structured, and in those areas they will be unable to communicate all of their experiences across the lexical divide. Though individuals may belong to several interrelated communities (thus, be multilinguals), they experience aspects of the world differently as they move from one to the next.

Remarks like these suggest that the world is somehow mind-dependent, perhaps an invention or construction of the creatures which inhabit it, and in recent years such suggestions have been widely pursued. But the metaphors of invention, construction, and mind-dependence are in two respects grossly misleading. First, the world is not invented or constructed. The creatures to whom this responsibility is imputed, in fact, find the world already in place, its rudiments at their birth and its increasingly full actuality during their educational socialization, a socialization in which examples of the way the world is play an essential part. That world, furthermore, has been experientially given, in part to the new inhabitants directly, and in part indirectly, by inheritance, embodying the experience of their forebears. As such, it is entirely solid: not in the least respectful of an observer's wishes and desires; quite capable of providing decisive evidence against invented hypotheses which fail to match its behavior. Creatures born into it must take it as they find it. They can, of course, interact with it, altering both it and

10. H. Putnam, *Meaning and the Moral Sciences* (London: Routledge, 1978), pp. 123–38.

themselves in the process, and the populated world thus altered is the one that will be found in place by the generation which follows. The point closely parallels the one made earlier about the nature of evaluation seen from a developmental perspective: there, what required evaluation was not belief but change in some aspects of belief, the rest held fixed in the process; here, what people can effect or invent is not the world but changes in some aspects of it, the balance remaining as before. In both cases, too, the changes that can be made are not introduced at will. Most proposals for change are rejected on the evidence; the nature of those that remain can rarely be foreseen, and the consequences of accepting one or another of them often prove to be undesired.

Can a world that alters with time and from one community to the next correspond to what is generally referred to as "the real world"? I do not see how its right to that title can be denied. It provides the environment, the stage, for all individual and social life. On such life it places rigid constraints; continued existence depends on adaptation to them; and in the modern world scientific activity has become a primary tool for adaptation. What more can reasonably be asked of a real world?

In the penultimate sentence, above, the word 'adaptation' is clearly problematic. Can the members of a group properly be said to adapt to an environment which they are constantly adjusting to fit their needs? Is it the creatures who adapt to the world, or does the world adapt to the creatures? Doesn't this whole way of talking imply a mutual plasticity incompatible with the rigidity of the constraints that make the world real and that made it appropriate to describe the creatures as adapted to it? These difficulties are genuine, but they necessarily inhere in any and all descriptions of undirected evolutionary processes. The identical problem is, for example, currently the subject of much discussion in evolutionary biology. On the one hand the evolutionary process gives rise to creatures more and more closely adapted to a narrower and narrower biological niche. On the other, the niche to which they are adapted is recognizable only in retrospect, with its population in place: it has no existence independent of the community which is adapted to it.[11] What actually evolves, therefore, are creatures and niches together: what creates the tensions inherent in talk of adaptation is the need, if discussion and analysis are to be possible, to draw a line between the

11. R. C. Lewontin, "Adaptation," *Scientific American* 239 (1978): 212–30.

creatures within the niche, on the one hand, and their "external" environment, on the other.

Niches may not seem to be worlds, but the difference is one of viewpoint. Niches are where *other* creatures live. We see them from outside and thus in physical interaction with their inhabitants. But the inhabitants of a niche see it from inside and their interactions with it are, to them, intentionally mediated through something like a mental representation. Biologically, that is, a niche is the world of the group which inhabits it, thus constituting it a niche. Conceptually, the world is *our* representation of *our* niche, the residence of the particular human community with whose members we are currently interacting.

The world-constitutive role assigned here to intentionality and mental representations recurs to a theme characteristic of my viewpoint throughout its long development: compare my earlier recourse to gestalt switches, seeing as understanding, and so on. This is the aspect of my work that, more than any other, has suggested that I took the world to be mind-dependent. But the metaphor of a mind-dependent world—like its cousin, the constructed or invented world—proves to be deeply misleading. It is groups and group practices that constitute worlds (and are constituted by them). And the practice-in-the-world of some of those groups *is* science. The primary unit through which the sciences develop is thus, as previously stressed, the group, and groups do not have minds. Under the unfortunate title "Are species individuals?" contemporary biological theory offers a significant parallel.[12] In one sense the procreating organisms which perpetuate a species are the units whose practice permits evolution to occur. But to understand the outcome of that process one must see the evolutionary unit (not to be confused with a unit of selection) as the gene pool shared by those organisms, the organisms which carry the gene pool serving only as the parts which, through bisexual reproduction, exchange genes within the population. Cognitive evolution depends, similarly, upon the exchange, through discourse, of statements within a community. Though the units which exchange those statements are individual scientists, understanding the advance of knowledge, the outcome of their practice, depends upon seeing them as atoms constitutive of a larger whole, the community of practitioners of some scientific specialty.

---

12. D. J. Hull provides an especially useful introduction to the literature in "Are Species Really Individual?" *Systematic Zoology* 25 (1976): 174–91.

The primacy of the community over its members is reflected also in the theory of the lexicon, the unit which embodies the shared conceptual or taxonomic structure that holds the community together and simultaneously isolates it from other groups. Conceive the lexicon as a module within the head of an individual group member. It can then be shown (though not here) that what characterizes members of the group is possession not of identical lexicons, but of mutually congruent ones, of lexicons with the same structure. The lexical structure which characterizes a group is more abstract than, different in kind from, the individual lexicons or mental modules which embody it. And it is only that structure, not its various individual embodiments, that members of the community must share. The mechanics of taxonomizing are in this respect like its function: neither can be fully understood except as grounded within the community it serves.

By now it may be clear that the position I'm developing is a sort of post-Darwinian Kantianism. Like the Kantian categories, the lexicon supplies preconditions of possible experience. But lexical categories, unlike their Kantian forebears, can and do change, both with time and with the passage from one community to another. None of those changes, of course, is ever vast. Whether the communities in question are displaced in time or in conceptual space, their lexical structures must overlap in major ways, or there could be no bridgeheads permitting a member of one to acquire the lexicon of the other. Nor, in the absence of major overlap, would it be possible for the members of a single community to evaluate proposed new theories when their acceptance required lexical change. Small changes, however, can have large-scale effects. The Copernican revolution provides especially well-known illustrations.

Underlying all these processes of differentiation and change, there must, of course, be something permanent, fixed, and stable. But, like Kant's *Ding an sich*, it is ineffable, undescribable, undiscussible. Located outside of space and time, this Kantian source of stability is the whole from which have been fabricated both creatures and their niches, both the "internal" and the "external" worlds. Experience and description are possible only with the described and describer separated, and the lexical structure which marks that separation can do so in different ways, each resulting in a different, though never wholly different, form of life. Some ways are better suited to some purposes, some to others. But none is to be accepted as true or rejected as false; none gives privileged access to a real, as against an invented, world. The ways of being-in-the-world which a lexicon provides are not candidates for true/false.

Chapter Five

The Trouble with the Historical Philosophy of Science

"The Trouble with the Historical Philosophy of Science" was the first lecture of the Robert and Maurine Rothschild Distinguished Lecture Series, delivered at Harvard University on November 19, 1991. It was published in booklet form the following year by the Department of History of Science, Harvard University.

———◦———

THE INVITATION TO INAUGURATE the Robert and Maurine Rothschild Lectures has given me great pleasure. For it I have, in the first instance, the Rothschilds to thank, and I join the Department and the University in doing so. Without this new example of their generosity, there would be no series. But I want also to thank the Department of the History of Science for asking me to lead off. Invitations to participate in a series like this one characteristically include a rather frightening list of the distinguished people who've previously graced the chair. Only the inaugurator of a series escapes, and it's been most helpful to me, in preparing for this occasion, that I've not had a roster of elder statesmen peering over my shoulder. Unfortunately, however, I found a substitute. Given my chosen title, few of you will be surprised to find that during much of this lecture the person peering over my shoulder will be me.

Turning to my topic, let me begin by telling you what I shall be trying to do. As many of you know, the image of science current both inside and, less completely, outside the academy has, during the last quarter century, been quite radically transformed. I was myself a contributor to that transformation, think it was badly needed, and have few significant regrets. The change has, I think, begun to yield a far more realistic understanding of what the scientific enterprise is, how it operates, and what it can and cannot achieve than was available before. But the transformation has had a by-product—centrally philosophical, but with implications also for the historical and sociological study of science—that frequently troubles me, not least because it was initially emphasized and developed by people who often called themselves Kuhnians. I think their viewpoint damagingly mistaken, have been pained to be associated with it, and have for years attributed that association to misunderstanding. Recently, however, I've increasingly recognized that something central to the new view of science was also involved, and I shall be trying this afternoon to confront it.

The talk that results has three parts. The first is a brief account of what I think went wrong and of some possible reasons for its having done so. The second sketches a route by which damage might be avoided and our understanding of the scientific enterprise further improved. In this more constructive portion of my talk, I'll be drawing on bits and pieces of a far larger project, the book on which I'm currently at work. Drastic condensation and simplification of even those pieces is going to be essential, and the central part of my project—a theory of what I once called incommensurability—will have to be omitted entirely. Finally, at the very end of this talk, I shall briefly suggest how the viewpoint I develop today fits the larger pattern of my past and future work.

The new approach that has so fundamentally altered the received image of science was historical in nature, but none of those who produced it was in the first instance a historian. Rather they were philosophers, mostly professionals, plus a few amateurs, the latter usually trained in science. I'm a case in point. Though most of my career has been devoted to history of science, I began as a theoretical physicist with a strong avocational interest in philosophy and almost none in history. Philosophical goals prompted my move to history; it's to philosophy that I've gone back in the last ten or fifteen years; and it's as a philosopher that I speak this afternoon. Like my fellow innovators I was primarily motivated by widely recognized difficulties in the then current philosophy of science, most prominently in positivism or logical empiricism but in other sorts of empiricism as well. What we mostly thought we were doing as we turned to history was building a philosophy of science on observations of scientific life, the historical record providing our data.

All of us had been brought up to believe, more or less strictly, in one or another version of a traditional set of beliefs about which I'll briefly and schematically remind you. Science proceeds from facts given by observation. Those facts are objective in the sense that they are interpersonal: they are, it was said, accessible to and indubitable for all normally equipped human observers. They had to be discovered, of course, before they could become data for science, and their discovery often required the invention of elaborate new instruments. But the need to search out the facts of observation was not seen as a threat to their authority once they were found. Their status as the objective starting point, available to all, remained secure. These facts, the older image of science continued, are prior to the scientific laws and theories for which they provide the foundation, and which are themselves, in turn, the basis for explanations of natural phenomena.

Unlike the facts on which they are based, these laws, theories, and explanations are not simply given. To find them one must interpret the facts—invent laws, theories, and explanations to fit them. And interpretation is a human process, by no means the same for all: different individuals can be expected to interpret the facts differently, to invent characteristically different laws and theories. But, again, observed facts were said to provide a court of final appeal. Two sets of laws and theories do not ordinarily have all the same consequences, and tests designed to see which set of consequences are observed will eliminate at least one of them.

Variously construed, these processes constituted something called the scientific method. Sometimes thought to have been invented in the seventeenth century, this was the method by which scientists discovered true generalizations about and explanations for natural phenomena. Or if not exactly true, at least approximations to the truth. And if not certain approximations, then at least highly probable ones. Something of this sort we had all been taught, and we all knew that attempts to refine that understanding of scientific method and what it produced had encountered deep, though isolated, difficulties that were not, after centuries of effort, responding to treatment. It was those difficulties which drove us to observations of scientific life and to history, and we were considerably disconcerted by what we found there.

In the first place, the supposedly solid facts of observations turned out to be pliable. The results achieved by different people apparently observing the same phenomena differed from one another, though never a great deal. And those differences—though contained in the same ball park—were often sufficient to affect crucial points of interpretation. In addition the so-called facts proved never to be mere facts, independent of existing belief and theory. Producing them required apparatus which itself depended on theory, often on the theory that the experiments were supposed to test. Even when the apparatus could be redesigned to eliminate or reduce these disagreements, the design process sometimes forced the revision of conceptions about what was being observed. And after that, disagreements, though reduced, were still present, and sometimes sufficient to have a bearing on interpretation. Observations, that is, including those designed as tests, always left room for disagreement about whether some particular law or theory should be accepted. That space for disagreement was often exploited: discrepancies that to an outsider looked trivial were frequently matters of deep import to those on whom the research impinged.

In these circumstances—a third aspect of what we found in the record—individuals committed to one interpretation or another sometimes defended their viewpoint in ways that violated their professed canons of professional behavior. I am not thinking primarily of fraud, which was relatively rare. But failure to acknowledge contrary findings, the substitution of personal innuendo for argument, and other techniques of the sort were not. Controversy about scientific matters sometimes looked much like a cat fight.

Philosophically, none of this need have been a problem. None of the things I've just said was quite new. Practitioners of traditional philosophy of science had been at least dimly aware of them. They were taken to be reminders that science was practiced by fallible humans in a less than ideal world. Traditional philosophy of science was concerned to provide methodological norms, and it supposed these were sufficiently powerful to withstand the effects of occasional violations. The behavior I've just described was acknowledged but left aside; it was taken to play no positive role in the formation of scientific doctrine. But the historically inclined philosophers of science looked at these observations differently. We were already dissatisfied with the prevailing tradition and were seeking behavioral clues with which to reform it. These aspects of scientific life provided a plausible starting point.

If observation and experiment were insufficient to bring different individuals to the same decision, the differences in what they took to be facts and in the decisions they based upon them must, we thought, be due to personal factors, unacknowledged by previous philosophy of science. Individuals might, for example, differ by virtue of the personal history and tastes which underlay their research agendas. Another likely source of difference was the estimated rewards and penalties, whether financial or in prestige, likely to result from the individual's decision. These and other individual interests could all be observed at work in the historical record, and they did not appear to be eliminable. Where observation itself was insufficient to compel even an individual's decision, only factors like these or else the toss of a coin could fill the gap.

Given this initial divergence between the conclusions of individuals, it became urgent to determine the process by which differences of belief were reconciled en route to an ultimate consensus within the group. What is the process, that is, by which the outcome of experiments is uniquely specified as fact and by which authoritative new beliefs—new scientific laws and theories—come to be based upon that outcome? Those are the questions central to the work of the generation that followed mine, and the principal contributions to them have come not from philosophy but from a new kind of historical and, more especially, of sociological studies to which the work of my generation helped give rise. These studies have dealt, in microscopic detail, with the processes within a scientific community or group from which an authoritative consensus finally emerges, a process this literature often refers to as "negotiation." Some of these studies seem to me brilliant, and all of them are revealing of aspects of the scientific process that we have badly needed to know. There can be no question, I think, about either their novelty or their importance. But their net effect, at least from a philosophical perspective, has been to deepen rather than to eliminate the very difficulty they were intended to resolve.

What negotiation, so-called, seeks to establish is the facts from which scientific conclusions should be drawn, together with the conclusions—the new laws or theories—which should be based upon them. These two aspects of the negotiation—the factual and the interpretive—are carried on concurrently, the conclusions shaping the description of facts just as the facts shape the conclusions drawn from them. Such a process is clearly circular, and it becomes very difficult to see what role experiment could have in determining its outcome. That difficulty is further aggravated because the need for negotiation appears to result from the sorts of individual differences generally described as mere matters of biographical fact. What causes the parties to the negotiation to reach different conclusions are, as I've indicated, things like differences in individual history, research agenda, personal interest. These are the sorts of differences that might be eliminated by retraining or brainwashing, but they are not in principle accessible to reasoned argument or negotiation.

The question therefore arose, how can a process so nearly circular and so largely dependent on individual contingencies be said to result in either true or probable conclusions about the nature of reality? I take that to be a serious question and think that inability to answer it is a grave loss in our understanding of the nature of scientific knowledge. But the question emerged during the 1960s, when distrust of all sorts of authority was widespread, and it was then a small step to regard that loss as gain. Negotiations in science, like those in politics, diplomacy, business, and many other aspects of societal life, were widely said— especially by sociologists and political scientists—to be governed by interest, their outcome determined by considerations of authority and power. That was the thesis of those who first applied the term 'negotiation' to the scientific process, and the term carried much of the thesis with it.

I do not think that either the term or the description of the activities it covered was merely mistaken. Interest, politics, power, and authority undoubtedly do play a significant role in scientific life and its development. But the form taken by studies of "negotiation" has, as I've indicated, made it hard to see what else may play a role as well. Indeed, the most extreme form of the movement, called by its proponents "the strong program," has been widely understood as claiming that power and interest are all there are. Nature itself, whatever that may be, has seemed to have no part in the development of beliefs about it. Talk of evidence, of the rationality of claims drawn from it, and of the truth or probability of those claims has been seen as simply the rhetoric behind which the victorious party cloaks its power. What passes for scientific knowledge becomes, then, simply the belief of the winners.

I am among those who have found the claims of the strong program absurd: an example of deconstruction gone mad. And the more qualified sociological and historical formulations that currently strive to replace it are, in my view, scarcely more satisfactory. These newer formulations freely acknowledge that observations of nature do play a role in scientific development. But they remain almost totally uninformative about that role—about the way, that is, in which nature enters the negotiations that produce beliefs about it.

The strong program and its descendants have repeatedly been dismissed as uncontrolled expressions of hostility to authority in general and science in particular. For some years I reacted somewhat that way myself. But I now think that easy evaluation ignores a real philosophical challenge. There's a continuous line (or continuous slippery slope) from the inescapable initial observations that underlie microsociological studies to their still entirely unacceptable conclusions. Much that ought not be abandoned has been learned by travelling that line. And it remains unclear how, without abandoning those lessons, the line can be deflected or interrupted, how its unacceptable conclusions can be avoided.

A remark recently made to me by Marcello Pera suggests a likely clue to these difficulties. The authors of microsociological studies are, he suggests, taking the traditional view of scientific knowledge too much for granted. They seem, that is, to feel that traditional philosophy of science was correct in its understanding of what *knowledge* must be. Facts must come first, and inescapable conclusions, at least about probabilities, must be based upon them. If science doesn't produce knowledge in that sense, they conclude, it cannot be producing knowledge at all. It is possible, however, that the tradition was wrong not simply about the methods by which knowledge was obtained, but about the nature of knowledge itself. Perhaps knowledge, properly understood, is the product of the very process these new studies describe. I think something of that sort is the case, and I shall try in the remainder of this lecture to make the point by sketching a few aspects of my current work.

—◦—

Early in this talk I suggested that my generation of philosopher/ historians saw ourselves as building a philosophy on observations of actual scientific behavior. Looking back now, I think that that image of what we were up to is misleading. Given what I shall call the historical perspective, one can reach many of the central conclusions we drew with scarcely a glance at the historical record itself. That historical perspective was, of course, initially foreign to all of us. The questions which led us to examine the historical record were products of a philosophical tradition that took science as a static body of knowledge and asked what rational warrant there was for taking one or another of its component beliefs to be true. Only gradually, as a by-product of our study of historical "facts," did we learn to replace that static image with a dynamic one, an image that made science an ever-developing enterprise or practice. And it is taking longer still to realize that, with that perspective achieved, many of the most central conclusions we drew from the historical record can be derived instead from first principles. Approaching them in that way reduces their apparent contingency, making them harder to dismiss as a product of muckraking investigation by those hostile to science. And the approach from principle yields, in addition, a very different view of what's at stake in the evaluative processes that have been taken to epitomize such concepts as reason, evidence, and truth. Both these changes are clear gains.

The characteristic concern of the historian is development over time, and the typical result of his or her activity is embodied in narrative. Whatever its subject, the narrative must always open by setting the stage, by describing, that is, the state of affairs in place at the beginning of the series of events that constitutes the narrative proper. If that narrative deals with beliefs about nature, then it must open with a description of what people believed at the time when it began. That description must make it plausible that the beliefs were held by human actors, for which purpose it must include a specification of the conceptual vocabulary in which natural phenomena were described and in which beliefs about phenomena were stated. With the stage thus set, the narrative proper begins, and it tells the story of change of belief over time and of the changing context within which those alterations occurred. By the end of the narrative those changes may be considerable, but they have occurred in small increments, each stage historically situated in a climate somewhat different from that of the one before. And at each of those stages except the first, the historian's problem is to understand, not why people held the beliefs that they did, but why they elected to change them, why the incremental change took place.

For the philosopher who adopts the historical perspective, the problem is the same: understanding small incremental *changes* of belief. When questions about rationality, objectivity, or evidence arise in that context, they are addressed not to the beliefs that were current either before or after the change, but simply to the change itself. Why, that is, given the body of belief with which they began, do the members of a scientific group elect to alter it, a process that is seldom a mere addition but ordinarily calls for the adjustment or abandonment of a few beliefs already in place? From the philosophical point of view, the difference between those two formulations—the rationality of belief versus the rationality of incremental change of belief—is vast. I shall touch on just three of a number of significant differences, each requiring discussion at greater length than any of them can be given here.

As I've already said, the tradition supposed that good reasons for belief could be supplied only by neutral observations, the sort of observations, that is, which are the same for all observers and also independent of all other beliefs and theories. These provided the stable Archimedean platform required to determine the truth or the probability of the particular belief, law, or theory to be evaluated. But observations that satisfy such criteria have proved, as I've indicated, to be few and far between. The traditional Archimedean platform provides an insufficient base for the rational evaluation of belief, a fact that the strong program and its relatives have exploited. From the historical perspective, however, where change of belief is what's at issue, the *rationality* of the conclusions requires only that the observations invoked be neutral for, or shared by, the members of the group making the decision, and for them only at the time the decision is being made. By the same token, the observations involved need no longer be independent of all prior beliefs, but only of those that would be modified as a result of the change. The very large body of beliefs unaffected by the change provides a basis on which discussion of the desirability of change can rest. It is simply irrelevant that some or all of those beliefs may be set aside at some future time. To provide a basis for rational discussion they, like the observations the discussion invokes, need only be shared by the discussants. There is no higher criterion of the rationality of discussion than that. The historical perspective, thus, also invokes an Archimedean platform, but it is not fixed. Rather, it moves with time and changes with community and sub-community, with culture and subculture. Neither of those sorts of change interferes with its providing a basis for reasoned discussions and evaluations of proposed changes in the body of belief current in a given community at a given time.

The second difference between the evaluation of belief and the evaluation of change of belief can be more briefly put. From the historical perspective the changes to be evaluated are always small. In retrospect, some of them seem gigantic, and these regularly affect a considerable body of beliefs. But all of them have been prepared gradually, step by step, leaving only a keystone to be put in place by the innovator whose name they bear. And that step too is small, clearly foreshadowed by the steps that have been taken before: only in retrospect, after it has been taken, does it gain the status of a keystone. No wonder the process of evaluating the desirability of change seems circular. Many of the considerations that suggested the nature of the change to the innovator are also the ones that supply reasons for accepting the proposal he or she has made. The question which came first, the idea or the observation, is like the question about the chicken and the egg, and *that* question has never raised doubts that one outcome of the process is chickens.

The third effect of shifting the evaluation from belief to change of belief is closely related and perhaps more striking. Within the main formulation of the previous tradition in philosophy of science, beliefs were to be evaluated for their truth or for their probability of being true, where truth meant something like corresponding to the real, the mind-independent external world. There was also a secondary formulation which held that beliefs were to be evaluated for their utility, but I must for lack of time omit that formulation here. A dogmatic assertion that it fails to account for essential aspects of scientific development will have to substitute for discussion.

Sticking therefore with the formulation that assumes truth to be the goal of evaluations, notice that it requires evaluation to be indirect. Seldom or never can one compare a newly proposed law or theory directly with reality. Rather, for purposes of evaluation, one must embed it in a relevant body of currently accepted beliefs—for example, those governing the instruments with which the relevant observations have been made—and then apply to the whole a set of secondary criteria. Accuracy is one of these, consistency with other accepted beliefs is another, breadth of applicability a third, simplicity a fourth, and there are others besides. All these criteria are equivocal, and they are rarely all satisfied at once. Accuracy is ordinarily approximate, and often unavailable. Consistency is at best local: it has not characterized the sciences as a whole since at least the seventeenth century. Breadth of applicability becomes narrower with time, a point to which I shall return. Simplicity is in the eye of the beholder. And so on.

These traditional criteria of evaluation have been scrutinized also by the microsociologists, who ask, not unreasonably, how, in the circumstances, they can be viewed as more than window dressing. But look what happens to these same criteria—upon which I cannot much improve—when applied to comparative evaluation, to change of belief rather than directly to belief itself. To ask which of two bodies of belief is *more* accurate, displays *fewer* inconsistencies, has a *wider* range of applications, or achieves these goals with the *simpler* machinery does not eliminate all ground for disagreement, but the comparative judgment is clearly far more tractable than the traditional one from which it derives.

Especially since what must be compared are only sets of beliefs actually in place in the historical situation. For that comparison, even a somewhat equivocal set of criteria may over time be adequate.

I take that change in the object of evaluation to be both clear and important. But there is a price to be paid for it, and it's again one that may help to explain the appeal of the microsociological point of view. A new body of belief could be *more* accurate, *more* consistent, broad*er* in its range of applicability, and also simpl*er* without for those reasons being any tru*er*. Indeed, even the term 'truer' has a vaguely ungrammatical ring: it is hard to know quite what those who use it have in mind. For 'truer' some people would therefore substitute 'more probable', but this leads to difficulties of another sort, ones emphasized in a slightly different context by Hilary Putnam. All past beliefs about nature have sooner or later turned out to be false. On the record, therefore, the probability that any currently proposed belief will fare better must be close to zero. What remains to be claimed is embodied in a standard locution developed within the tradition: successive scientific law and theories grow closer and closer to the truth. That could, of course, be the case, but at present it's not even clear what is being claimed. Only a fixed, rigid Archimedean platform could supply a base from which to measure the distance between current belief and true belief. In the absence of that platform, it's hard to imagine what such a measurement would be, what the phrase 'closer to the truth' can mean.

Lacking time to carry this part of my argument further, I shall simply assert or reassert a tripartite conviction. First, the Archimedean platform outside of history, outside of time and space, is gone beyond recall. Second, in its absence, comparative evaluation is all there is. Scientific development is like Darwinian evolution, a process driven from behind rather than pulled toward some fixed goal to which it grows ever closer. And third, if the notion of truth has a role to play in scientific development, which I shall elsewhere argue that it does, then truth cannot be anything quite like correspondence to reality. I am not suggesting, let me emphasize, that there is a reality which science fails to get at. My point is rather that no sense can be made of the notion of reality as it has ordinarily functioned in philosophy of science.

Notice now that, to this point, my position is much like that of the strong program—facts are not prior to conclusions drawn from them, and those conclusions cannot claim truth. But I've reached that position from principles that must govern all developmental processes, without, that is, needing to call upon actual examples of scientific behavior. Nothing along that route has suggested replacing evidence and reason by power and interest. Of course power and interest play a role in scientific development, but there's room for a great deal else besides.

————◦————

To clarify the way in which other determinants of scientific development enter, let me turn to some even more excessively condensed remarks about a second aspect of the historical or developmental perspective. This one, unlike the last, is not a necessary or an a priori characteristic, but must be suggested by observations. The observations involved are not, however, restricted to the sciences and they require, in any case, no more than a glance. What I have in mind is the apparently inexorable (albeit ultimately self-limiting) growth in the number of distinct human practices or specialties over the course of human history. I shall use the term 'speciation' to describe this aspect of development, though the parallel to biological evolution is by no means so precise in this case as it was in the last. In my closing remarks I'll advert to one particularly important disanalogy.

I know the proliferation of specialties best in the sciences, where it is, perhaps, especially prominent. But it's clearly been present in all realms of human activity. Kings and chieftains dispensed justice before there were judges and lawyers. There were wars before there was a military, and a military before there was an army, navy, and air force. Or, within the area of religion, think just of the Pauline church and the number of churches that have sprung and are still springing from it.

In the sciences the pattern is, if anything, more obvious. In antiquity there is mathematics—which includes astronomy, optics, mechanics, geography, and music—as well as medicine and natural philosophy, neither of which you may want to call sciences, but which are recognized practices that will later be a major source of sciences. During the later seventeenth century, the various components of mathematics separate from their parent and from each other. Simultaneously the speculative chemistry within natural philosophy begins to become a field in its own right through involvement with problems from medicine and the crafts. The specialties that are going to be physics begin to separate from natural philosophy, and the same sort of proliferation produces the early biological sciences from medicine. During the nineteenth century these individual specialties that collectively constitute science rapidly acquire their own special societies, journals, university departments, and special chairs.

The same pattern continues even more rapidly today, something I can most easily document from personal experience. When I left Harvard in 1957, the life sciences were the preserve of a single department, biology. That institutionalization was, I thought, a natural division of knowledge, written in stone, or at Harvard in brick. And I was correspondingly shocked to discover on arrival in California that my new home, Berkeley, required three departments to cover the topics covered in Cambridge by one. Now on returning to Cambridge, I find that there are four departments for the life sciences, and it would surprise me if there were not by now an even larger number at Berkeley. While this has been going on, something only slightly less dramatic had been happening to my old field, physics. When I took my degree a single journal, the *Physical Review*, published most of the contributions to knowledge made by U.S. physicists. All professionals subscribed to it, though only a few of them could (and fewer did) read all the articles in any given issue. Now that journal has been split in four and few individuals subscribe to more than one or two of them. Though departments have not been divided, there's been a considerable elaboration of the substructure of the profession: more subgroups with their own societies and their own special journals. What results is the elaborate and rather ramshackle structure of separate fields, specialities, and subspecialties within which the business of producing scientific knowledge is carried forward.

Knowledge production is the particular business of the subspecialties, whose practitioners struggle to improve *incrementally* the accuracy, consistency, breadth of applicability, and simplicity of the set of beliefs they acquired during their education, their initiation into the practice. It is the beliefs modified in this process that they transmit to their successors who carry on from there, working with and modifying scientific knowledge as they go. Occasionally the process runs aground, and the proliferation and reorganization of specialties is usually part of the required remedy. What I am thus suggesting, in an excessively compressed way, is that human practices in general and scientific practices in particular have evolved over a very long time span, and their development forms something very roughly like an evolutionary tree.

Some characteristics of the various practices entered early in this evolutionary development and are shared by all human practices. I take power, authority, interest, and other "political" characteristics to be in this early set. With respect to these, scientists are no more immune than anyone else, a fact that need not have occasioned surprise. Other characteristics enter later, at some developmental branching point, and they are thereafter characteristic only of the group of practices formed by further episodes of proliferation among descendants from that branch. The sciences constitute one such group, though their development involved several branch points and much recombination. The characteristics of the members of this group are, in addition to their concern with the study of natural phenomena, the evaluative procedures I've already described and others like them. I again have in mind such characteristics as accuracy, consistency, breadth of application, simplicity, and so on—characteristics that are passed, together with illustrations, from one generation of practitioners to the next.

These characteristics are somewhat differently understood in the different scientific specialties and subspecialties. And in none of them are they by any means always observed. Nevertheless, in the fields where they have once taken hold, they account for the continuous emergence of more and more refined—and also more specialized—tools for the accurate, consistent, comprehensive, and simple descriptions of nature. Which is only to say that in such fields they're sufficient to account for the continued development of scientific knowledge. What else is scientific knowledge, and what else would you expect practices characterized by these evaluative tools to produce?

————◄o►————

With these remarks I conclude the presentation of the subject announced in my title. I shall shortly add a very short coda for those who know my earlier work. But first let me summarize the point we've reached. The trouble with the historical philosophy of science has been, I've suggested, that by basing itself upon observations of the historical record it has undermined the pillars on which the authority of scientific knowledge was formerly thought to rest without supplying anything to replace them. The most central of the pillars I have in mind were two: first, that facts are prior to and independent of the beliefs for which they are said to supply the evidence, and, second, that what emerges from the practice of science are truths, probable truths, or approximations to the truth about a mind- and culture-independent external world.

What's gone on since the undermining occurred has been efforts either to shore up those pillars or else to erase all vestige of them by showing that even in its own domain science has no special authority whatsoever. I've tried to suggest another approach. The difficulties that have seemed to undermine the authority of science should not be seen simply as observed facts about its practice. Rather they are necessary characteristics of any developmental or evolutionary process. That change makes it possible to reconceive what it is that scientists produce and how it is that they produce it.

Sketching the needed reconceptualization, I've indicated three of its main aspects. First, that what scientists produce and evaluate is not belief *tout court* but change of belief, a process which I've argued has intrinsic elements of circularity, but of a circularity that is not vicious. Second, that what evaluation aims to select is not beliefs that correspond to a so-called real external world, but simply the better or best of the bodies of belief actually present to the evaluators at the time their judgments are reached. The criteria with respect to which evaluation is made are the standard philosopher's set: accuracy, breadth of application, consistency, simplicity, and so on. And, finally, I've suggested that the plausibility of this view depends upon abandoning the view of science as a single monolithic enterprise, bound by a unique method. Rather, it should be seen as a complex but unsystematic structure of distinct specialties or species, each responsible for a different domain of phenomena, and each dedicated to changing current beliefs about its domain in ways that increase its accuracy and the other standard criteria I've mentioned. For that enterprise, I suggest, the sciences, which must then be viewed as plural, can be seen to retain a very considerable authority.

———◦———

So much for my summary: let me now turn to a three-minute coda. Those of you who know of me are likely to know me primarily as the author of *The Structure of Scientific Revolutions*. That's a book in which the most central notions are "revolutionary change," on the one hand, and something called "incommensurability," on the other. Explicating those notions, especially incommensurability, is at the heart of the project from which the ideas I've presented here are abstracted. But they've not been mentioned here, and some of you are likely to be wondering how they can still fit in. Let me offer three components of an answer, each presented so briefly and dogmatically that it's unlikely to make much sense. I give them in the order of apparently increasing absurdity.

First, the episodes that I once described as scientific revolutions are intimately associated with the ones I've here compared with speciation.

It's at this point that the previously mentioned disanalogy enters, for revolutions directly displace some of the concepts basic to the earlier practice in a field in favor of others, a destructive element not nearly so directly present in biological speciation. But in addition to the destructive element in revolutions, there's also a narrowing of focus. The mode of practice permitted by the new concepts never covers all of the field for which the earlier one took responsibility. There's always a residue (sometimes a very large one), the pursuit of which continues as an increasingly distinct specialty. Though the process of proliferation is often more complex than my reference to speciation suggests, there are regularly more specialities after a revolutionary change than there were before. The older, more encompassing modes of practice simply die off: they are the fossils whose paleontologists are historians of science.

The second component of my return to my past is the specification of what makes these specialties distinct, what keeps them apart and leaves the ground between them as apparently empty space. To that the answer is incommensurability, a growing conceptual disparity between the tools deployed in the two specialties. Once the two specialties have grown apart, that disparity makes it impossible for the practitioners of one to communicate fully with the practitioners of the other. And those communication problems reduce, though they never altogether eliminate, the likelihood that the two will produce fertile offspring.

Finally, what replaces the one big mind-independent world about which scientists were once said to discover the truth is the variety of niches within which the practitioners of these various specialties practice their trade. Those niches, which both create and are created by the conceptual and instrumental tools with which their inhabitants practice upon them, are as solid, real, resistant to arbitrary change as the external world was once said to be. But, unlike the so-called external world, they are not independent of mind and culture, and they do not sum to a single coherent whole of which we and the practitioners of all the individual scientific specialties are inhabitants.

That, far too briefly, is the context from which the ideas I've been developing this afternoon are mostly abstracted. My coda is therefore at an end. For those members of the orchestra who desire it, I conclude with the standard instruction, *da capo al fine*.

Comments and Replies

Chapter Six

Reflections on My Critics

"Reflections on My Critics" is a lengthy reply to seven essays—by John Watkins, Stephen Toulmin, L. Pearce Williams, Karl Popper, Margaret Masterman, Imre Lakatos, and Paul Feyerabend—each more or less critical of ideas put forward by Kuhn, especially in The Structure of Scientific Revolutions. *The first four of those essays were presented, following an introductory paper by Kuhn, titled "Logic of Discovery or Psychology of Research?" at a symposium entitled "Criticism and the Growth of Knowledge" at the Fourth International Colloquium in the Philosophy of Science, held in London in July 1965. The fifth essay was completed a year later, but the last two, and Kuhn's reply, were not completed until 1969. All of these were then published together as* Criticism and the Growth of Knowledge, *edited by Imre Lakatos and Alan Musgrave (London: Cambridge University Press, 1970). Reprinted with the permission of Cambridge University Press.*

———◦———

IT IS NOW FOUR YEARS since Professor Watkins and I exchanged mutually impenetrable views at the International Colloquium in the Philosophy of Science held at Bedford College, London. Rereading our contributions together with those that have since accreted to them, I am tempted to posit the existence of two Thomas Kuhns. Kuhn$_1$ is the author of this essay and of an earlier piece in this volume.[1] He also

---

Though my battle with a publication deadline allowed them almost no time for it, my colleagues C. G. Hempel and R. E. Grandy both managed to read my first manuscript and offer useful suggestions for its improvement, conceptual and stylistic. I am most grateful to them, but they should not be blamed for my views.

1. T. S. Kuhn, "Logic of Discovery or Psychology of Research?" in *Criticism and the Growth of Knowledge: Proceedings of the International Colloquium in the Philosophy of Science, London*

published in 1962 a book called *The Structure of Scientific Revolutions,* the one which he and Miss Masterman discuss above. Kuhn₂ is the author of another book with the same title. It is the one here cited repeatedly by Sir Karl Popper as well as by Professors Feyerabend, Lakatos, Toulmin, and Watkins. That both books bear the same title cannot be altogether accidental, for the views they present often overlap and are, in any case, expressed in the same words. But their central concerns are, I conclude, usually very different. As reported by his critics (his original has unfortunately been unavailable to me), Kuhn₂ seems on occasion to make points that subvert essential aspects of the position outlined by his namesake.

Lacking the wit to extend this introductory fantasy, I will instead explain why I have embarked upon it. Much in this volume testifies to what I described above as the gestalt switch that divides readers of *Structure* into two groups. Together with that book, this collection of essays therefore provides an extended example of what I have elsewhere called partial or incomplete communication—the talking-through-each-other that regularly characterizes discourse between participants in incommensurable points of view.

Such communication breakdown is important and needs much study. Unlike Paul Feyerabend (at least as I and others are reading him), I do not believe that it is ever total or beyond recourse. Where he talks of incommensurability *tout court,* I have regularly spoken also of partial communication, and I believe it can be improved upon to whatever extent circumstances may demand and patience permit, a point to be elaborated below. But neither do I believe, as Sir Karl does, that the sense in which "we are prisoners caught in the framework of our theories; our expectations; our past experiences; our language" is merely "Pickwickian." Nor do I suppose that "we can break out of our framework at any time . . . [into] a better and roomier one . . . [from which] we can at any moment break out . . . again."[2] If that possibility were routinely available, there ought to be no very special difficulties about stepping into someone else's framework in order to evaluate it. My critics' attempts to step into mine suggest, however, that changes of framework, of theory, of language, or of paradigm pose deeper problems of both principle and practice than the preceding quotations recog-

---

2. K. R. Popper, "Normal Science and Its Dangers," in *Criticism and the Growth of Knowledge,* p. 56.

nize. These problems are not simply those of ordinary discourse, nor will they be resolved by quite the same techniques. If they could be, or if changes of framework were normal, occurring at will and at any moment, they would not be comparable, in Sir Karl's phrase, to "the culture clash[es] which [have] stimulated some of the greatest intellectual revolutions" (p. 57). The very possibility of that comparison is what makes them so very important.

One especially interesting aspect of this volume is, then, that it provides a developed example of a minor culture clash, of the severe communication difficulties which characterize such clashes, and of the linguistic techniques deployed in the attempt to end them. Read as an example, it could be an object for study and analysis, providing concrete information concerning a type of developmental episode about which we know very little. For some readers, I suspect, the recurrent failure of these essays to intersect on intellectual issues will provide this book's greatest interest. Indeed, because those failures illustrate a phenomenon at the heart of my own point of view, the book has that interest for me. I am, however, too much a participant, too deeply involved, to provide the analysis which the breakdown of communication warrants. Instead, though I remain convinced that their fire is frequently misplaced and that it often obscures the deeper differences between Sir Karl's views and my own, I must here speak primarily to the points raised by my present critics.

Those points, excepting for the moment the ones raised in Miss Masterman's stimulating paper, fall into three coherent categories, each of which illustrates what I have just called the failure of our discussion to intersect on issues. The first, for purposes of my discussion, is the perceived difference in our methods: logic versus history and social psychology; normative versus descriptive. These, as I shall shortly try to show, are odd contrasts with which to discriminate among the contributors to this volume. All of us, unlike the members of what has until recently been the main movement in philosophy of science, do historical research and rely both on it and on observation of contemporary scientists in developing our viewpoints. In those viewpoints, furthermore, the descriptive and the normative are inextricably mixed. Though we may differ in our standards and surely differ about some matters of substance, we are scarcely to be distinguished by our methods. The title of my earlier paper, "Logic of Discovery or Psychology of Research?" was not chosen to suggest what Sir Karl *ought* to do but rather to describe *what he does*. When Lakatos writes, "But Kuhn's conceptual framework . . . is socio-psychological: mine is normative,"[3] I can only think that he is employing a sleight of hand to reserve the philosophical mantle for himself. Surely Feyerabend is right in claiming that my work repeatedly makes normative claims. Equally surely, though the point will require more discussion, Lakatos's position is social-psychological in its repeated reliance on decisions governed not by logical rules but by the mature sensibility of the trained scientist. If I differ from Lakatos (or Sir Karl, Feyerabend, Toulmin, or Watkins), it is with respect to substance rather than method.

As to substance, our most apparent difference is about normal science, the topic to which I shall turn immediately after discussing method. A disproportionate part of this volume is devoted to normal science, and it calls forth some of the oddest rhetoric: normal science does not exist *and* is uninteresting. On this issue we do disagree, but not, I think, either consequentially or in the ways my critics suppose. When I take it up, I shall deal in part with the real difficulties in retrieving normal scientific traditions from history, but my first and more central point will be a logical one. The existence of normal science is a corollary of the existence of revolutions, a point implicit in Sir Karl's paper and explicit in Lakatos's. If it did not exist (or if it were nonessential, dispensable for science), then revolutions would be in jeopardy also. But, about the latter, I and my critics (excepting Toulmin) agree. Revolutions through criticism demand normal science no less than revolutions through crisis. Inevitably, the term 'cross-purposes' better catches the nature of our discourse than 'disagreement'.

Discussion of normal science raises the third set of issues about which criticism has here clustered: the nature of the change from one normal-scientific tradition to another and of the techniques by which the resulting conflicts are resolved. My critics respond to my views on this subject with charges of irrationality, relativism, and the defense of mob rule. These are all labels which I categorically reject, even when they are used in my defense by Feyerabend. To say that, in matters of theory choice, the force of logic and observation cannot in principle be compelling is neither to discard logic and observation nor to suggest that there are not good reasons for favoring one theory over another. To say that trained scientists are, in such matters, the highest court of appeal is neither to defend mob rule nor to suggest that scientists could have

---

3. I. Lakatos, "Falsification and the Methodology of Scientific Research Programmes," in *Criticism and the Growth of Knowledge*, p. 177.

decided to accept any theory at all. In this area, too, my critics and I differ, but our points of difference have yet to be seen for what they are.

These three sets of issues—method, normal science, and mob rule— are the ones which bulk largest in this volume and, for that reason, in my response. But my reply cannot close without going one step beyond them to consider the problem of paradigms to which Miss Masterman's essay is devoted. I concur in her judgment that the term 'paradigm' points to the central philosophical aspect of my book but that its treatment there is badly confused. No aspect of my viewpoint has evolved more since the book was written, and her paper has helped in that development. Though my present position differs from hers in many details, we approach the problem in the same spirit including a common conviction of the relevance of the philosophy of language and of metaphor.

I shall not here be able to deal at all fully with the problems presented by my initial treatment of paradigms, but two considerations necessitate my touching upon them. Even brief discussion should permit the isolation of two quite different ways in which the term is deployed in my book and thus eliminate a constellation of confusions which has handicapped me as well as my critics. The resulting clarification will, in addition, permit me to suggest what I take to be the root of my single most fundamental difference from Sir Karl.

He and his followers share with more traditional philosophers of science the assumption that the problem of theory choice can be resolved by techniques which are semantically neutral. The observational consequences of both theories are first stated in a shared basic vocabulary (not necessarily complete or permanent). Some comparative measure of their truth/falsity count then provides the basis for a choice between them. For Sir Karl and his school, no less than for Carnap and Reichenbach, canons of rationality thus derive exclusively from those of logical and linguistic syntax. Paul Feyerabend provides the exception which proves that rule. Denying the existence of a vocabulary adequate to neutral observation reports, he at once concludes to the intrinsic irrationality of theory choice.

That conclusion is surely Pickwickian. No process essential to scientific development can be labeled 'irrational' without vast violence to the term. It is therefore fortunate that the conclusion is unnecessary. One can deny, as Feyerabend and I do, the existence of an observation language shared in its entirety by two theories and still hope to preserve good reasons for choosing between them. To achieve that goal, however, philosophers of science will need to follow other contemporary philosophers in examining, to a previously unprecedented depth, the manner in which language fits the world, asking how terms attach to nature, how those attachments are learned, and how they are transmitted from one generation to another by the members of a language community. Because paradigms, in one of the two separable senses of the term, are fundamental to my own attempts to answer questions of that sort, they must also find a place in this essay.

Methodology: The Role of History and Sociology

Doubts about the appropriateness of my methods to my conclusions unite many of the essays in this volume. History and social psychology are not, my critics claim, a proper basis for philosophical conclusions. Their reservations are not, however, all of a piece. I shall therefore consider seriatim the somewhat different forms they take in the essays by Sir Karl, Watkins, Feyerabend, and Lakatos.

Sir Karl concludes his paper by pointing out that to him "the idea of turning for enlightenment concerning the aims of science, and its possible progress, to sociology or psychology (or . . . to the history of science) is surprising and disappointing . . . . how," he asks, "can the regress to these often spurious sciences help us in this particular difficulty?"[4] I am puzzled to know what these remarks intend, for in this area I think there are no differences between Sir Karl and myself. If he means that the generalizations which constitute received theories in sociology and psychology (and history?) are weak reeds from which to weave a philosophy of science, I could not agree more heartily. My work relies on them no more than his. If, on the other hand, he is challenging the relevance to philosophy of science of the sorts of observations collected by historians and sociologists, I wonder how his own work is to be understood. His writings are crowded with historical examples and with generalizations about scientific behavior, some of them discussed in my earlier essay. He does write on historical themes, and he cites those papers in his central philosophical works. A consistent interest in historical problems and a willingness to engage in original historical research distinguishes the men he has trained from the members of any other current school in philosophy of science. On these points I am an unrepentant Popperian.

4. Popper, "Normal Science," pp. 57–58.

John Watkins voices a different sort of doubt. Early in his paper he writes that "methodology . . . is concerned with science at its best, or with science as it should be conducted, rather than with hack science,"[5] a point with which, at least in a more careful formulation, I fully agree. Later he argues that what I have called normal science is hack science, and he then asks why I am so "concerned to up-value Normal Science and down-value Extraordinary Science" (p. 31). In so far as that question is about normal science in particular, I reserve my response until later (at which point I shall attempt also to unravel Watkins's extraordinary distortion of my position). But Watkins seems also to be asking a more general question, one that relates closely to an issue raised by Feyerabend. Both grant, at least for the sake of their argument, that scientists do behave as I have said they do (I shall later consider their qualifications of that concession). Why should the philosopher or methodologist, they then ask, take the facts seriously? He is, after all, concerned not with a full description of science but with the discovery of the essentials of the enterprise, i.e., with rational reconstruction. By what right and what criteria does the historian/observer or sociologist/observer tell the philosopher which facts of scientific life he must include in his reconstruction, which he may ignore?

To avoid lengthy disquisitions on the philosophy of history and of sociology, I restrict myself to a personal response. I am no less concerned with rational reconstruction, with the discovery of essentials, than are philosophers of science. My objective, too, is an understanding of science, of the reasons for its special efficacy, of the cognitive status of its theories. But unlike most philosophers of science, I began as a historian of science, examining closely the facts of scientific life. Having discovered in the process that much scientific behavior, including that of the very greatest scientists, persistently violated accepted methodological canons, I had to ask why those failures to conform did not seem at all to inhibit the success of the enterprise. When I later discovered that an altered view of the nature of science transformed what had previously seemed aberrant behavior into an essential part of an explanation for science's success, the discovery was a source of confidence in that new explanation. My criterion for emphasizing any particular aspect of scientific behavior is therefore not simply that it occurs, nor merely that it occurs frequently, but rather that it fits a theory of scientific knowledge.

5. J. W. N. Watkins, "Against 'Normal Science'," in *Criticism and the Growth of Knowledge*, p. 27.

Conversely, my confidence in that theory derives from its ability to make coherent sense of many facts which, on an older view, had been either aberrant or irrelevant. Readers will observe a circularity in the argument, but it is not vicious, and its presence does not at all distinguish my view from those of my present critics. Here, too, I am behaving as they do.

That my criteria for discriminating between the essential and nonessential elements of observed scientific behavior are to a significant extent theoretical provides also an answer to what Feyerabend calls the ambiguity of my presentation. Are Kuhn's remarks about scientific development, he asks, to be read as descriptions or prescriptions?[6] The answer, of course, is that they should be read in both ways at once. If I have a theory of how and why science works, it must necessarily have implications for the way in which scientists should behave if their enterprise is to flourish. The structure of my argument is simple and, I think, unexceptionable: scientists behave in the following ways; those modes of behavior have (here theory enters) the following essential functions; in the absence of an alternate mode *that would serve similar functions*, scientists should behave essentially as they do if their concern is to improve scientific knowledge.

Note that nothing in that argument sets the value of science itself, and that Feyerabend's "plea for hedonism" (p. 209) is correspondingly irrelevant. Partly because they have misconstrued my prescription (a point to which I shall return), both Sir Karl and Feyerabend find menace in the enterprise I have described. It is "liable to corrupt our understanding and diminish our pleasure" (Feyerabend, p. 209); it is "a danger . . . indeed to our civilization" (Sir Karl, p. 53). I am not led to that evaluation nor are many of my readers, but nothing in my argument depends on its being wrong. To explain why an enterprise works is not to approve or disapprove it.

Lakatos's paper raises a fourth problem about method, and it is the most fundamental of all. I have already confessed my inability to understand what he means when he says things like, "Kuhn's conceptual framework . . . is socio-psychological: mine is normative." If I ask, however, not what he intends, but why he finds this sort of rhetoric appropriate, an important point emerges, one that is almost explicit in

6. P. K. Feyerabend, "Consolations for the Specialist," in *Criticism and the Growth of Knowledge*, p. 198. For a far deeper and more careful examination of some contexts in which the descriptive and normative merge, see S. Cavell, "Must We Mean What We Say?" in *Must We Mean What We Say? A Book of Essays* (New York: Scribner, 1969), pp. 1–42.

the first paragraph of his section 4. Some of the principles deployed in my explanation of science are irreducibly sociological, at least at this time. In particular, confronted with the problem of theory choice, the structure of my response runs roughly as follows: take a *group* of the ablest available people with the most appropriate motivation; train them in some science and in the specialties relevant to the choice at hand; imbue them with the value system, the ideology, current in their discipline (and to a great extent in other scientific fields as well); and, finally, *let them make the choice*. If that technique does not account for scientific development as we know it, then no other will. There can be no set of rules of choice adequate to dictate desired *individual* behavior in the concrete cases that scientists will meet in the course of their careers. Whatever scientific progress may be, we must account for it by examining the nature of the scientific group, discovering what it values, what it tolerates, and what it disdains.

That position is intrinsically sociological and, as such, a major retreat from the canons of explanation licensed by the traditions which Lakatos labels justificationism and falsificationism, both dogmatic and naive. I shall later specify it further and defend it. But my present concern is simply with its structure, which both Lakatos and Sir Karl find unacceptable in principle. My question is, why should they? Both repeatedly use arguments of the same structure themselves.

Sir Karl does not, it is true, do so all the time. That part of his writing which seeks an algorithm for verisimilitude would, if successful, eliminate all need for recourse to group values, to judgments made by minds prepared in a particular way. But, as I pointed out at the end of my previous essay, there are many passages throughout Sir Karl's writings which can only be read as descriptions of the values and attitudes which scientists must possess if, when the chips are down, they are to succeed in advancing their enterprise. Lakatos's sophisticated falsificationism goes even further. In all but a few respects, only two of them essential, his position is now very close to my own. Among the respects in which we agree, though he has not yet seen it, is our common use of explanatory principles that are ultimately sociological or ideological in structure.

Lakatos's sophisticated falsificationism isolates a number of issues about which scientists employing the method must make decisions, individually or collectively. (I distrust the term 'decision' in this context since it implies conscious deliberation on each issue prior to the assumption of a research stance. For the moment, however, I shall use it. Until the last section of this paper very little will depend upon the distinction between making a decision and finding oneself in the position that would have resulted from making it.) Scientists must, for example, *decide* which statements to make "unfalsifiable by *fiat*" and which not.[7] Or, dealing with a probabilistic theory, they must *decide* on a probability threshold below which statistical evidence will be held "inconsistent" with that theory (p. 109). Above all, viewing theories as research programs to be evaluated over time, scientists must *decide* whether a given program at a given time is "progressive" (whence scientific) or "degenerative" (whence pseudo-scientific) (p. 118 ff.). If the first, it is to be pursued; if the latter, rejected.

Notice now that a call for decisions like these may be read in two ways. It may be taken to name or describe decision points for which procedures applicable in concrete cases must still be supplied. On this reading Lakatos has yet to tell us how scientists are to select the particular statements that are to be unfalsifiable by their *fiat;* he must also still specify criteria which can be used at the time to distinguish a degenerative from a progressive research program; and so on. Otherwise, he has told us nothing at all. Alternatively, his remarks about the need for particular decisions may be read as already complete descriptions (at least in form—their particular content may be preliminary) of directives, or maxims which the scientist is required to follow. On this interpretation, the third decision directive would read: "As a scientist, you may not refrain from deciding whether your research program is progressive or degenerative, and you must take the consequences of your decision, abandoning the program in one case, pursuing it in the other." Correspondingly, the second directive would read: "Working with a probabilistic theory, you must constantly ask yourself whether the result of some particular experiment is not so improbable as to be inconsistent with your theory, and you must, as a scientist, also answer." Finally, the first directive would read: "As a scientist, you will have to take risks, choosing certain statements as the basis for your work and ignoring, at least until your research program has developed, all actual and potential attacks upon them."

The second reading is, of course, far weaker than the first. It demands the same decisions, but it neither supplies nor promises to supply rules which would dictate their outcomes. Instead, it assimilates these deci-

7. Lakatos, "Falsification," p. 106.

sions to judgments of value (a subject about which I shall have more to say) rather than to measurements or computations, say, of weight. Nevertheless, conceived merely as imperatives which commit the scientist to making certain sorts of decisions, these directives are strong enough to affect scientific development profoundly. A group whose members felt no obligations to wrestle with such decisions (but which instead emphasized others, or none at all) would behave in notably different ways, and their discipline would change accordingly. Though Lakatos's discussion of his decision directives is often equivocal, I believe that it is just this second sort of efficacy upon which his methodology depends. Certainly he does little to specify algorithms by which the decisions he demands are to be made, and the tenor of his discussion of naive and dogmatic falsificationism suggests that he no longer thinks such specification possible. In that case, however, his decision imperatives are, in form though not always in content, identical to my own. They specify ideological commitments which scientists must share if their enterprise is to succeed. They are therefore irreducibly sociological in the same sense and to the same extent as my explanatory principles.

Under these circumstances I am not sure what Lakatos is criticizing or what, in this area, he thinks we disagree about. A strange footnote late in his paper may, however, provide a clue:

> There are *two kinds of psychologistic philosophies of science*. According to one kind there can be no philosophy of science: only a psychology of individual scientists. According to the other kind there is a psychology of the "scientific," "ideal," or "normal" mind: this turns philosophy of science into a psychology of this ideal mind. . . . Kuhn does not seem to have noticed this distinction. (p. 180, n.3)

If I understand him correctly, Lakatos identifies the first kind of psychologistic philosophy of science with me, the second with himself. But he is misunderstanding me. We are not nearly so far apart as his description would suggest, and, where we do differ, his literal position would demand a renunciation of our common goal.

Part of what Lakatos is rejecting is explanations that demand recourse to the factors which individuate particular scientists ("the psychology of the individual scientist" versus "the psychology of the . . . 'normal' mind"). But that does not separate us. My recourse has been exclusively to social psychology (I prefer 'sociology'), a field quite different from individual psychology reiterated n times. Correspondingly, my unit for purposes of explanation is the normal (i.e., nonpathological) scientific group, account being taken of the fact that its members differ but not of what makes any given individual unique. In addition, Lakatos would like to reject those characteristics of even normal scientific minds which make them the minds of human beings. Apparently he sees no other way to retain the methodology of an ideal science in explaining the observed success of actual science. But his way will not do if he hopes to explain an enterprise practiced by people. There are no ideal minds, and the "psychology of this ideal mind" is therefore unavailable as a basis for explanation. Nor is Lakatos's manner of introducing the ideal needed to achieve what he aims at. Shared ideals affect behavior without making those who hold them ideal. The type of question I ask has therefore been: how will a particular constellation of beliefs, values, and imperatives affect group behavior? My explanations follow from the answer. I am not sure Lakatos means anything else, but, if he does not, there is nothing in this area for us to disagree about.

Having misconstrued the sociological base of my position, Lakatos and my other critics inevitably fail to note a special feature which follows from taking the normal group rather than the normal mind as unit. Given a shared algorithm adequate, let us say, to individual choice between competing theories or to the identification of severe anomaly, all members of a scientific group will reach the same decision. That would be the case even if the algorithm were probabilistic, for all those who used it would evaluate the evidence in the same way. The effects of a shared ideology, however, are less uniform, for its mode of application is of a different sort. Given a group all the members of which are committed to choosing between alternative theories and also to considering such values as accuracy, simplicity, scope, and so on while making their choice, the concrete decisions of individual members in individual cases will nevertheless vary. Group behavior will be affected decisively by the shared commitments, but individual choice will be a function also of personality, education, and the prior pattern of professional research. (These variables *are* the province of individual psychology.) To many of my critics this variability seems a weakness of my position. When considering the problems of crisis and of theory choice I shall want, however, to argue that it is instead a strength. If a decision must be made under circumstances in which even the most deliberate and considered judgment may be wrong, it may be vitally important that different individuals decide in different ways. How else could the group as a whole hedge its bets?[8]

Normal Science: Its Nature and Functions

As to methods, then, the ones I employ are not significantly different from those of my Popperian critics. Applying those methods, we, of course, draw somewhat different conclusions, but even they are not so far apart as several of my critics believe. In particular, all of us excepting Toulmin share the conviction that the central episodes in scientific advance—those which make the game worth playing and the play worth studying—are revolutions. Watkins is constructing an opponent from his own straw when he describes me as having "down-valued" scientific revolutions, taken a "philosophical dislike" to them, or suggested that they "can hardly be called science at all."[9] Discovering the puzzling nature of revolutions was what drew me to history and philosophy of science in the first place. Almost everything I have written since deals with them, a fact which Watkins points out and then ignores.

If, however, we agree about this much, we cannot altogether disagree about normal science, the aspect of my work which most disturbs my present critics. By their nature revolutions cannot be the whole of science: something different must necessarily go on in between. Sir Karl sets up the point admirably. Underlining what I have always recognized as one of our principal areas of agreement, he stresses that "scientists *necessarily* develop their ideas within a definite theoretical framework".[10] For him, as for me, furthermore, revolutions demand such frameworks, since they always involve the rejection and replacement of a framework or of some of its integral parts. Since the science which I call normal is precisely research within a framework, it can only be the opposite side of a coin the face of which is revolutions. No wonder Sir Karl has been "dimly aware of the distinction" between normal science and revolutions (p. 52). It follows from his premises.

8. If human motivation were not at issue, the same effect could be achieved by first computing a probability and then *assigning* a certain fraction of the profession to each of the competing theories, the exact fraction to depend on the result of the probabilistic computation. Somehow that alternative makes my point by reductio ad absurdum.

9. Watkins, "Against 'Normal Science'," pp. 31, 32, and 29.

10. Popper, "Normal Science," p. 51, italics added. Unless explicitly noted, all italic passages in the quotations in this paper are in the originals.

Something else follows as well. If frameworks are necessary to scientists, if to break with one is inevitably to break into another—points which Sir Karl embraces explicitly—then the hold of a framework on a scientist's mind may not be accounted for *merely* as the result of his having "been badly taught, . . . a victim of indoctrination" (p. 53). Nor may it, as Watkins supposes, be explained *entirely* by reference to the prevalence of third-rate minds, fit only for "plodding, uncritical" work.[11] Those things do exist, and most of them do damage. Nevertheless, if frameworks are the prerequisite of research, their grip on the mind is not merely "Pickwickian," nor can it be quite right to say that, "if we try, we can break out of our framework at any time."[12] To be simultaneously essential and freely dispensible is very nearly a contradiction in terms. My critics become incoherent when they embrace it.

None of that is said in an effort to show that my critics really agree with me, if only they knew it. They do not! Rather I am trying, by eliminating irrelevancies, to discover what we disagree about. I have so far argued that Sir Karl's phrase 'revolutions in permanence' does not, any more than 'square circle', describe a phenomenon that could exist. Frameworks must be lived with and explored before they can be broken. But that does not imply that scientists ought not aim at perpetual framework-breaking, however unobtainable that goal. 'Revolutions in permanence' could name an important ideological imperative. If Sir Karl and I disagree at all about normal science, it is over this point. He and his group argue that the scientist should try at all times to be a critic and a proliferator of alternate theories. I urge the desirability of an alternate strategy which reserves such behavior for special occasions.

That disagreement, being restricted to research strategy, is already narrower than the one my critics have envisaged. To see what is at stake it must be narrowed further. Everything that has been said so far, though phrased for science and scientists, applies equally to a number of other fields. My methodological prescription is, however, directed exclusively to the sciences and, among them, to those fields which display the special developmental pattern known as progress. Sir Karl neatly catches the distinction I have in mind. At the start of his paper he writes: " 'A scientist engaged in a piece of research . . . can go at once to the heart of . . . an organized structure . . . [and of] a generally accepted problem-situation . . . [leaving] it to others to fit his contribu-

11. Watkins, "Against 'Normal Science'," p. 32.
12. Popper, "Normal Science," p. 56.

tion into the framework of scientific knowledge.' . . . the philosopher,"
he continues, "finds himself in a different position."[13] Nevertheless, having pointed to the difference, Sir Karl thereafter ignores it, recommend-
ing the same strategy to both scientists and philosophers. In the process
he misses the consequences for research design of the special detail and
precision with which, as he says, the framework of a mature science
informs its practitioners what to do. In the absence of that detailed
guidance, Sir Karl's critical strategy seems to me the very best available.
It will not induce the special developmental pattern which characterizes,
say, physics, but neither will any other methodological prescription.
Given a framework which does provide such guidance, however, then
I do intend my methodological recommendations to apply.

Consider for a moment the evolution of philosophy or of the arts
since the end of the Renaissance. These are fields often contrasted with
the established sciences as ones which do not progress. That contrast
cannot be due to the absence of revolutions or of an intervening mode
of normal practice. On the contrary, long before the similar structure
of scientific development was noticed, historians portrayed these fields
as developing through a succession of traditions punctuated by revolutionary alterations of artistic style and taste or of philosophical viewpoint and goal. Nor can the contrast be due to the absence from philosophy and the arts of a Popperian methodology. As Miss Masterman
observes for philosophy,[14] these are just the fields in which it is best
exemplified, in which practitioners do find current tradition stifling, do
struggle to break with it, and do regularly seek a style or a philosophical
viewpoint of their own. In the arts, in particular, the work of men who
do not succeed in innovation is described as 'derivative', a term of dero-
gation significantly absent from scientific discourse, which does, on the
other hand, repeatedly refer to 'fads'. In none of these fields, whether
arts or philosophy, does the practitioner who fails to alter traditional
practice have significant impact on the discipline's development.[15] These

13. Popper, "Normal Science," p. 51. Readers who know my *Structure of Scientific Revolutions*
(Chicago: University of Chicago Press, 1962) will recognize how closely Sir Karl's phrase "leaving it to others to fit his contribution into the framework of scientific knowledge" catches the
essential implications of my description of normal science.
14. M. Masterman, "The Nature of a Paradigm," in *Criticism and the Growth of Knowledge*, p. 69 ff.
15. For a fuller discussion of differences between scientific and artistic communities and between
the corresponding developmental patterns, see my "Comment" [on the Relations of Science and
Art], *Comparative Studies in Society and History* 11 (1969) : 403–12; reprinted as "Comment on
the Relations of Science and Art," in *The Essential Tension: Selected Studies in Scientific Tradition
and Change* (Chicago: University of Chicago Press, 1977), pp. 340–51.

are, in short, fields to which Sir Karl's method is essential because without constant criticism and the proliferation of new modes of practice there would be no revolutions. Substituting my own methodology for Sir Karl's would induce stagnation for exactly the reasons my critics underscore. In no obvious sense, however, does his methodology produce progress. The relation of pre- to post-revolutionary practice in these fields is not what we have learned to expect from the developed sciences.

My critics will suggest that the reasons for that difference are obvious. Fields like philosophy and the arts do not claim to be sciences, nor do they satisfy Sir Karl's demarcation criterion. They do not, that is, generate results which can in principle be tested through a point-by-point comparison with nature. But that argument seems to me mistaken. Without satisfying Sir Karl's criterion these fields could not be sciences, but they could nevertheless progress as the sciences do. In antiquity and during the Renaissance, the arts rather than the sciences provided the accepted paradigms of progress.[16] Few philosophers find reasons of principle why their field should not move steadily ahead, though many bemoan its failure to do so. In any case, there are many fields—I shall call them proto-sciences—in which practice does generate testable conclusions but which nonetheless resemble philosophy and the arts rather than the established sciences in their developmental patterns. I think, for example, of fields like chemistry and electricity before the mid-eighteenth century, of the study of heredity and phylogeny before the mid-nineteenth, or of many of the social sciences today. In these fields, too, though they satisfy Sir Karl's demarcation criterion, incessant criticism and continual striving for a fresh start are primary forces, and need to be. No more than in philosophy and the arts, however, do they result in clear-cut progress.

I conclude, in short, that the proto-sciences, like the arts and philosophy, lack some element which, in the mature sciences, permits the more obvious forms of progress. It is not, however, anything that a methodological prescription can provide. Unlike my present critics, Lakatos at this point included, I claim no therapy to assist the transformation of a proto-science to a science, nor do I suppose that anything of the sort is to be had. If, as Feyerabend suggests, some social scientists take from me the view that they can improve the status of their field by first

16. E. H. Gombrich, *Art and Illusion: A Study in the Psychology of Pictorial Representation* (New York: Pantheon, 1960), pp. 11 ff.

legislating agreement on fundamentals and then turning to puzzle-solving, they are badly misconstruing my point.[17] A sentence I once used when discussing the special efficacy of mathematical theories applies equally here: "As in individual development, so in the scientific group, maturity comes most surely to those who know how to wait."[18] Fortunately, though no prescription will force it, the transition to maturity does come to many fields, and it is well worth waiting and struggling to attain. Each of the currently established sciences has emerged from a previously more speculative branch of natural philosophy, medicine, or the crafts at some relatively well defined period in the past. Other fields will surely experience the same transition in the future. Only after it occurs does progress become an obvious characteristic of a field. And only then do those prescriptions of mine which my critics decry come into play.

About the nature of that change I have written at length in *Structure* and more briefly when discussing demarcation criteria in my earlier contribution to this volume. Here I shall be content with an abstract descriptive summary. Confine attention first to fields which aim to explain in detail some range of natural phenomena. (If, as my critics point out, my further description fits theology and bank robbery as well, no problems are thereby created.) Such a field first gains maturity when provided with theory and technique which satisfy the four following conditions. First is Sir Karl's demarcation criterion, without which no field is potentially a science: for some range of natural phenomena, concrete predictions must emerge from the practice of the field. Second, for some interesting subclass of phenomena, whatever passes for predictive success must be consistently achieved. (Ptolemaic astronomy always predicted planetary position within widely recognized limits of error. The companion astrological tradition could not, except for the tides and the average menstrual cycle, specify in advance which prediction would succeed, which fail.) Third, predictive techniques must have roots in a theory which, however metaphysical, simultaneously justifies them, explains their limited success, and suggests means for their improvement in both precision and scope. Finally, the improvement of predictive technique must be a challenging task, demanding on occasions the very highest measure of talent and devotion.

17. Feyerabend, "Consolations for the Specialist," p. 198. Note, however, that the passage Feyerabend quotes in note 3 does not say at all what he reports.
18. See T. S. Kuhn, "The Function of Measurement in Modern Physical Science," *Isis* 52 (1962), p. 190.

These conditions are, of course, tantamount to the description of a good scientific theory. But once hope for a therapeutic prescription is abandoned, there is no reason to expect anything less. My claim has been—it is my single genuine disagreement with Sir Karl about normal science—that with such a theory in hand the time for steady criticism and theory proliferation has passed. Scientists for the first time have an alternative which is not merely aping what has gone before. They can instead apply their talents to the puzzles which lie in what Lakatos now calls the "protective belt." One of their objectives then is to extend the range and precision of existing experiment and theory as well as to improve the match between them. Another is to eliminate conflicts both between the different theories employed in their work and between the ways in which a single theory is used in different applications. (Watkins is right, I now think, in charging that my book gives too small a role to these inter- and intra-theoretic puzzles, but Lakatos's attempt to reduce science to mathematics, leaving no significant role to experiment, goes vastly too far. He could not, for example, be more mistaken about the irrelevance of the Balmer formula to the development of Bohr's atom model.[19]) These puzzles and others like them constitute the main activity of normal science. Though I cannot argue the point again, they are not, *pace* Watkins, for hacks, nor do they, *pace* Sir Karl, resemble the problems of applied science and engineering. Of course the men fascinated by them are a special breed, but so are philosophers or artists.

Even given a theory which permits normal science, however, scientists need not engage the puzzles it supplies. They could instead behave as practitioners of the proto-sciences must; they could, that is, seek potential weak spots, of which there are always large numbers, and endeavor to erect alternate theories around them. Most of my present critics believe they should do so. I disagree but exclusively on strategic grounds. Feyerabend mispresents me in a way I particularly regret when he reports, for example, that I "criticized Bohm for disturbing the uniformity of the contemporary quantum theory."[20] My record as a troublemaker should be hard to reconcile with that report. In fact, I confessed

19. Lakatos, "Falsification," p. 147. This attitude toward the role of experiment is found throughout much of Lakatos's paper. For the actual role of the Balmer formula in Bohr's work, see J. L. Heilbron and T. S. Kuhn, "The Genesis of the Bohr Atom," *Historical Studies in the Physical Sciences* 1 (1969): 211–90.

20. Feyerabend, "Consolations for the Specialist," p. 206. An implicit answer to the contrast Feyerabend draws between my attitudes toward Bohm and Einstein as critics will be found below.

to Feyerabend that I shared Bohm's discontent but thought his exclusive attention to it almost certain to fail. No one, I suggested, was likely to resolve the paradoxes of the quantum theory until he could relate them to some concrete technical puzzle of current physics. In the developed sciences, unlike philosophy, it is technical puzzles that provide the usual occasion and often the concrete materials for revolution. Their availability together with the information and signals they provide account in large part for the special nature of scientific progress. Because they can ordinarily take current theory for granted, exploiting rather than criticizing it, the practitioners of mature sciences are freed to explore nature to an esoteric depth and detail otherwise unimaginable. Because that exploration will ultimately isolate severe trouble spots, they can be confident that the pursuit of normal science will inform them when and where they can most usefully become Popperian critics. Even in the developed sciences, there is an essential role for Sir Karl's methodology. It is the strategy appropriate to those occasions when something goes wrong with normal science, when the discipline encounters crisis.

I have discussed those points at great length elsewhere and shall not elaborate them here. Let me instead conclude this section by returning to the generalization with which it began. Despite the energy and space which my critics have devoted to it, I do not think the position just outlined departs very greatly from Sir Karl's. On this set of questions our differences are over nuances. I hold that in the developed sciences occasions for criticism need not, and by most practitioners ought not, deliberately be sought. When they are found, a decent restraint is the appropriate first response. Sir Karl, though he sees the need to defend a theory when first attacked, gives more emphasis than I to the purposeful search for weak points. There is not a great deal to choose between us.

Why is it, then, that my present critics see our crucial differences here? One reason I have already suggested: their sense—which I do not share but which is in any case irrelevant—that my strategic prescription violates a higher morality. A second reason, which I shall discuss in the next section, is their apparent inability to see in historical examples the detailed functions of the breakdown of normal science in setting the stage for revolutions. Lakatos's case histories are in this respect particularly interesting, for he describes clearly the transition from the progressive to the degenerative phase of a research program (the transition from normal science to crisis) and then appears to deny the critical importance of what results. With a third reason, however, I must deal at this point. It emerges from a criticism voiced by Watkins, which, however, in the present context serves a purpose he by no means intends.

"By contrast with the relatively sharp idea of testability," Watkins writes, "the notion of [normal science's] 'ceasing adequately to support a puzzle-solving tradition' is essentially vague."[21] With the charge of vagueness I agree, but it is a mistake to suppose that it differentiates my position from Sir Karl's. What is precise about Sir Karl's position is, as Watkins also points out, the idea of testability in principle. On that much I rely too, for no theory that was not *in principle* testable could function or cease to function adequately when applied to scientific puzzle-solving. I do, despite Watkins's strange failure to see it, take Sir Karl's notion of the asymmetry of falsification and confirmation very seriously indeed. What is vague, however, about my position is the actual criteria (if that is what is called for) to be applied when deciding whether a particular failure in puzzle-solving is or is not to be attributed to fundamental theory and thus to become an occasion for deep concern. That decision is, however, identical in kind with the decision whether or not the result of a particular test actually falsifies a particular theory, and on that subject Sir Karl is necessarily as vague as I. To drive a wedge between us on this issue, Watkins transfers the sharpness of testability-in-principle to the shady area of testability-in-practice without even hinting how the transfer is to be effected. It is not an unprecedented mistake, and it regularly makes Sir Karl's methodology appear more a logic, less an ideology, than it is.

Besides, reverting to a point made at the end of the last section, one may legitimately ask whether what Watkins calls vagueness is a disadvantage. All scientists must be taught—it is a vital element in their ideology—to be alert for and responsible to theory breakdown, whether it be described as severe anomaly or falsification. In addition, they must be supplied with examples of what their theories can, with sufficient care and skill, be expected to do. Given only that much, they will, of course, often reach different judgments in concrete cases, one man seeing a cause of crisis where another sees only evidence of limited talent for research. But they do reach judgments, and their lack of unanimity may then be what saves their profession. Most judgments that a theory has ceased adequately to support a puzzle-solving tradition prove to be wrong. If everyone agreed in such judgments, no one would

21. Watkins, "Against 'Normal Science'," p. 30.

be left to show how existing theory could account for the apparent anomaly, as it usually does. If, on the other hand, no one were willing to take the risk and then seek an alternate theory, there would be none of the revolutionary transformations on which scientific development depends. As Watkins says, "there must be a critical level at which a tolerable turns into an intolerable amount of anomaly" (p. 30). But that level ought not be the same for everyone, nor need any individual specify his own tolerance level in advance. He need only be certain that he has one and aware of some sorts of discrepancies which would drive him toward it.

Normal Science: Its Retrieval from History

I have so far argued that, if there are revolutions, then there must be normal science. One may, however, legitimately ask whether either exists. Toulmin has done so, and my Popperian critics have difficulties in retrieving from history a significant normal science upon the existence of which that of revolutions depends. Toulmin's questions are of particular value, for a response to them will require me to confront some genuine difficulties presented by *Structure* and to modify my original presentation accordingly. Unfortunately, however, those difficulties are not the ones Toulmin sees. Before they can be isolated, the dust he has imported must be swept away.

Though there have been important changes in my position during the seven years since my book was published, the retreat from a concern with macro- to a concentration on micro-revolutions is not among them. Part of that retreat Toulmin finds by contrasting a paper *read* in 1961 with a book *published* in 1962.[22] The paper was, however, both written and published after the book, and its first footnote specifies the relationship which Toulmin inverts. Other evidence of retreat Toulmin retrieves from a comparison of the book with the manuscript of my first essay in this volume.[23] But no one else has, to my knowledge, even noticed the differences which he underlines, and the book is in any case quite explicit about the centrality of the concern which Toulmin finds

---

22. S. E. Toulmin, "Does the Distinction between Normal and Revolutionary Science Hold Water?" in *Criticism and the Growth of Knowledge*, pp. 39 ff.
23. See also S. E. Toulmin, "The Evolutionary Development of Natural Science," *American Scientist* 55 (1967): 456–71, especially p. 471, n. 8. The publication of this biographical canard in advance of the article on which it claims to be based has given me much trouble.

only in my more recent work. Among the revolutions discussed in the body of the book are, for example, discoveries like those of X rays and of the planet Uranus. "Admittedly," the preface states, "the extension [of the term 'revolution' to episodes like these] strains customary usage. Nevertheless, I shall continue to speak even of discoveries as revolutionary, because it is just the possibility of relating their structure to that of, say, the Copernican revolution that makes the extended conception seem to me so important."[24] My concern, in short, has never been with scientific revolutions as "something that tended to happen in a given branch of science only once every two hundred years or so."[25] Rather it has been throughout what Toulmin now takes it to have become: a little-studied type of conceptual change which occurs frequently in science and is fundamental to its advance.

To that concern Toulmin's geological analogy is entirely appropriate, but not in the way he uses it. He emphasizes the aspect of the uniformitarian-catastrophist debate which dealt with the possibility of attributing catastrophes to natural causes, and he suggests that once that issue had been resolved "'catastrophes' became *uniform* and law-governed just like any other geological and palaeontological phenomena" (p. 43, my italics). But his insertion of the term 'uniform' is gratuitous. Besides the issue of natural causes, the debate had a second central aspect: the question whether catastrophes existed, whether a major role in geological evolution should be attributed to phenomena like earthquakes and volcanic action which acted more suddenly and destructively than erosion and sedimentary deposition. This part of the debate the uniformitarians lost. When it was over, geologists recognized two sorts of geological change, no less distinct because both due to natural causes; one acted gradually and uniformly, the other suddenly and catastrophically. Even today we do not treat tidal waves as special cases of erosion.

Correspondingly, my claim has been, not that revolutions were inscrutable unit events, but that in science as in geology there are two sorts of change. One of them, normal science, is the generally cumulative process by which the accepted beliefs of a scientific community are fleshed out, articulated, and extended. It is what scientists are trained to do, and the main tradition in English-speaking philosophy of science derives from the examination of the exemplary works in which that

---

24. Cf. *Structure*, pp. 7 f. On p. 6 the possibility of extending the conception to micro-revolutions is described as a "fundamental thesis" of the book.
25. Toulmin, "Does the Distinction," p. 44.

training is embodied. Unfortunately, as indicated in my previous essay, proponents of that philosophical tradition generally choose their examples from changes of another sort which are then tailored to fit. The result is a failure to recognize the prevalence of changes in which conceptual commitments fundamental to the practice of some scientific specialty must be jettisoned and replaced. Of course, as Toulmin says, the two sorts of change interpenetrate: revolutions are no more total in science than in other aspects of life, but recognizing continuity through revolutions has not led historians or anyone else to abandon the notion. It was a weakness of *Structure* that it could only name, not analyze, the phenomenon it repeatedly referred to as 'partial communication'. But partial communication was never, as Toulmin would have it, "complete [mutual] incomprehension" (p. 43). It named a problem to be worked on, not elevated to inscrutability. Unless we can learn more about it (I shall offer some hints in the next section), we shall continue to mistake the nature of scientific progress and thus perhaps of knowledge. Nothing in Toulmin's essay begins to persuade me that we shall succeed if we continue to treat all scientific change as one.

The fundamental challenge of his paper, however, remains. Can we distinguish mere articulations and extensions of shared belief from changes which involve reconstruction? The answer in extreme cases is obviously "Yes." Bohr's theory of the hydrogen spectrum was revolutionary as Sommerfeld's theory of the hydrogen fine structure was not; Copernican astronomical theory was revolutionary but the caloric theory of adiabatic compression was not. These examples are, however, too extreme to be fully informative: there are too many differences between the theories contrasted, and the revolutionary changes affected too many people. Fortunately, however, we are not restricted to them: Ampère's theory of the electric circuit was revolutionary (at least among French electricians), because it severed electric-current and electrostatic effects which had previously been conceptually united. Ohm's law was again revolutionary, and was resisted accordingly, because it demanded a reintegration of concepts previously applied separately to current and charge.[26] On the other hand, the Joule-Lenz law relating the heat generated in a wire to the resistance and current was a product of normal science, for both the qualitative effects and the concepts required for

---

26. On these topics, see T. M. Brown, "The Electric Current in Early Nineteenth-Century French Physics," *Historical Studies in the Physical Sciences* 1 (1969): 61–103; M. L. Schagrin, "Resistance to Ohm's Law," *American Journal of Physics* 31 (1963) : 536–37.

quantification were in hand. Again, at a level which is not so obviously theoretical, Lavoisier's discovery of oxygen (though perhaps not Scheele's and surely not Priestley's) was revolutionary, for it was inseparable from a new theory of combustion and acidity. The discovery of neon, however, was not, for helium had supplied both the notion of an inert gas and the needed column of the periodic table.

One may question, however, how far and how universally this process of discrimination can be pressed. I am repeatedly asked whether such-and-such a development was "normal or revolutionary," and I usually have to answer that I do not know. Nothing depends upon my, or anyone else's, being able to respond in every conceivable case, but much depends on the discrimination's being applicable to a far larger number of cases than have been supplied so far. Part of the difficulty in answering is that the discrimination of normal from revolutionary episodes demands close historical study, and few parts of the history of science have received it. One must know not simply the name of the change, but the nature and structure of group commitments before and after it occurred. Often, to determine these, one must also know the manner in which the change was received when first proposed. (There is no area in which I am more deeply conscious of the need for additional historical research, though I dissent from the conclusions Pearce Williams draws from that need and doubt that the results of investigation will draw Sir Karl and me closer.) My difficulty, however, has a deeper aspect. Though much depends upon more research, the investigations required are not simply of the sort indicated above. Furthermore, the structure of the argument in *Structure* somewhat obscures the nature of what is missing. If I were rewriting the book now I would significantly change its organization.

The gist of the problem is that to answer the question "normal or revolutionary?" one must first ask, "for whom?" Sometimes the answer is easy: Copernican astronomy was a revolution for everyone; oxygen was a revolution for chemists but not for, say, mathematical astronomers unless, like Laplace, they were interested in chemical and thermal subjects too. For the latter group oxygen was simply another gas, and its discovery was merely an increment to their knowledge; nothing essential to them as astronomers had to be changed in the discovery's assimilation. It is not, however, usually possible to identify groups which share cognitive commitments simply by naming a scientific subject matter—astronomy, chemistry, mathematics, or the like. That is, however, what I have just done here and did earlier in my book. Some scientific subjects—for example, the study of heat—have belonged to different scientific communities at different times, sometimes to several at once, without becoming the special province of any. In addition, though scientists are much more nearly unanimous in their commitments than practitioners of, say, philosophy and the arts, there are such things as schools in science, communities which approach the same subject from very different points of view. French electricians in the first decades of the nineteenth century were members of a school which included almost none of the British electricians of the day, and so on. If I were writing my book again now, I would therefore begin by discussing the community structure of science, and I would not rely exclusively on shared subject matter in doing so. Community structure is a topic about which we have very little information at present, but it has recently become a major concern for sociologists, and historians are now increasingly concerned with it as well.[27]

The research problems involved are by no means trivial. Historians of science who engage in them must cease to rely exclusively on the techniques of the intellectual historian and use those of the social and cultural historian as well. Even though work has scarcely begun, there is every reason to expect it to succeed, particularly for the developed sciences, those which have severed their historical roots in the philosophical or medical communities. What one would then have would be a roster of the different specialists' groups through which science was advanced at various periods of time. The analytic unit would be the practitioners of a given specialty, men bound together by common elements in their education and apprenticeship, aware of each other's work, and characterized by the relative fullness of their professional communication and the relative unanimity of their professional judgment. In the mature sciences the members of such communities would ordinarily see themselves and be seen by others as the men exclusively responsible for a given subject matter and a given set of goals, including the training of their successors. Research would, however, disclose the existence of rival schools as well. Typical communities, at least on the contemporary scientific scene, may consist of a hundred members, sometimes significantly fewer. Individuals, particularly the ablest, may belong to several such groups, either simultaneously or in succession, and they will

27. A somewhat more detailed discussion of this reorganization together with some preliminary bibliography is included in my "Second Thoughts on Paradigms," in *The Structure of Scientific Theories*, ed. F. Suppe (Urbana: University of Illinois Press, 1974), pp. 459–82; reprinted in *The Essential Tension*, pp. 293–319.

change or at least adjust their thinking caps as they go from one to another.

Groups like these should, I suggest, be regarded as the units which produce scientific knowledge. They could not, of course, function without individuals as members, but the very idea of scientific knowledge as a private product presents the same intrinsic problems as the notion of a private language, a parallel to which I shall return. Neither knowledge nor language remains the same when conceived as something an individual can possess and develop alone. It is, therefore, with respect to groups like these that the question "normal or revolutionary?" should be asked. Many episodes will then be revolutionary for no communities, many others for only a single small group, still others for several communities together, a few for all of science. Posed in that way, the question will, I believe, have answers as precise as my distinction requires. One reason for thinking so I shall illustrate in a moment by applying this approach to some of the concrete cases used by my critics to raise doubts about the existence and role of normal science. First, however, I must point out one aspect of my present position which, far more clearly than normal science, represents a deep divide between my viewpoint and Sir Karl's.

The program just outlined makes even clearer than it has been before the sociological base of my position. More important, it highlights what has perhaps not been clear before, the extent to which I regard scientific knowledge as intrinsically a product of a congeries of specialists' communities. Sir Karl sees "a great danger in . . . specialization," and the context in which he provides this evaluation suggests that the danger is the same one he sees in normal science.[28] But with respect to the former, at least, the battle has clearly been lost from the start. Not that one might not wish for good reasons to oppose specialization and even succeed in doing so, but that the effort would necessarily be to oppose science as well. Whenever Sir Karl contrasts science with philosophy, as he does at the start of his paper, or physics with sociology, psychology, and history, as he does at the end, he is contrasting an esoteric, isolated, and largely self-contained discipline with one that still aims to communicate with and persuade an audience larger than their own profession. (Science is not the only activity the practitioners of which can be grouped into communities, but it is the only one in which each

28. Popper, "Normal Science," p. 53.

community is its own exclusive audience and judge.[29]) The contrast is not a new one, characteristic, say, of Big Science and the contemporary scene. Mathematics and astronomy were esoteric subjects in antiquity; mechanics became so after Galileo and Newton; electricity after Coulomb and Poisson; and so on until economics today. For the most part that transition to a closed specialists' group was part of the transition to maturity that I discussed above when considering the emergence of puzzle-solving. It is hard to believe that it is a dispensable characteristic. Perhaps science could again become like philosophy, as Sir Karl wishes, but I suspect that he would then admire it less.

To conclude this part of my discussion, I turn to some concrete cases by means of which my critics illustrate their difficulties in finding normal science and its functions in history, taking up first a problem raised by Sir Karl and Watkins. Both point out that nothing like a consensus over fundamentals "emerged during the long history of the theory of *matter:* here from the pre-Socratics to the present day there has been an unending *debate* between continuous and discontinuous concepts of matter, between various atomic theories on the one hand, and ether, wave and field theories on the other."[30] Feyerabend makes a very similar point for the second half of the nineteenth century by contrasting the mechanical, phenomenological, and field-theoretic approaches to problems of physics.[31] With all of their descriptions of what went on I agree. But the term 'theories of matter' does not, at least until the last thirty years, even differentiate the concerns of science from those of philosophy, much less single out a community or small group of communities responsible for and expert in the subject.

I am not suggesting that scientists do not have and use theories of matter, nor that their work is unaffected by such theories, nor that their research results have no role in the theories of matter held by others. But until this century theories of matter have been a tool for scientists rather than a subject matter. That different specialties have chosen different tools and sometimes criticized each others' choices does not mean that they have not each been practicing normal science. The frequently

29. See my "Comment" [on the Relations of Science and Art].
30. Watkins, "Against 'Normal Science'," pp. 34 ff., 54–55. As Watkins notes, Dudley Shapere has made a similar point in his review of *Structure* (*Philosophical Review* 73 [1964]: 383–94), in connection with the role of atomism in chemistry in the first half of the nineteenth century. I deal with that case immediately below.
31. Feyerabend, "Consolations for the Specialist," p. 207.

heard generalization that, before the advent of wave mechanics, physicists and chemists deployed characteristic and irreconcilable theories of
matter is too simplistic (partly because it can equally well be said about
different chemical specialties even today). But the very possibility of
such a generalization suggests the way in which the issue raised by
Watkins and Sir Karl must be approached. For that matter, the practitioners of a given community or school need not always share a theory
of matter. Chemistry during the first half of the nineteenth century is
a case in point. Though many of its fundamental tools—constant proportion, multiple proportion, combining weights, and so on—had been
developed and become common property through Dalton's atomic theory, the men who used them could, after the event, adopt widely varying attitudes about the nature and even the existence of atoms. Their
discipline, or at least many parts of it, did not depend upon a shared
model for matter.

Even where they admit the existence of normal science, my critics
regularly have difficulty discovering crisis and its role. Watkins provides an example, and its resolution follows at once from the sort of
analysis deployed above. Kepler's laws, Watkins reminds us, were incompatible with Newton's planetary theory, but astronomers had not
previously been dissatisfied with them. Newton's revolutionary treatment of planetary motions was not, Watkins therefore asserts, preceded
by astronomical crisis. But why should it have been? In the first place,
the transition from Keplerian to Newtonian orbits need not have been
(I lack the evidence to be certain) a revolution *for astronomers*. Most of
them followed Kepler and explained the shape of the planetary orbits
in mechanical rather than geometrical terms. (Their explanation did not,
that is, make use of the ellipse's 'geometric perfection', if any, or of
some other characteristic of which the orbit was deprived by Newtonian
perturbations.) Though the transition from circle to ellipse had been
part of a revolution for them, a minor adjustment of mechanism would
account, as it did with Newton, for departure from ellipticity. More
important, Newton's adjustment of Keplerian orbits was a by-product of
his work in mechanics, a field to which the community of mathematical
astronomers made passing reference in their prefaces but which thereafter played only the most global role in their work. In mechanics, however, where Newton did induce a revolution, there had been a widely
recognized crisis since the acceptance of Copernicanism. Watkins's
counterexample is the best sort of grist for my mill.

I turn finally to one of Lakatos's extended case histories, that of the

Bohr research program, for it illustrates what most puzzles me about his often admirable paper and suggests how deep even residual Popperianism can be. Though his terminology is different, his analytic apparatus is as close to mine as need be: hard core, work in the protective belt, and degenerative phase are close parallels for my paradigms, normal science, and crisis. Yet in important ways Lakatos fails to see how these shared notions function even when applying them to what is for me an ideal case. Let me illustrate some of the things he could have seen and might have said. My version, like his or like any other bit of historical narrative, will be a rational reconstruction. But I shall not ask my readers to apply 'tons of salt' nor add footnotes pointing out that what is said in my text is false.[32]

Consider Lakatos's account of the origin of the Bohr atom. "The background problem," he writes, "was the riddle of how Rutherford atoms . . . can remain stable; for, according to the well-corroborated Maxwell-Lorentz theory of electromagnetism they should collapse."[33] That is a genuine Popperian problem (not a Kuhnian puzzle) arising from the conflict between two increasingly well established parts of physics. It had, in addition, been available for some time as a potential focal point for criticism. It did not originate with Rutherford's model in 1911; radiative instability was equally a difficulty for most older atom models, including both Thomson's and Nagaoka's. Furthermore, it is the problem which Bohr (in some sense) solved in his famous three-part paper of 1913, thereby inaugurating a revolution. No wonder Lakatos would like it to be the "background problem" for the research program that produced the revolution, but it emphatically is not.[34]

Instead, the background was an entirely normal puzzle. Bohr set out to improve the physical approximations in a paper by C. G. Darwin

---

32. Lakatos, "Falsification," pp. 138, 140, 146, and elsewhere. One may reasonably ask about the evidential force of examples that call for this sort of qualification (and is 'qualification' quite the right word?). I shall, however, in another context be very grateful for these 'case histories' of Lakatos's. More clearly, because more explicitly, than any other examples I know, they illustrate the differences between the way philosophers and historians usually do history. The problem is not that philosophers are likely to make errors—Lakatos knows the facts better than many historians who have written on these subjects, and historians do make egregious errors. But a historian would not *include in his narrative* a factual report which he *knew* to be false. If he had done so, he would be so sensitive to the offense that he could not conceivably compose a footnote calling attention to it. Both groups are scrupulous, but they differ in what they are scrupulous about. I have discussed some differences of this sort in "The Relations between History and Philosophy of Science," in *The Essential Tension*, pp. 1–20.

33. Lakatos, "Falsification," p. 141.

34. For what follows, see Heilbron and Kuhn, "The Genesis of the Bohr Atom."

on the energy lost by charged particles passing through matter. In the process he made what was to him the surprising discovery that the Rutherford atom, unlike other current models, was mechanically unstable and that a Planck-like ad hoc device for stabilizing it provided a promising explanation of the periodicities in Mendeleev's table, something else for which he had not been looking. At that point his model still had no excited states, nor was Bohr yet concerned to apply it to atomic spectra. Those steps followed, however, as he attempted to reconcile his model with the apparently incompatible one developed by J. W. Nicholson and, in the process, encountered Balmer's formula. Like much of the research that produces revolutions, Bohr's biggest achievements in 1913 were products, therefore, of a research program directed to goals very different from those obtained. Though he could not have stabilized the Rutherford model by quantization if unaware of the crisis which Planck's work had introduced to physics, his own work illustrates with particular clarity the revolutionary efficacy of normal research puzzles.

Examine, finally, the concluding portion of Lakatos's case history, the degenerative phase of the old quantum theory. Most of the story he tells well, and I shall simply point it up. From 1900 on it was increasingly widely recognized among physicists that Planck's quantum had introduced a fundamental inconsistency into physics. At first many of them tried to eliminate it, but, after 1911 and particularly after the invention of Bohr's atom, those critical efforts were increasingly abandoned. Einstein was, for more than a decade, the only physicist of note who continued to direct his energies toward the search for a consistent physics. Others learned to live with inconsistency and tried instead to solve technical puzzles with the tools at hand. Particularly in the areas of atomic spectra, atomic structure, and specific heats, their achievements were unprecedented. Though the inconsistency of physical theory was widely acknowledged, physicists could nevertheless exploit it and by doing so made fundamental discoveries at an extraordinary rate between 1913 and 1921. Quite suddenly, however, beginning in 1922, these very successes were seen to have isolated three obdurate problems—the helium model, the anomalous Zeeman effect, and optical dispersion—which could not, physicists were increasingly convinced, be resolved by anything quite like existing technique. As a result, many of them changed their research stance, proliferating more and wilder versions of the old quantum theory than before, designing and testing each attempt against the three recognized trouble spots.

It is this last phase, 1922 and after, which Lakatos calls the degenerative stage of Bohr's program. For me it is a casebook example of crisis, clearly documented in publications, correspondence, and anecdote. We see it in very nearly the same way. Lakatos might therefore have told the rest of the story. To those who were experiencing this crisis, two of the three problems which had provoked it proved immensely informative, dispersion and the anomalous Zeeman effect. By a series of connected steps too complex to be outlined here, their pursuit led first to the adoption in Copenhagen of an atom model in which so-called virtual oscillators coupled discrete quantum states, then to a formula for quantum-theoretical dispersion, and finally to matrix mechanics which terminated the crisis barely three years after it had begun. For that first formulation of quantum mechanics, the degenerative phase of the old quantum theory provided both occasion and much detailed technical substance. History of science, to my knowledge, offers no equally clear, detailed, and cogent example of the creative functions of normal science and crisis.

Lakatos, however, ignores this chapter and jumps instead to wave mechanics, the second and at first quite different formulation of a new quantum theory. First, he describes the degenerative phase of the old quantum theory as filled with "ever more sterile inconsistencies and ever more *ad hoc* hypotheses" ("*ad hoc*" and "inconsistencies" are right; "sterile" could not be more wrong; not only did these hypotheses lead to matrix mechanics but also to electron spin). Then, he produces the crisis-resolving innovation like a magician pulling a rabbit from a hat: "A rival research programme soon appeared: wave mechanics . . . [which] soon caught up with, vanquished and replaced Bohr's programme. De Broglie's paper came at a time when Bohr's programme was degenerating. *But this was mere coincidence.* One wonders what would have happened if de Broglie had published his paper in 1914 instead of 1924."[35]

To the closing rhetorical question, the answer is clear: nothing at all. Both de Broglie's paper and the route from it to the Schrödinger wave equation depend in detail on developments which occurred after 1914: on work by Einstein and by Schrödinger himself as well as on the discovery of the Compton effect in 1922.[36] Even if that point could

35. Lakatos, "Falsification," p. 154; italics added.
36. See M. J. Klein, "Einstein and the Wave-Particle Duality," *The Natural Philosopher* 3 (1964): 1–49; V. V. Raman and P. Forman, "Why Was It Schrödinger Who Developed de Broglie's Ideas?" *Historical Studies in the Physical Sciences* 1 (1969) : 291–314.

not be documented in detail, however, is not coincidence strained beyond recognition when used to explain the simultaneous emergence of two independent and at first quite different theories, both capable of resolving a crisis that had been visible for only three years?

Let me be scrupulous. Though Lakatos entirely misses the essential creative functions of the crisis of the old quantum theory, he is not altogether wrong about its relevance to the invention of wave mechanics. The wave equation was not a response to the crisis which began in 1922, but to the one which dates from Planck's work in 1900 and on which most physicists had turned their backs after 1911. If Einstein had not tenaciously refused to set aside his deep dissatisfaction with the fundamental inconsistencies of the old quantum theory (and if he had not been able to attach that discontent to the concrete technical puzzles of electromagnetic fluctuation phenomena—something for which he found no equivalent after 1925), the wave equation would not have emerged when and as it did. The research route which leads to it is not the same as the route to matrix mechanics.

But neither are the two independent, nor is the simultaneity of their termination due merely to coincidence. Among the several research episodes which tie them together is, for example, Compton's convincing demonstration in 1922 of the particulate properties of light, the by-product of a very high-class piece of normal research on X-ray scattering. Before physicists could consider the idea of matter waves, they had first to take the idea of the photon seriously, and this few of them had done before 1922. De Broglie's work started as photon theory, its main thrust being to reconcile Planck's radiation law with the particulate structure of light; matter waves entered along the way. De Broglie himself may not have needed Compton's discovery in order to take the photon seriously, but his audience, French and foreign, certainly did. Though wave mechanics in no sense follows from the Compton effect, there are historical ties between the two. On the road to matrix mechanics the role of the Compton effect is even clearer. The first use of the virtual oscillator model in Copenhagen was to show how that effect could be explained *without* recourse to Einstein's photon, a concept that Bohr had been notoriously reluctant to accept. The same model was next applied to dispersion and the clues to matrix mechanics found. The Compton effect is therefore one bridge across the gap which Lakatos hides under "coincidence."

Having provided elsewhere many other examples of the significant roles of normal science and crisis, I shall not multiply them further here.

For lack of additional research I could not, in any case, provide enough. When completed, that research need not bear me out, but what has been done so far surely fails to support my critics. They must look further for counterexamples.

Irrationality and Theory Choice

I consider now one last set of concerns voiced by my present critics, in this case one they share with a number of other philosophers. It arises mainly from my description of the procedures by which scientists choose between competing theories, and it results in charges which cluster about such terms as 'irrationality', 'mob rule', and 'relativism'. In this section I aim to eliminate misunderstandings for which my own past rhetoric is doubtless partially responsible. In my concluding section, which follows, I shall touch upon some deeper issues raised by the problem of theory choice. At that point the terms 'paradigm' and 'incommensurability', which I have so far almost entirely avoided, will necessarily reenter the discussion.

In *Structure*, normal science is at one point described as "a strenuous and devoted attempt to force nature into the conceptual boxes supplied by professional education" (p. 5). Later, discussing the problems which surround the choice between competing sets of boxes, theories, or paradigms, I described them as:

> about techniques of persuasion, or about argument and counter argument in a situation in which . . . neither proof nor error is at issue. The transfer of allegiance from paradigm to paradigm is a conversion experience that cannot be forced. Lifelong resistance . . . is not a violation of scientific standards but an index to the nature of scientific research itself. . . . Though the historian can always find men—Priestley, for instance—who were unreasonable to resist for as long as they did, he will not find a point at which resistance becomes illogical or unscientific. At most he may wish to say that the man who continues to resist after his whole profession has been converted has *ipso facto* ceased to be a scientist. (p. 151)

Not surprisingly (though I have myself been very much surprised), passages like these are in some quarters read as implying that, in the developed sciences, might makes right. Members of a scientific community can, I am held to have claimed, believe anything they please if only they will first decide what they agree about and then enforce it both on their colleagues and on nature. The factors which determine what they do choose to believe are fundamentally irrational, matters of accident and personal taste. Neither logic nor observation nor good reason is implicated in theory choice. Whatever scientific truth may be, it is through-and-through relativistic.

These are all damaging misinterpretations, whatever my responsibility may be for making them possible. Though their elimination will still leave a deep divide between my critics and me, it is prerequisite even to discovering our disagreement. Before treating them individually, however, one general remark should be helpful. The sorts of misinterpretations just outlined are voiced only by philosophers, a group already familiar with the points at which I aim in passages like the above. Unlike readers to whom the point is less familiar, they sometimes suppose that I intend more than I do. What I mean to be saying, however, is only the following.

In a debate over choice of theory, neither party has access to an argument which resembles a proof in logic or formal mathematics. In the latter, both premises and rules of inference are stipulated in advance. If there is disagreement about conclusions, the parties to the debate can retrace their steps one by one, checking each against prior stipulation. At the end of that process, one or the other must concede that at an isolable point in the argument he has made a mistake, violated or misapplied a previously accepted rule. After that concession he has no recourse, and his opponent's proof is then compelling. Only if the two discover instead that they differ about the meaning or applicability of a stipulated rule, that their prior agreement does not provide a sufficient basis for proof, does the ensuing debate resemble what inevitably occurs in science.

Nothing about this relatively familiar thesis should suggest that scientists do not *use* logic (and mathematics) in their arguments, including those which aim to persuade a colleague to renounce a favored theory and embrace another. I am dumbfounded by Sir Karl's attempt to convict me of self-contradiction because I employ logical arguments myself.[37] What might better be said is that I do not expect that, merely because my arguments are logical, they will be compelling. Sir Karl underscores my point, not his, when he describes them as logical but mistaken, and then makes no attempt to isolate the mistake or to display its logical character. What he means is that, though my arguments are

37. Popper, "Normal Science," pp. 55, 57.

logical, he disagrees with my conclusion. Our disagreement must be about premises or the manner in which they are to be applied, a situation which is standard among scientists debating theory choice. When it occurs, their recourse is to persuasion as a prelude to the possibility of proof.

To name persuasion as the scientist's recourse is not to suggest that there are not many good reasons for choosing one theory rather than another.[38] It is emphatically *not* my view that "adoption of a new scientific theory is an intuitive or mystical affair, a matter for psychological description rather than logical or methodological codification."[39] On the contrary, the chapter of *Structure* from which the preceding quotation was abstracted explicitly denies "that new paradigms triumph ultimately through some mystical aesthetic," and the pages which precede that denial contain a preliminary codification of good reasons for theory choice.[40] These are, furthermore, reasons of exactly the kind standard in philosophy of science: accuracy, scope, simplicity, fruitfulness, and the like. It is vitally important that scientists be taught to value these characteristics and that they be provided with examples that illustrate them in practice. If they did not hold values like these, their disciplines would develop very differently. Note, for example, that the periods in which the history of art was a history of progress were also the periods in which the artist's aim was accuracy of representation. With the abandonment of that value, the developmental pattern changed drastically though very significant development continued.[41]

What I am denying then is neither the existence of good reasons nor that these reasons are of the sort usually described. I am, however, insisting that such reasons constitute values to be used in making choices rather than rules of choice. Scientists who share them may nevertheless make different choices in the same concrete situation. Two factors are deeply involved. First, in many concrete situations, different values, though all constitutive of good reasons, dictate different conclusions, different choices. In such cases of value conflict (e.g., one theory is

38. For one version of the view that Kuhn insists that "the decisions of a scientific group to adopt a new paradigm cannot be based on good reasons of any kind, factual or otherwise," see D. Shapere, "Meaning and Scientific Change," in *Mind and Cosmos: Essays in Contemporary Science and Philosophy*, ed. R. G. Colodny, University of Pittsburgh Series in the Philosophy of Science, vol. 3 (Pittsburgh: University of Pittsburgh Press, 1966), pp. 41–85, especially p. 67.
39. Cf. I. Scheffler, *Science and Subjectivity* (Indianapolis: Bobbs-Merrill, 1967), p. 18.
40. Cf. *Structure*, p. 157.
41. Gombrich, *Art and Illusion*, p. 11 f.

simpler but the other is more accurate) the relative weight placed on different values by different individuals can play a decisive role in individual choice. More important, though scientists share these values and must continue to do so if science is to survive, they do not all apply them in the same way. Simplicity, scope, fruitfulness, and even accuracy can be judged quite differently (which is not to say they may be judged arbitrarily) by different people. Again, they may differ in their conclusions without violating any accepted rule.

That variability of judgment may, as I suggested above in connection with the recognition of crises, even be essential to scientific advance. The choice of a theory, which is, as Lakatos says, equally the choice of a research program, involves major risks, particularly in its early stages. Some scientists must, by virtue of a value system differing in its applicability from the average, choose it early, or it will not be developed to the point of general persuasiveness. The choices dictated by these atypical value systems are, however, generally wrong. If all members of the community applied values in the same high-risk way, the group's enterprise would cease. This last point, I think, Lakatos misses, and with it the essential role of individual variability in what is only belatedly the unanimous decision of the group. As Feyerabend also emphasizes, to give these decisions a '*historical character*' or to suggest that they are made only '*with hindsight*' deprives them of their function.[42] The scientific community cannot wait for history, though some individual members do. The needed results are instead achieved by distributing the risk that must be taken among the group's members.

Does anything in this argument suggest the appropriateness of phrases like decision by 'mob psychology'?[43] I think not. On the contrary, one characteristic of a mob is its rejection of values which its members ordinarily share. Done by scientists, the result should be the end of their science, and the Lysenko case suggests that it would be. My argument, however, goes even further, for it emphasizes that, unlike most disciplines, the responsibility for applying shared scientific values must be left to the specialists' group.[44] It may not even be extended to all scientists, much less to all educated laymen, much less to the mob. If the specialists' group behaves as a mob, renouncing its normal values, then science is already past saving.

---

42. Lakatos, "Falsification," p. 120; Feyerabend, "Consolations for the Specialist," pp. 215 ff.
43. Lakatos, "Falsification," p. 140, n. 3, and p. 178.
44. Cf. *Structure*, p. 167.

By the same token, no part of the argument here or in my book implies that scientists may choose any theory they like so long as they agree in their choice and thereafter enforce it.[45] Most of the puzzles of normal science are directly presented by nature, and all involve nature indirectly. Though different solutions have been received as valid at different times, nature cannot be forced into an arbitrary set of conceptual boxes. On the contrary, the history of proto-science shows that normal science is possible only with very special boxes, and the history of developed science shows that nature will not indefinitely be confined in any set which scientists have constructed so far. If I sometimes say that any choice made by scientists on the basis of their past experience and in conformity with their traditional values is ipso facto valid science for its time, I am only underscoring a tautology. Decisions made in other ways or decisions that could not be made in this way provide no basis for science and would not be scientific.

The charges of irrationality and relativism remain. To the first, however, I have already spoken, for I have discussed the issues, excepting incommensurability, from which it seems to arise. I am not sanguine in this matter, however, for I have not previously and do not now understand quite what my critics mean when they employ terms like 'irrational' and 'irrationality' to characterize my views. These labels seem to me mere shibboleths, barriers to a joint enterprise whether conversation or research. My difficulties in understanding are, however, even clearer and more acute when these terms are used not to criticize my position but in its defense. Obviously there is much in the last part of Feyerabend's paper with which I agree, but to describe the argument as a defense of irrationality in science seems to me not only absurd but vaguely obscene. I would describe it, together with my own, as an attempt to show that existing theories of rationality are not quite right and that we must readjust or change them to explain why science works as it does. To suppose, instead, that we possess criteria of rationality which are independent of our understanding of the essentials of the scientific process is to open the door to cloud-cuckoo land.

An answer to the charge of relativism must be more complex than

45. Some sense of my surprise and chagrin over this and related ways of reading my book may be generated by the following anecdote. During a meeting, I was talking to a usually far-distant friend and colleague whom I knew, from a published review, to be enthusiastic about my book. She turned to me and said, "Well, Tom, it seems to me that your biggest problem now is showing in what sense science can be empirical." My jaw dropped and still sags slightly. I have total visual recall of that scene and of no other since de Gaulle's entry into Paris in 1944.

those which precede, for the charge arises from more than misunderstanding. In one sense of the term I may be a relativist; in a more essential one I am not. What I can hope to do here is separate the two. It must already be clear that my view of scientific development is fundamentally evolutionary. Imagine, therefore, an evolutionary tree representing the development of the scientific specialties from their common origin in, say, primitive natural philosophy. Imagine, in addition, a line drawn up that tree from the base of the trunk to the tip of some limb without doubling back on itself. Any two theories found along this line are related to each other by descent. Now consider two such theories, each chosen from a point not too near its origin. I believe it would be easy to design a set of criteria—including maximum accuracy of predictions, degree of specialization, number (but not scope) of concrete problem solutions—which would enable any observer involved with neither theory to tell which was the older, which the descendant. For me, therefore, scientific development is, like biological evolution, unidirectional and irreversible. One scientific theory is not as good as another for doing what scientists normally do. In that sense I am not a relativist.

But there are reasons why I get called one, and they relate to the contexts in which I am wary about applying the label 'truth'. In the present context, its intra-theoretic uses seem to me unproblematic. Members of a given scientific community will generally agree which consequences of a shared theory sustain the test of experiment and are therefore true, which are false as theory is currently applied, and which are as yet untested. Dealing with the comparison of theories designed to cover the same range of natural phenomena, I am more cautious. If they are historical theories, like those considered above, I can join Sir Karl in saying that each was believed to be true in its time but was later abandoned as false. In addition, I can say that the later theory was the better of the two as a tool for the practice of normal science, and I can hope to add enough about the senses in which it was better to account for the main developmental characteristics of the sciences. Being able to go that far, I do not myself feel that I am a relativist. Nevertheless, there is another step, or kind of step, which many philosophers of science wish to take and which I refuse. They wish, that is to compare theories as representations of nature, as statements about "what is really out there." Granting that neither theory of a historical pair is true, they nonetheless seek a sense in which the later is a better approximation to the truth. I believe nothing of that sort can be found. On the other hand, I no longer feel that anything is lost, least of all the ability to explain scientific progress, by taking this position.

What I am rejecting will be clarified by reference to Sir Karl's paper and to his other writings. He has proposed a criterion of verisimilitude which permits him to write that "a later theory . . . t_2 has superseded t_1 . . . by approaching more closely to the truth than t_1." Also, when discussing a succession of frameworks, he speaks of each later member of the series as "better *and roomier*" than its predecessors; and he implies that the limit of the series, at least if carried to infinity, is "'absolute' or 'objective' truth, in Tarski's sense."[46] Those positions present, however, two problems, about the first of which I am uncertain of Sir Karl's position. To say, for example, of a field theory that it "approach[es] more closely to the truth" than an older matter-and-force theory should mean, unless words are being oddly used, that the ultimate constituents of nature are more like fields than like matter and force. But in this ontological context it is far from clear how the phrase 'more like' is to be applied. Comparison of historical theories gives no sense that their ontologies are approaching a limit: in some fundamental ways Einstein's general relativity resembles Aristotle's physics more than Newton's. In any case, the evidence from which conclusions about an ontological limit are to be drawn is the comparison not of whole theories but of their empirical consequences. That is a major leap, particularly in the face of a theorem that any finite set of consequences of a given theory can be derived from another incompatible one.

The other difficulty is highlighted by Sir Karl's reference to Tarski and is more fundamental. The semantic conception of truth is regularly epitomized in the example: 'Snow is white' is true if and only if snow is white. To apply that conception in the comparison of two theories, one must therefore suppose that their proponents agree about technical equivalents of such matters of fact as whether snow is white. If that supposition were exclusively about objective observation of nature, it would present no insuperable problems, but it involves as well the assumption that the objective observers in question understand 'snow is white' in the same way, a matter which may not be obvious if the sentence reads 'elements combine in constant proportion by weight'. Sir Karl takes it for granted that the proponents of competing theories do share a neutral language adequate to the comparison of such observa-

46. K. R. Popper, *Conjectures and Refutations: The Growth of Scientific Knowledge* (London: Routledge & Kegan Paul, 1963); Popper, "Normal Science," p. 56; italics added.

tion reports. I am about to argue that they do not. If I am right, then 'truth' may, like 'proof', be a term with only intra-theoretical applications. Until this problem of a neutral observation language is resolved, confusion will only be perpetuated by those who point out (as Watkins does when responding to my closely parallel remarks about 'mistakes'[47]) that the term is regularly used as though the transfer from intra- to inter-theoretical contexts made no difference.

Incommensurability and Paradigms

At last we arrive at the central constellation of issues which separate me from most of my critics. I regret the length of the journey to this point but accept only partial responsibility for the brush that has had to be cleared from the path. Unfortunately, the necessity of relegating these issues to my concluding section results in a relatively cursory and dogmatic treatment. I can hope only to isolate some aspects of my viewpoint which my critics have generally missed or dismissed and to provide motives for further reading and discussion.

The point-by-point comparison of two successive theories demands a language into which at least the empirical consequences of both can be translated without loss or change. That such a language lies ready to hand has been widely assumed since at least the seventeenth century, when philosophers took the neutrality of pure sensation reports for granted and sought a 'universal character' which would display all languages for expressing them as one. Ideally the primitive vocabulary of such a language would consist of pure sense-datum terms plus syntactic connectives. Philosophers have now abandoned hope of achieving any such ideal, but many of them continue to assume that theories can be compared by recourse to a basic vocabulary consisting entirely of words which are attached to nature in ways that are unproblematic and, to the extent necessary, independent of theory. That is the vocabulary in which Sir Karl's basic statements are framed. He requires it in order to compare the verisimilitude of alternate theories or to show that one is 'roomier' than (or includes) its predecessor. Feyerabend and I have argued at length that no such vocabulary is available. In the transition from one theory to the next words change their meanings or conditions

---

47. Watkins, "Against 'Normal Science'," p. 26, n. 3.

of applicability in subtle ways.[48] Though most of the same signs are used before and after a revolution—e.g., force, mass, element, compound, cell—the ways in which some of them attach to nature has somehow changed. Successive theories are thus, we say, incommensurable.

Our choice of the term 'incommensurable' has bothered a number of readers. Though it does not mean 'incomparable' in the field from which it was borrowed, critics have regularly insisted that we cannot mean it literally since men who hold different theories do communicate and sometimes change each others' views.[49] More important, critics often slide from the observed existence of such communication, which I have underscored myself, to the conclusion that it can present no essential problems. Toulmin seems content to admit 'conceptual incongruities' and then go on as before (p. 44). Lakatos inserts parenthetically the phrase "or from semantical reinterpretations" when telling us how to compare successive theories and thereafter treats the comparison as purely logical.[50] Sir Karl exorcises the difficulty in a way that has particular interest: "It is just a dogma—a dangerous dogma—that the different frameworks are like mutually untranslatable languages. The fact is that even totally different languages (like English and Hopi, or Chinese) are not untranslatable, and that there are many Hopis or Chinese who have learnt to master English very well."[51]

I accept the utility, indeed the importance, of the linguistic parallel, and shall therefore dwell for a bit upon it. Presumably Sir Karl accepts it too since he uses it. If he does, the dogma to which he objects is not that frameworks are like languages but that languages are untranslatable. But no one ever believed they were! What people have believed, and what makes the parallel important, is that the difficulties of learning a second language are different from and far less problematic than the difficulties of translation. Though one must know two languages in order to translate at all, and though translation can then always be managed up to a point, it can present grave difficulties to even the most

48. In his review of *Structure*, Shapere criticizes, in part quite properly, the way I discuss meaning change in my book. In the process he challenges me to specify the "cash difference" between a change in meaning and an alteration in the application of a term. Need I say that, in the present state of the theory of meaning, there is none. The identical point can be made using either term.

49. See, for example, Toulmin, "Does the Distinction," pp. 43–44.

50. Lakatos, "Falsification," p. 118. Perhaps only because of its excessive brevity, Lakatos's other reference to this problem on p. 179, n. 1, is equally little helpful.

51. Popper, "Normal Science," p. 56.

adept bilingual. He must find the best available compromises between incompatible objectives. Nuances must be preserved but not at the price of sentences so long that communication breaks down. Literalness is desirable but not if it demands introducing too many foreign words which must be separately discussed in a glossary or appendix. People deeply committed both to accuracy and to felicity of expression find translation painful, and some cannot do it at all.

Translation, in short, always involves compromises which alter communication. The translator must decide what alterations are acceptable. To do that he needs to know what aspects of the original it is most important to preserve and also something about the prior education and experience of those who will read his work. Not surprisingly, therefore, it is today a deep and open question what a perfect translation would be and how nearly an actual translation can approach the ideal. Quine has recently concluded "that rival systems of analytic hypotheses [for the preparation of translations] can conform to all speech dispositions within each of the languages concerned and yet dictate, in countless cases, utterly disparate translation. . . . Two such translations might even be patently contrary in truth value."[52] One need not go that far to recognize that reference to translation only isolates but does not resolve the problems which have led Feyerabend and me to talk of incommensurability. To me at least, what the existence of translations suggests is that recourse is available to scientists who hold incommensurable theories. That recourse need not, however, be to full restatement in a neutral language of even the theories' consequences. The problem of theory comparison remains.

Why is translation, whether between theories or languages, so difficult? Because, as has often been remarked, languages cut up the world in different ways, and we have no access to a neutral sublinguistic means of reporting. Quine points out that, though the linguist engaged in radical translation can readily discover that his native informant utters 'Gavagai' because he has seen a rabbit, it is more difficult to discover how 'Gavagai' should be translated. Should the linguist render it as 'rabbit', 'rabbit-kind', 'rabbit-part', 'rabbit-occurrence', or by some other phrase he may not even have thought to formulate? I extend the example by supposing that, in the community under examination, rabbits change color, length of hair, characteristic gait, and so on during the rainy

52. W. V. O. Quine, *Word and Object* (Cambridge, MA: Technology Press of the Massachusetts Institute of Technology, 1960), pp. 73 ff.

season, and that their appearance then elicits the term 'Bavagai'. Should 'Bavagai' be translated 'wet rabbit', 'shaggy rabbit', 'limping rabbit', all of these together, or should the linguist conclude that the native community has not recognized that 'Bavagai' and 'Gavagai' refer to the same animal? Evidence relevant to a choice among these alternatives will emerge from further investigation, and the result will be a reasonable analytic hypothesis with implication for the translation of other terms as well. But it will be only a hypothesis (none of the alternatives considered above need be right); the result of any error may be later difficulties in communication; when it occurs, it will be far from clear whether the problem is with translation and, if so, where the root difficulty lies.

These examples suggest that a translation manual inevitably embodies a theory, which offers the same sorts of reward, but also is prone to the same hazards, as other theories. To me they also suggest that the class of translators includes both the historian of science and the scientist trying to communicate with a colleague who embraces a different theory.[53] (Note, however, that the motives and correlated sensitivities of the scientists and historian are very different, which accounts for many systematic differences in their results.) They often have the inestimable advantage that the signs used in the two languages are identical or nearly so, that most of them function the same way in both languages, and that, where function has changed, there are nevertheless informative reasons for retaining the same sign. But those advantages bring with them penalties illustrated in both scientific discourse and history of science. They make it excessively easy to ignore functional changes that would be apparent if they had been accompanied by a change of sign.

The parallel between the task of the historian and the linguist highlights an aspect of translation with which Quine does not deal (he need not) and that has made trouble for linguists.[54] Teaching Aristotelian physics to students, I regularly point out that matter (in the *Physics*, not the *Metaphysics*), just because of its omnipresence and qualitative neutrality, is a physically dispensable concept. What populates the Aris-

53. A number of these ideas about translation were developed in my Princeton seminar. I cannot now distinguish my contributions from those of the students and colleagues who attended. A paper by Tyler Burge was, however, particularly helpful.

54. See particularly E. A. Nida, "Linguistics and Ethnology in Translation-Problems," in *Language and Culture in Society: A Reader in Linguistics and Anthropology*, ed. D. H. Hymes (New York: Harper and Row, 1964), pp. 90–97. I am much indebted to Sarah Kuhn for calling this paper to my attention.

totelian universe, accounting for both its diversity and regularity, is immaterial 'natures' or 'essences'; the appropriate parallel for the contemporary periodic table is not the four Aristotelian elements, but the quadrangle of four fundamental forms. Similarly, when teaching the development of Dalton's atomic theory, I point out that it implied a new view of chemical combination with the result that the line separating the referents of the terms 'mixture' and 'compound' shifted; alloys were compounds before Dalton, mixtures after.[55] Those remarks are part and parcel of my attempt to translate older theories into modern terms, and my students characteristically read source materials, though already rendered into English, differently after I have made them than they did before. By the same token, a good translation manual, particularly for the language of another region and culture, should include or be accompanied by discursive paragraphs explaining how native speakers view the world, what sorts of ontological categories they deploy. Part of learning to translate a language or a theory is learning to describe the world with which the language or theory functions.

Having introduced translation to illustrate the illumination that can be had by regarding scientific communities as language communities, I now leave it for a time in order to examine a particularly important aspect of the parallelism. In learning either a science or a language, vocabulary is generally acquired together with at least a minimal battery of generalizations which exhibit it applied to nature. In neither case, however, do the generalizations embody more than a fraction of the knowledge of nature which has been acquired in the learning process. Much of it is embodied instead in the mechanism, whatever it may be, which is used to attach terms to nature.[56] Both natural and scientific language are designed to describe the world as it is, not any conceivable

55. This example makes particularly clear the inadequacy of Scheffler's suggestion that the problems raised by Feyerabend and me vanish if one substitutes sameness-of-reference for sameness-of-meaning (*Science and Subjectivity*, chapter 3). Whatever the reference of 'compound' may be, in this example it changes. But, as the following discussion will indicate, sameness-of-reference is no more free of difficulty than sameness-of-meaning in any of the applications that concern me and Feyerabend. Is the referent of 'rabbit' the same as that of 'rabbit-kind' or of 'rabbit-occurrence'? Consider the criteria of individuation and of self-identity which fit each of the terms.

56. For an extended example, see my "A Function for Thought Experiments," in *Mélanges Alexandre Koyré*, vol. 2, *L'aventure de l'esprit*, ed. I. B. Cohen and R. Taton (Paris: Hermann, 1964), pp. 307–34; reprinted in *The Essential Tension*, pp. 240–65. A more analytic discussion will be found in my "Second Thoughts on Paradigms."

world. The former, it is true, adapts to the unexpected occurrence more easily than the latter, but often at the price of long sentences and dubious syntax. Things which cannot *readily* be said in a language are things that its speakers do not expect to have occasion to say. If we forget this or underestimate its importance, that is probably because its converse does not hold. We can readily describe many things (unicorns, for example) which we do not expect to see.

How, then, do we acquire the knowledge of nature that is built into language? For the most part by the same techniques and at the same time as we acquire language itself, whether everyday or scientific. Parts of the process are well known. The definitions in a dictionary tell us something about what words mean and simultaneously inform us of the objects and situations about which we may need to read or speak. About some of these words we learn more, and about others everything we know, by encountering them in a variety of sentences. Under those circumstances, as Carnap has shown, we acquire laws of nature together with a knowledge of meanings. Given a verbal definition of two tests, each definitive, for the presence of an electric charge, we learn both about the term 'charge' and also that a body which passes one test will also pass the other. These procedures for language-nature learning are, however, purely linguistic. They relate words to other words and thus can function only if we already possess some vocabulary acquired by a nonverbal or incompletely verbal process. Presumably that part of learning is by ostension or some elaboration of it, the direct matching of whole words or phrases to nature. If Sir Karl and I have a fundamental philosophical dispute, it is about the relevance of this last mode of language-nature learning to philosophy of science. Though he knows that many words needed by scientists, particularly for the formulation of basic sentences, are learned by a process not fully linguistic, he treats those terms and the knowledge acquired with them as unproblematic, at least in the context of theory choice. I believe he misses a central point, the one which led me to introduce the notion of paradigms in *Structure*.

When I speak of knowledge embedded in terms and phrases learned by some nonlinguistic process like ostension, I am making the same point that my book aimed to make by repeated reference to the role of paradigms as concrete problem solutions, the exemplary objects of an ostension. When I speak of that knowledge as consequential for science and for theory construction, I am identifying what Miss Masterman underscores about paradigms by saying that they "can function when the theory is not there."[57] These ties are not, however, likely to be apparent to anyone who has taken the notion of paradigm less seriously than Miss Masterman, for, as she quite properly emphasizes, I have used the term in a number of different ways. To discover what is presently the issue, I must briefly digress to unravel confusions, in this case ones that are entirely of my own making.

Above, I remarked that a new version of *Structure* would open with a discussion of community structure. Having isolated an individual specialists' group, I would next ask what its members shared that enabled them to solve puzzles and that accounted for their relative unanimity in problem choice and in the evaluation of problem solutions. One answer which my book licenses to that question is "a paradigm" or "a set of paradigms." (This is Miss Masterman's sociological sense of the term.) For it I should now like some other phrase, perhaps 'disciplinary matrix': 'disciplinary', because it is common to the practitioners of a specified discipline; 'matrix', because it consists of ordered elements which require individual specification. All of the objects of commitment described in my book as paradigms, parts of paradigms, or paradigmatic would find a place in the disciplinary matrix, but they would not be lumped together as paradigms, individually or collectively. Among them would be: shared symbolic generalizations, like '$f = ma$', or 'elements combine in constant proportion by weight'; shared models, whether metaphysical, like atomism, or heuristic, like the hydrodynamic model of the electric circuit; shared values, like the emphasis on accuracy of prediction, discussed above; and other elements of the sort. Among the latter I would particularly emphasize concrete problem solutions, the sorts of standard examples of solved problems which scientists encounter first in student laboratories, in the problems at the ends of chapters in science texts, and on examinations. If I could, I would call these problem solutions paradigms, for they are what led me to the choice of the term in the first place. Having lost control of the word, however, I shall henceforth describe them as exemplars.[58]

---

57. Masterman, "The Nature of a Paradigm," p. 66.

58. This modification and almost everything else in the remainder of this paper is discussed in far more detail and with more evidence in my "Second Thoughts on Paradigms." I refer readers to it even for bibliographical references. One additional remark is, however, in place here. The change just outlined in my text deprives me of recourse to the phrases 'pre-paradigm period' and 'post-paradigm period' when describing the maturation of a scientific specialty. In retrospect that seems to me all to the good, for, in both senses of the term, paradigms have throughout been possessed by any scientific community, including the schools of what I previously called

Ordinarily problem solutions of this sort are viewed as mere applications of theory that has already been learned. The student does them for practice, to gain facility in the use of what he already knows. Undoubtedly that description is correct after enough problems have been done, but never, I think, at the start. Rather, doing problems is learning the language of a theory and acquiring the knowledge of nature embedded in that language. In mechanics, for example, many problems involve applications of Newton's second law, usually stated as '$f = ma$'. That symbolic expression is, however, a law-sketch rather than a law. It must be rewritten in a different symbolic form for each physical problem before logical and mathematical deduction are applied to it. For free fall it becomes

$$mg = \frac{md^2 s}{dt^2};$$

for the pendulum it is

$$mg\ \mathrm{Sin}\ \theta = -ml\frac{d^2 \theta}{dt^2};$$

for coupled harmonic oscillators it becomes two equations, the first of which may be written

$$m_1 \frac{d^2 s_1}{dt^2} + k_1 s_1 = k_2(d + s_2 - s_1);$$

and so on.

Lacking space to develop an argument, I shall simply assert that physicists share few rules, explicit or implicit, by which they make the transition from law-sketch to the specific symbolic forms demanded by individual problems. Instead, exposure to a series of exemplary problem solutions teaches them to see different physical situations as like each

---

the "pre-paradigm period." My failure to see that point earlier has certainly helped to make a paradigm seem a quasi-mystical entity or property that, like charisma, transforms those infected by it. Note, however, as indicated above, that this alteration in terminology does not at all alter my description of the maturation process. The early stages in the development of most sciences are characterized by the presence of a number of competing schools. Later, usually in the aftermath of a notable scientific achievement, all or most of these schools vanish, a change which permits a far more powerful professional behavior to the members of the remaining community. On this whole problem, Miss Masterman's remarks ("The Nature of a Paradigm," pp. 70–72) seem very telling.

other; they are, if you will, seen in a Newtonian gestalt. Once students have acquired the ability to see a number of problem situations in that way, they can write down ad lib the symbolic forms demanded by other such situations as they arise. Before that acquisition, however, Newton's second law was to them little or no more than a string of uninterpreted symbols. Though they shared it, they did not know what it meant and it therefore told them little about nature. What they had yet to learn was not, however, embodied in additional symbolic formulations. Rather it was gained by a process like ostension, the direct exposure to a series of situations each of which, they were told, were Newtonian.

Seeing problem situations as like each other, as subjects for the application of similar techniques, is also an important part of normal scientific work. One example may both illustrate the point and drive it home. Galileo found that a ball rolling down an incline acquires just enough velocity to return it to the same vertical height on a second incline of any slope, and he learned to see that experimental situation as like the pendulum with a point-mass for a bob. Huyghens then solved the problem of the center of oscillation of a physical pendulum by imagining that the extended body of the latter was composed of Galilean point-pendula, the bonds between which could be released at any point in the swing. After the bonds were released, the individual point-pendula would swing freely, but their collective center of gravity, when each was at its highest point, would be only at the height from which the center of gravity of the extended pendulum had begun to fall. Finally, Daniel Bernoulli, still with no aid from Newton's laws, discovered how to make the flow of water from an orifice in a storage tank resemble Huyghens's pendulum. Determine the descent of the center of gravity of the water in tank and jet during an infinitesimal period of time. Next imagine that each particle of water afterward moves separately upward to the maximum height obtainable with the velocity it possessed at the end of the interval of descent. The ascent of the center of gravity of the separate particles must then equal the descent of the center of gravity of the water in tank and jet. From that view of the problem the long-sought speed of efflux followed at once. These examples display what Miss Masterman has in mind when she speaks of a paradigm as fundamentally an artefact which transforms problems to puzzles and enables them to be solved even in the absence of an adequate body of theory.

Is it clear that we are back to language and its attachment to nature? Only one law was used in all of the preceding examples. Known as the principle of *vis viva*, it was generally stated as 'Actual descent equals potential ascent'. Contemplating the examples is an essential part (though only part) of learning what the words in that law mean individually and collectively, or in learning how they attach to nature. Equally, it is part of learning how the world behaves. The two cannot be separated. The same double role is played by the textbook problems from which students learn, for example, to discover forces, masses, accelerations in nature and in the process find out what '$f = ma$' means and how it attaches to and legislates for nature. In none of these cases do the examples function alone, of course. The student must know mathematics, some logic, and above all natural language and the world to which it applies. But the latter pair has to a considerable extent been learned in the same way, by a series of ostensions which have taught him to see mother as always like herself and different from father and sister, which have taught him to see dogs as similar to each other and unlike cats, and so on. These learned similarity-dissimilarity relationships are ones that we all deploy every day, unproblematically, yet without being able to name the characteristics by which we make the identifications and discriminations. They are prior, that is, to a list of criteria which, joined in a symbolic generalization, would enable us to define our terms. Rather, they are parts of a language-conditioned or language-correlated way of seeing the world. Until we have acquired them, we do not see a world at all.

For a more leisurely and developed account of this aspect of the language-theory parallel, I shall have to refer readers to the previously cited paper from which much in the last few paragraphs is abstracted. Before returning to the problem of theory choice, however, I must at least state the point which that paper primarily aims to defend. When I speak of learning language and nature together by ostension, and particularly when I speak of learning to cluster the objects of perception into similarity sets without answering questions like, 'similar with respect to what?' I am not calling upon some mystic process to be covered by the label 'intuition' and thereafter left alone. On the contrary, the sort of process I have in mind can perfectly well be modeled on a computer and thus compared with the more familiar mode of learning which resorts to criteria rather than to a learned similarity relationship. I am currently in the early stages of such a comparison, hoping, among other things, to discover something about the circumstances under which each of the two strategies works more effectively. In both programs the computer will be given a series of stimuli (modeled as ordered sets of integers) together with the name of the class from which each stimulus was selected. In the criterion-learning program the machine is instructed to abstract criteria which will permit the classification of additional stimuli, and it may thereafter discard the original set from which it learned to do the job. In the similarity-learning program, the machine is instead instructed to retain all stimuli and to classify each new one by a global comparison with the clustered exemplars it has already encountered. Both programs will work, but they do not give identical results. They differ in many of the same ways and for many of the same reasons as case law and codified law.

One of my claims is, then, that we have too long ignored the manner in which knowledge of nature can be tacitly embodied in whole experiences without intervening abstraction of criteria or generalizations. Those experiences are presented to us during education and professional initiation by a generation which already knows what they are exemplars of. By assimilating a sufficient number of exemplars, we learn to recognize and work with the world our teachers already know. My main past applications of that claim have, of course, been to normal science and the manner in which it is altered by revolutions, but an additional application is worth noting here. Recognizing the cognitive function of examples may also remove the taint of irrationality from my earlier remarks about the decisions I described as ideologically based. Given examples of what a scientific theory does and being bound by shared values to keep doing science, one need not also have criteria in order to discover that something has gone wrong or to make choices in case of conflict. On the contrary, though I have as yet no hard evidence, I believe that one of the differences between my similarity and criteria programs will be the special effectiveness with which the former deals with situations of this sort.

Against that background I return finally to the problem of theory choice and the recourse offered by translation. One of the things upon which the practice of normal science depends is a learned ability to group objects and situations into similarity classes which are primitive in the sense that the grouping is done without an answer to the question, 'similar with respect to what?' One aspect of every revolution is, then, that some of the similarity relations change. Objects which were grouped in the same set before are grouped in different sets afterward and vice versa. Think of the sun, moon, Mars, and Earth before and after Copernicus; of free fall, pendular, and planetary motion before and after Galileo; or of salts, alloys, and a sulfur/iron-filing mix before and after Dalton. Since most objects within even the altered sets con-

tinue to be grouped together, the names of the sets are generally preserved. Nevertheless, the transfer of a subset can crucially affect the
network of interrelations among sets. Transferring the metals from the
set of compounds to the set of elements was part of a new theory of
combustion, of acidity, and of the difference between physical and
chemical combination. In short order, those changes had spread through
all of chemistry. When such a redistribution of objects among similarity
sets occurs, two men whose discourse had proceeded for some time
with apparently full understanding may suddenly find themselves responding to the same stimulus with incompatible descriptions or generalizations. Just because neither can then say, "I use the word 'element'
(or 'mixture', or 'planet', or 'unconstrained motion') in ways governed
by such and such criteria," the source of the breakdown in their communication may be extraordinarily difficult to isolate and bypass.

I do not claim that there is no recourse in such situations, but before
asking what it is, let me emphasize just how deep differences of this
sort go. They are not simply about names or language but equally and
inseparably about nature. We cannot say with any assurance that the
two men even see the same thing, possess the same data, but identify
or interpret it differently. What they are responding to differently is
stimuli, and stimuli receive much neural processing before anything is
seen or any data are given to the sense. Since we now know (as Descartes did not) that the stimulus-sensation correlation is neither one-
to-one nor independent of education, we may reasonably suspect that
it varies somewhat from community to community, the variation being
correlated with the corresponding differences in the language-nature
interaction. The sorts of communication breakdowns now being considered are likely evidence that the men involved are processing certain
stimuli differently, receiving different data from them, seeing different
things or the same things differently. I think it likely myself that much
or all of the clustering of stimuli into similarity sets takes place in the
stimulus-to-sensation portion of our neural processing apparatus; that
the educational programming of that apparatus takes place when we are
presented with stimuli that we are told emanate from members of the
same similarity class; and that, after programming has been completed,
we recognize, say, cats and dogs (or pick out forces, masses, and constraints) because they (or the situations in which they appear) then do,
for the first time, look like the examples we have seen before.

Nevertheless, there must be recourse. Though they have no direct
access to it, the stimuli to which the participants in a communication

breakdown respond are, under pain of solipsism, the same. So is their general neural apparatus, however different the programming. Furthermore, except in a small, if all-important, area of experience, the programming must be the same, for the men involved share a history (except the immediate past), a language, an everyday world, and most of a scientific one. Given what they share, they can find out much about how they differ. At least they can do so if they have sufficient will, patience, and tolerance of threatening ambiguity, characteristics which, in matters of this sort, cannot be taken for granted. Indeed, the sorts of therapeutic efforts to which I now turn are rarely carried far by scientists.

First and foremost, men experiencing communication breakdown can discover by experiment—sometimes by thought experiment, armchair science—the area within which it occurs. Often the linguistic center of the difficulty will involve a set of terms, like element and compound, which both men deploy unproblematically but which it can now be seen they attach to nature in different ways. For each, these are terms in a basic vocabulary, at least in the sense that their normal intragroup use elicits no discussion, request for explication, or disagreement. Having discovered, however, that for intergroup discussion, these words are the locus of special difficulties, our men may resort to their shared everyday vocabularies in a further attempt to elucidate their troubles. Each may, that is, try to discover what the other would see and say when presented with a stimulus to which his visual and verbal response would be different. With time and skill, they may become very good predictors of each other's behavior, something that the historian regularly learns to do (or should) when dealing with older scientific theories.

What each participant in a communication breakdown has then found is, of course, a way to translate the other's theory into his own language and simultaneously to describe the world in which that theory or language applies. Without at least preliminary steps in that direction, there would be no process that one were even tempted to describe as theory *choice*. Arbitrary conversion (except that I doubt the existence of such a thing in any aspect of life) would be all that was involved. Note, however, that the possibility of translation does not make the term 'conversion' inappropriate. In the absence of a neutral language, the choice of a new theory is a decision to adopt a different native language and to deploy it in a correspondingly different world. That sort of transition is, however, not one which the terms 'choice' and 'decision' quite fit, though the reasons for wanting to apply them after the event are clear.

Exploring an alternative theory by techniques like those outlined above, one is likely to find that one is already using it (as one suddenly notes that one is thinking in, not translating out of, a foreign language). At no point was one aware of having reached a decision, made a choice. That sort of change is, however, conversion, and the techniques which induce it may well be described as therapeutic, if only because, when they succeed, one learns one had been sick before. No wonder the techniques are resisted and the nature of the change disguised in later reports.

Chapter Seven

Theory Change as Structure Change: Comments on the Sneed Formalism

"Theory Change as Structure Change: Comments on the Sneed Formalism" first appeared in Erkenntnis *10 (1976): 179–99. Reprinted with the kind permission of Kluwer Academic Publishers.*

———◦———

IT IS NOW MORE THAN a year and a half since Professor Stegmüller kindly sent me a copy of his *Theorie und Erfahrung,*[1] thus drawing my attention for the first time to the existence of Dr. Sneed's new formalism and its likely relevance to my own work. At that time set theory was to me an unknown and altogether forbidding language, but I was quickly persuaded that I must somehow find time to acquire it. Even now I cannot claim entire success: I shall here sometimes refer to, but never attempt to speak, set theory. Nevertheless, I have learned enough to embrace with enthusiasm the two major conclusions of Stegmüller's book. First, though still at an early stage of its development, the new formalism makes important new territory accessible to analytic philosophy of science. Second, though sketched with a pen I can still scarcely hold, preliminary charts of the new terrain display remarkable resem-

1. W. Stegmüller, *Probleme und Resultate der Wissenschaftstheorie und analytischen Philosophie,* vol. 2, *Theorie und Erfahrung,* part 2, *Theorienstrukturen und Theoriendynamik* (Berlin: Springer-Verlag, 1973); reprinted as *The Structure and Dynamics of Theories,* trans. W. Wohlhueter (New York: Springer-Verlag, 1976).

blance to a map I had previously sketched from scattered travelers' reports brought back by itinerant historians of science.

The resemblance is firmly underscored in the closing chapter of Sneed's book[2]; its detailed elaboration is a primary contribution of Stegmüller's. That the rapprochement both see is genuine should be sufficiently indicated by the fact that Stegmüller, approaching my work through Sneed's, has understood it better than any other philosopher who has made more than passing reference to it. From these developments I take great encouragement. Whatever its limitations (I take them to be severe), formal representation provides a primary technique for exploring and clarifying ideas. But traditional formalisms, whether set-theoretical or propositional, have made no contact whatsoever with mine. Dr. Sneed's formalism does, and at a few especially strategic points. Though neither he, nor Stegmüller, nor I suppose that it can solve all the outstanding problems in philosophy of science, we are united in regarding it as an important tool, thoroughly worth much additional development.

Just because the new formalism does illuminate some of my own characteristic heresies, my evaluation of it is unlikely to be free from bias. But I shall not pause merely to deplore the inevitable. Instead, I turn to my subject proper, beginning with a cursory sketch of some aspects of the new formalism that seem to me particularly appealing. Premising them, I shall next explore two aspects of the Sneed-Stegmüller position that, in their present form, seem to me significantly incomplete. Finally, I shall examine one central difficulty that will not be resolved within the formalism but presumably requires resort to philosophy of language. Before turning to that program, however, let me avoid misunderstanding by indicating an area in which this paper makes no claims at all. What has excited me about the Sneed formalism is the issues it makes it possible to explore with precision, not the particular apparatus developed for that purpose. About such questions as whether or not those achievements demand the use of set and model theory, I have no basis for opinions. Or rather, I have a basis for only one: those who think set theory an illegitimate tool for analyzing the logical structure of scientific theories are now challenged to produce similar results in another way.

2. J. D. Sneed, *The Logical Structure of Mathematical Physics* (Dordrecht, Boston: D. Reidel, 1971), esp. pp. 288–307.

Appraising the Formalism

What has struck me from the start about the Sneed formalism is that even its elementary structural form captures significant features of scientific theory and practice notably absent from the earlier formalisms known to me. That is perhaps not surprising, for Sneed has repeatedly inquired, while preparing his book, how theories are presented to students of science and then used by them (e.g., pp. 3 f., 28, 33, 110–14). One result of this procedure is the elimination of artificialities that have in the past often made philosophical formalisms seem irrelevant to both practitioners and historians of science. The one physicist with whom I have to date discussed Sneed's views has been fascinated by them. As a historian, I shall myself mention below one way in which the formalism has already begun to influence my work. Though even guesses about the future are premature, I shall risk one. If only simpler and more palatable ways of representing the essentials of Sneed's position can be found, philosophers, practitioners, and historians of science may, for the first time in years, find fruitful channels for interdisciplinary communication.

To make this global claim more concrete, consider the three classes of models required by Sneed's presentation. The second, his potential partial models or M_{pp}'s, are (or include) the entities to which a given theory might be applied by virtue of their description in the nontheoretical vocabulary of the theory. The third, his models or M's, derive from the subset of the M_{pp}'s to which, after suitable theoretical extension, the laws of the theory actually do apply. Both find obvious parallels in traditional formal treatments. But Sneed's partial models, his M_p's, do not. They are the set of models obtained by adding theoretical functions to all the suitable members of M_{pp}, thus completing or extending them prior to the application of the theory's fundamental laws. It is, in part, by giving them a central place in the reconstruction of theory that Sneed adds significant verisimilitude to the structures that result.

Lacking time for an extended argument, I shall be content here with three assertions. First, teaching a student to make the transition from partial potential models to partial models is a large part of what scientific, or at least physics, education is about. That is what student laboratories and the problems at the ends of chapters of textbooks are for. The familiar student who can solve problems which are stated in equations but cannot produce equations for problems exhibited in the laboratory or stated in words has not begun to acquire this essential talent.

Second, almost a corollary, the creative imagination required to find an M_p corresponding to a nonstandard M_{pp} (say, a vibrating membrane or string before these were normal applications of Newtonian mechanics) is among the criteria by which great scientists may sometimes be distinguished from mediocre.[3] Third, failure to pay attention to the manner in which this task is done has for years disguised the nature of the problem presented by the meaning of theoretical terms.

Except in the case of fully mathematized theories, neither Stegmüller nor Sneed has much to say about how M_{pp}'s are, in fact, extended to M_p's. But the view Sneed develops with precision for his special case is strikingly like the one I had earlier articulated vaguely for the general one, and the two may henceforth fruitfully interact, a point to which I shall return. In both cases the process of extension depends upon assuming that the theory has been correctly deployed in one or more previous applications and on then using those applications as guides to the specification of theoretical functions or concepts when transforming a new M_{pp} to an M_p.[4] For fully mathematical theories that guidance is supplied by what Sneed calls constraints, lawlike restrictions that limit the structure of pairs or sets of partial models rather than of individual ones. (The values assumed by theoretical functions in one application must, for example, be compatible with those assumed in others.) Together with the correlated notion of applications, that of constraints constitutes what I take to be the central conceptual innovation of Sneed's formalism, and another especially striking one follows from it. For him as for me, the adequate specification of a theory must include specification of some set of exemplary applications. Stegmüller's subsection, "Was ist ein Paradigma?" is a splendid elaboration of this point (pp. 195–207).

So far I have mentioned aspects of Sneed's formalism which cohere particularly closely with views I have developed in other places. Shortly I shall return to some others of the same sort. But I am not sure that closely associating our views will prove a favor to him, and there are

---

3. The absence from traditional reconstructions of any step like that from a member of M_{pp} to its extension, the corresponding member of M_p, may help to explain my lack of success in persuading philosophers that normal science might be anything but a totally routine enterprise.
4. Sneed and Stegmüller consider only theories of mathematical physics (only the mathematical parts of theories of mathematical physics, would be a better way to describe their subject). Therefore, they refer only to the role of constraints in the specification of theoretical *functions*. I add "or concepts" in anticipation of a needed generalization of the Sneed formalism. That Sneed himself believes that concepts are specified at least in part by mathematical structures which include constraints will appear below.

other reasons to take his seriously. Let me mention just a few closely related ones before returning to my main theme.

Roughly speaking, Sneed represents a theory as a set of distinct applications. In the case of classical particle mechanics, these might be the problems of planetary motion, of pendula, of free fall, of levers and balances, and so on. (Need I emphasize that learning a theory is learning successive applications in some appropriate order and that using it is designing still others?) Considered individually, each application might be reconstructed by a standard axiom system in a predicate calculus (thus raising the standard problem of theoretical terms). But the individual axiom systems would then ordinarily be somewhat different from each other.[5] What in Sneed's view supplies their unity, enables a sufficient set collectively to determine a theory, is partly the basic law or laws which all share (say, Newton's second law of motion) and partly the set of constraints which bind the applications together in pairs or at least in connected chains.

With such a set-theoretic structure, individual applications play a double role, one previously familiar at a pre-theoretic level from discussions of reduction sentences. Taken singly, individual applications, like individual reduction sentences, are vacuous, either because their theoretical terms are uninterpretable or because the interpretation they permit is circular. But when applications are tied together by constraints, as reduction sentences are tied together by the recurrence of a theoretical term, they prove capable simultaneously of specifying, on the one hand, the manner in which theoretical concepts or terms must be applied and, on the other, some empirical content of the theory itself. Introduced, like reduction sentences, to solve the problem of theoretical terms, constraints prove also, again like reduction sentences, to be a vehicle for empirical content.[6]

Numerous interesting consequences follow, of which I shall here mention three. Since the discovery that theoretical terms could not

5. Compare Kuhn, *The Structure of Scientific Revolutions*, 2d ed., rev. (Chicago: University of Chicago Press, 1970), pp. 187–91.

6. A third example of the process (this time operating at the level of observation terms) which introduces language and empirical content in an inextricably mixed form is sketched in the last pages of T. S. Kuhn, "Second Thoughts on Paradigms," in *The Structure of Scientific Theories*, ed. F. Suppe (Urbana: University of Illinois Press, 1974), pp. 459–82. Its reappearance at all three traditional levels (observation terms, theoretical terms, and whole theories) seems to me of likely significance.

readily be eliminated by strict definition, one has been puzzled how to distinguish the conventional from the empirical elements within the process by which they are introduced. The Sneed formalism clarifies the puzzle by giving it additional structure. If a theory, like Newtonian mechanics, had only a single application (for example, the determination of mass ratios for two bodies connected by a spring), then the specification of the theoretical functions it supplies would be literally circular and the application correspondingly vacuous. But, from Sneed's viewpoint, no single application yet constitutes a theory, and, when several applications are conjoined, the potential circularity ceases to be vacuous because distributed by constraints over the whole set of applications. As a result, certain other, sometimes nagging, problems change their form or disappear. Within the Sneed formalism, there is no temptation to ask the, to physicists, artificial question whether mass or, alternatively, force should be treated as a primitive in terms of which the other should be defined. Both, for Sneed, are theoretical and in most respects on a par, because neither can be learned or given meaning except within the theory, some applications of which must be presupposed. Finally, and perhaps of greatest long-range importance, is the new form taken by Ramsey sentences within Dr. Sneed's formalism. Just because constraints as well as laws take on empirical consequences, there are important new things to be said about the function and the eliminability of theoretical terms.[7]

These and other aspects of the Sneed formalism deserve and will presumably receive much additional attention, but for me their importance is dwarfed by that of another, with which this section of my paper concludes. To a far greater extent and also far more naturally than any previous mode of formalization, Sneed's lends itself to the reconstruction of theory dynamics, the process by which theories change and grow. Particularly striking to me, of course, is that its manner of doing so appears to demand the existence of (at least) two quite distinct sorts of alteration over time. In the first, what Sneed calls a theory-core remains fixed, as do at least some of a theory's exemplary applications. Progress then occurs either by discovering new applications which can be identified extensionally as members of the set of intended applications, *I,* or else by constructing a new theory-core-net (a new set of

7. On these subjects see Sneed, *Logical Structure,* pp. 31–37, 48–51, 65–86, 117–38, 150–51; Stegmüller, *Theorienstrukturen,* 45–103.

expansions of the core in Sneed's older vocabulary) which more precisely specifies the conditions for membership in I.[8] Both Stegmüller and Sneed emphasize that changes of this sort correspond to much of the theoretical part of what I have elsewhere called normal science,[9] and I entirely accept their identification. Since by its nature a theory-core is virtually immune to direct falsification, Sneed also suggests and Stegmüller elaborates the possibility that at least some cases of change of core correspond to what I have called scientific revolutions.[10]

Much of the rest of this paper is devoted to identifying difficulties with that second identification. Though the Sneed formalism does permit the existence of revolutions, it currently does virtually nothing to clarify the nature of revolutionary change. I see, however, no reason why it cannot be made to do so, and I mean here to be making a contribution toward that end. Even in its absence, furthermore, both my historical and my more philosophical work are illuminated by the attempt to view revolutions as changes of core. In particular, I find that much of my still unpublished research concerning both the genesis of the quantum theory and its transformation during the years 1925–26 discloses changes that can be well represented as juxtapositions of elements from a traditional core with others drawn from one of its recent expansions.[11] That way of regarding revolutions seems to me especially promising because it may shortly permit me for the first time to say something worthwhile about the continuities which endure through them.[12] Work

8. Stegmüller, who rejects what he calls "Sneed's Platonism," would put this point differently, and I find myself somewhat more comfortable with his approach. But its introduction here would call for additional symbolic apparatus irrelevant to this paper's main purposes.

9. Sneed, *Logical Structure*, pp. 284–88; Stegmüller, *Theorienstrukturen*, pp. 219–31.

10. Sneed, *Logical Structure*, pp. 296–306; Stegmüller, *Theorienstrukturen*, pp. 231–47.

11. I could, for example, paraphrase a central theme of my forthcoming book on the history of the black-body problem in the following way. From 1900 through the publication of his *Wärmestrahlung* in 1906, the basic equations of mechanics and electromagnetic theory were in the core of Planck's black-body theory; the equation for the energy element, $\varepsilon = h\nu$, was part of its expansion. In 1908, however, the equation defining the energy element became part of a new core; equations selected ad hoc from mechanics and electromagnetic theory were in its expansion. Though there was sizable overlap between the equations included in the two *expanded* cores (whence much continuity), the structures of the theories determined by the two cores were radically distinct.

12. Stegmüller (pp. 14, 182) suggests that my inability to resolve a cluster of difficulties presented by my position is due to my having accepted the traditional view of a theory as a set of statements. I shall express reservations about some of his illustrations of that suggestion below, but it is thoroughly relevant to the problem of continuity. Noticing that an equation or statement essential to a theory's success in a given application need not be a determinant of that theory's structure makes it possible to say much more about how new theories may be built from elements generated by their incompatible predecessors.

must be done first, however. I shall now begin to suggest what some of it may be.

Two Problems of Demarcation

I have already suggested that the central novelty of Sneed's approach is probably his concept of constraints. Let me now add that it might usefully be awarded a position even more fundamental than the one he attributes to it. In developing his formalism, Sneed begins by selecting a theory, like classical particle mechanics, for which, he emphasizes, strict identity criteria must be presupposed.[13] Examining that theory, he next distinguishes between the nontheoretical and the theoretical functions it deploys, the latter being those which cannot be specified, in *any* of that theory's applications, without resort to the theory's fundamental laws. Finally, in a third step, constraints are introduced to permit the specification of theoretical functions. That third step seems to me just right. But I am far less confident about the two it presupposes, and I therefore wonder about the possibility of inverting the order of their introduction. Could one not, that is, introduce applications and constraints between them as primitive notions, allowing subsequent investigation to reveal the extent to which criteria for theory identity and for a theoretical/nontheoretical distinction would follow?

Consider, for example, the classical formulations of mechanics and electromagnetic theory. Most applications of either theory can be carried through without recourse to the other, a sufficient reason for describing them as two theories rather than one. But the two have never been absolutely distinct. Both entered together, and thus constrained each other, in such applications as ether mechanics, stellar aberration, the electron theory of metals, X-rays, or the photoelectric effect. In such applications, furthermore, neither theory was ordinarily conceived as a mere tool to be presupposed while creatively manipulating the other. Instead, the two were deployed together, almost as a single theory of which most other applications were either purely mechanical, on the one hand, or purely electromagnetic, on the other.[14]

I think nothing of importance is lost by recognizing that what we

13. Sneed, *Logical Structure*, p. 35; Stegmüller, *Theorienstrukturen*, p. 50.
14. A further sense in which one theory constrains another is indicated by the traditional view that the compatibility of a new theory with others currently accepted is among the criteria legitimate to its evaluation.

ordinarily refer to as distinct theories do overlap in occasional important applications. But that opinion depends upon my being prepared simultaneously to surrender any criterion quite so strict as Sneed's for distinguishing theoretical from nontheoretical functions and concepts. What is involved can be illustrated by considering his discussion of classical particle mechanics. Because they can be learned only when some applications of that theory are presupposed, the mass and force functions are declared theoretical with respect to particle mechanics, and they are thus contrasted with the variables space and time, acquired independently of that theory. Something about that result seems to me profoundly right, but I am troubled that the argument appears to depend essentially on conceiving statics, the science of mechanical equilibria, as unproblematically a part of the more general theory that treats of matter in motion. Textbooks of advanced mechanics lend plausibility to that identification of the theory, but both history and elementary pedagogy suggest that statics might instead be considered a separate theory, the acquisition of which is prerequisite to that of dynamics, just as the acquisition of geometry is prerequisite to that of statics. If, however, mechanics were split in that way, then the force function would be theoretical only with respect to statics, from which it would enter dynamics with the aid of constraints. Newton's second law would be required only to permit the specification of mass, not of force.[15]

My point is not that this way of subdividing mechanics is right, Sneed's wrong. Rather I am suggesting that what is illuminating about

15. That a pan balance can be used to measure (inertial) mass can, of course, be justified only by resort to Newtonian theory. Presumably that is what Sneed has in mind when he argues (p. 117) that mass must be theoretical because Newtonian theory may be used to determine whether the design of a given balance is suitable for mass determination. That criterion (validation of a measuring instrument by a theory) is, I think, relevant to judgments of theoreticity, but it also illustrates the difficulties in making them categorical. Newtonian mechanics was, as a matter of course, used to check on the suitability of instruments for measuring time, and the ultimate result was the recognition of standards more precise than the diurnal rotation of the stars. I am not suggesting that Sneed's arguments for labeling time nontheoretical lack cogency. On the contrary, as already indicated, both they and their results fit my intuitions quite well. But I do think that the efforts to preserve a sharp distinction between theoretical and nontheoretical terms may by now be a dispensable aspect of a traditional mode of analysis.

My reservations about the full enforceability of Sneed's theoretical/nontheoretical distinction owe much to a conversation with my colleague C. G. Hempel. They were, however, initially stimulated by Stegmüller's repeated hints (pp. 60, 231–43) that the distinction would require the construction of a strict hierarchy of theories. Terms and functions established by theory at one level would then be nontheoretical at the next higher one. Again, I find the intuition illuminating, but I see neither much likelihood of making it precise nor much reason for trying to do so.

his argument may be independent of a choice between the two. My intuition of what it is to be theoretical would be satisfied by the suggestion that a function or concept is theoretical with respect to a given application if constraints are required to introduce it there. That a function like force may also seem theoretical relative to an entire theory would then be explained by its manner of entry into *most* of that theory's applications. A given function or concept might then be theoretical in some applications of a theory, nontheoretical in others, a result that does not seem to me troublesome. What it may seem to threaten was, in fact, surrendered long ago, with the abandonment of hope for a neutral observation language.

To this point I have been suggesting that much of what is most valuable in Sneed's approach can be preserved without solving a problem of demarcation raised by his present way of introducing his formalism. But other significant uses of the formalism presuppose distinctions of another sort, and the criteria relevant to them appear to require much additional specification. In discussing the development of a theory over time, both Sneed and Stegmüller make repeated reference to the difference between a theory-core and an expanded-theory-core. The first supplies the basic mathematical structure of the theory—Newton's second law in the case of classical particle mechanics—together with the constraints that govern all the theory's applications. An expanded core contains, in addition, some special laws required for special applications— for example, Hooke's law of elasticity—and it may also contain special constraints that apply only when those laws are invoked. Two men who subscribe to different cores ipso facto hold different theories. If, however, they share belief in a core and in certain of its exemplary applications, they are adherents of the same theory even though their beliefs about its permissible expansions differ widely. The same criteria for subscribing to one and the same theory apply to a single individual at different times.[16]

A core, in short, is a structure that cannot, unlike an expanded core, be abandoned without abandoning the corresponding theory. Since a theory's applications, excepting perhaps those which originated with it, depend on specially designed expansions, the failure of an empirical claim made for a theory can infirm only the expansion, not the core, and thus not the theory itself. The manner in which Sneed and Steg-

16. Sneed, *Logical Structure*, pp. 171–84, 266 f., 292 f.; Stegmüller, *Theorienstrukturen*, pp. 120–34, 189–95.

müller apply this insight to the explication of my views should be obvious. Also apparent, I take it, are their reasons for suggesting that at least some changes of core correspond to the episodes I have labeled scientific revolutions. As already indicated, I hope and am inclined to believe that claims of this sort can be made out, but in their present form they have an unfortunate air of circularity. To eliminate it, far more will need to be said about how to determine whether some particular element of structure, used when applying a theory, is to be attributed to that theory's core or to some of its expansions.

Though I have only intuitions to offer on this subject, its importance may justify my briefly exploring them, beginning with a pair that Stegmüller and Sneed clearly share. Suppose that gravitational attraction varied as the inverse cube of distance or that the force of elasticity were a quadratic function of displacement. In those cases, the world would be different, but Newtonian mechanics would still be both Newtonian and mechanics. Hooke's law of elasticity and Newton's law of gravity therefore belong within the expansions of classical particle mechanics, not within the core which determines that theory's identity. Newton's second law of motion, on the other hand, must be located in the theory's core, for it plays an essential role in giving content to the particular concepts of mass and force without which no particle mechanics would be Newtonian. Somehow the second law is constitutive of the entire mechanical tradition which descends from Newton's work.

What, however, is to be said of Newton's third law, the equality of action and reaction? Sneed, followed by Stegmüller, places it in an expanded core, apparently because, from the late nineteenth century, it was irreconcilable with electrodynamic theories of the interactions between charged particles and fields. That reason, however, if I have identified it correctly, only illustrates what I previously referred to as an "air of circularity." The necessity of abandoning the third law was one of a number of recognized conflicts between mechanics and electromagnetic theory in the late nineteenth century. To some physicists, at least, the third law as well as the second thus seemed constitutive of classical mechanics. We may not conclude that they were wrong simply because relativistic and quantum mechanics had not yet been invented to take classical mechanics' place. If we did proceed in that way, insisting that the core of classical mechanics must contain all and only those elements common to all theories called Newtonian mechanics during the entire period that theory endured, then the equation of change-of-core with change-of-theory would be literally circular. The analyst who felt, as some physicists have, that special relativity was the culmination of classical mechanics, not its overthrow, might prove his case by definition alone, supplying, that is, a core restricted to elements common to both theories.

I conclude, in short, that before the Sneed formalism can be used effectively to identify and analyze episodes in which theory change occurs by replacement, rather than simply by growth, some other techniques must be found to distinguish the elements in a core from those in its expansions. No problems of principle appear to block the way, for discussion of Sneed's formalism has already supplied important clues to their pursuit. What is needed, I take it, is an explicit and general articulation, within the formalism, of some widely shared intuitions, two of which were expressed above. Why is Newton's second law clearly constitutive of mechanics, his law of gravitation not? What underlies our conviction that relativistic mechanics differs conceptually from Newtonian in a way that, say, Lagrangian and Hamiltonian mechanics do not?[17]

In a letter responding to an earlier expression of these difficulties, Stegmüller has supplied some further clues. Perhaps, he suggests, a core must be rich enough to permit the evaluation of theoretical functions. Newton's second law is required for that purpose, he continues, but the third law and the law of gravity are not. That suggestion is precisely of the sort that is needed, for it begins to supply minimum conditions for the *adequacy* or *completeness* of a core. Even in so preliminary a form, furthermore, it is by no means trivial, for its systematic development may force the transfer of Newton's third law from the expansion of classical particle mechanics to its core. Though no expert in these matters, I see no way to distinguish inertial from gravitational mass (and thus mass from weight or force) without recourse to the third law. As to the distinction between classical and relativistic mechanics, remarks in Stegmüller's letter lead me to the following tentative formulation. Perhaps symbolically identical cores for the two theories could be

17. As the discussion to follow may indicate, the problem of distinguishing between a core and an extended core has a close counterpart in my own work: the problem of distinguishing between normal and revolutionary change. I have here and there used the term 'constitutive' in discussing that problem too, suggesting that what must be discarded during a revolutionary change is somehow a constitutive, rather than simply a contingent, part of the previous theory. The difficulty, then, is to find ways of unpacking the term 'constitutive'. My closest approach to a solution, still a mere *aperçu*, is the suggestion that constitutive elements are in some sense quasi-analytic, i.e., partially determined by the language in which nature is discussed rather than by nature *tout court* (Kuhn, *Structure*, pp. 183 f.; "Second Thoughts," p. 469n.).

found, but their identity would be only apparent. The two would, that is, make use of different theories of space-time for the specification of their nontheoretical functions. Obviously suggestions of this sort need work, but their ready accessibility is already reason to suspect that work will succeed.

Reduction and Revolutions

Suppose now that techniques adequate to distinguish a core from its expansions were developed. What might it then be possible to say about the relation between changes of core and the episodes I have labeled scientific revolutions? Answers to that question will ultimately depend on the application of Sneed's reduction relation to theory pairs in which one member at some time replaced the other as the accepted basis for research. No one to my knowledge has yet applied the new formalism to a pair of that sort,[18] but Sneed does tentatively suggest what such an application might try to show. Perhaps, he writes, the "new theory must be such that the old theory reduces to (a special case of) the new theory" (p. 305).

In his book, more clearly than in his contribution to this symposium, Stegmüller unequivocally endorses that relatively traditional suggestion, and he immediately employs it to eliminate what he calls the *Rationalitätslücken* in my viewpoint. For him as for many others these rationality gaps are found in my remarks on the incommensurability of pairs of theories separated by a revolution, in my consequent emphasis on the communication problems that confront adherents of the two, and in my insistence that those problems prevent any fully systematic, point-by-point comparison between them.[19] Turning to these issues, I concede

18. Sneed's examples are the reduction of rigid body mechanics by classical particle mechanics as well as the relations (more nearly equivalence than reduction) between the Newtonian, Lagrangian, and Hamiltonian formulations of particle mechanics. About them all, he has interesting things to say. But rigid body mechanics is, to a historical first approximation, younger than the theory by which it was reduced, and its conceptual structure is therefore straightforwardly related to that of the reducing theory. The relations between the three formulations of classical particle mechanics are more complex, but they coexisted without felt incompatibility. There is no apparent reason to suppose that the introduction of any but Newtonian mechanics constituted a revolution.

19. Stegmüller, *Theorienstrukturen*, pp. 14, 24, 165–69, 182 f., 247–52. See also W. Stegmüller, "Accidental ('Non-substantial') Theory Change and Theory Dislodgement," *Erkenntnis* 10 (1976): 147–78.

at once that, if a reduction relation could be used to show that a later theory resolved all problems solved by its predecessor and more besides, then nothing one might reasonably ask of a technique for comparing theories would be lacking. In fact, however, the Sneed formalism supplies no basis for Stegmüller's counterrevolutionary formulation. On the contrary, one of the formalism's main merits seems to me to be the specificity with which it can be made to localize the problem of incommensurability.

To show what is at issue, I begin by restating my position in a form somewhat more refined than the original. Most readers of my text have supposed that when I spoke of theories as incommensurable, I meant that they could not be compared. But 'incommensurability' is a term borrowed from mathematics, and it there has no such implication. The hypotenuse of an isosceles right triangle is incommensurable with its side, but the two can be compared to any required degree of precision. What is lacking is not comparability, but a unit of length in terms of which both can be measured directly and exactly. In applying the term 'incommensurability' to theories, I had intended only to insist that there was no common language within which both could be fully expressed and which could therefore be used in a point-by-point comparison between them.[20]

Seen in this way, the problem of comparing theories becomes in part a problem of translation, and my attitude toward it may be briefly indicated by reference to the related position developed by Quine in *Word and Object* and in subsequent publications. Unlike Quine, I do not believe that reference in natural or in scientific languages is ultimately inscrutable, only that it is very difficult to discover and that one may never be absolutely certain one has succeeded. But identifying reference in a foreign language is not equivalent to producing a systematic transla-

---

20. When I first made use of the term 'incommensurability', I conceived the hypothetical neutral language as one in which any theory at all might be described. Since then I have recognized that comparison requires only a language neutral with respect to the two theories at issue, but I doubt that anything of even that more limited neutrality can be designed. Conversation discloses that it is on this central point that Stegmüller and I most clearly disagree. Consider, for example, the comparison of classical and relativistic mechanics. He supposes that as one descends through the hierarchy from classical (relativistic) particle mechanics, to the more general mechanics that lack Newton's second law, to particle kinematics, *and so on*, one will at last reach a level at which the nontheoretical terms are neutral with respect to classical and relativistic theory. I doubt the availability of any such level, find his "and so on" unilluminating, and therefore suppose that systematic theory comparison requires determination of the referents of incommensurable terms.

tion manual for that language. Reference and translation are two problems, not one, and the two will not be resolved together. Translation always and necessarily involves imperfection and compromise; the best compromise for one purpose may not be the best for another; the able translator, moving through a single text, does not proceed fully systematically, but must repeatedly shift his choice of word and phrase, depending on which aspect of the original it seems most important to preserve. The translation of one theory into the language of another depends, I believe, upon compromises of the same sort, whence incommensurability. Comparing theories, however, demands only the identification of reference, a problem made more difficult, but not in principle impossible, by the intrinsic imperfections of translations.

Against this background, what I want first to suggest is that Stegmüller's use of the reduction relation is damagingly circular. Sneed's discussion of reduction depends upon an undiscussed early premise which I take to be equivalent to full translatability. A necessary condition for the reduction of a theory T by a theory T' is a similar reducibility relation between the corresponding cores, K and K'. It in turn requires a reducibility relation between the partial potential models characterizing these cores. One requires, that is, a relation ρ which uniquely associates each member of the set M'_{pp} with a single member of the generally smaller set M_{pp}. Both Sneed and Stegmüller emphasize that the members of the two sets may be very differently described and that they may thus exhibit very different structures.[21] Nevertheless, they take for granted the existence of a relation ρ sufficiently powerful to identify by its structure the member of M_{pp} which corresponds to a member of M'_{pp} with a different structure, described in different terms. That assumption is the one I take to be tantamount to unproblematic translation. Of course it eliminates the problems which, for me, cluster around incommensurability. But, in the present state of the literature, can the existence of any such relation simply be taken for granted?

In the case of qualitative theories, I think it clear that no relation of the sort ordinarily exists. Consider, for example, just one of the many counterexamples I have developed elsewhere.[22] The basic vocabulary of eighteenth-century chemistry was predominantly one of qualities, and the chemist's central problem was then to trace qualities through reactions. Bodies were identified as earthy, oily, metalline, and so forth.

---

21. Sneed, *Logical Structure*, p. 219 f.; Stegmüller, *Theorienstrukturen*, p. 145.
22. Kuhn, *Structure*, 2d ed., p. 107.

Phlogiston was a substance which, added to a variety of strikingly different earths, endowed them all with the luster, ductility, and so on common to the known metals. In the nineteenth century chemists largely abandoned such secondary qualities in favor of characteristics like combining proportions and combining weights. Knowing these for a given element or compound provided no clues to the qualities which had in the preceding century made it a distinct chemical species. That the metals had common properties could no longer be explained at all.[23] A sample identified as copper in the eighteenth century was still copper in the nineteenth, but the structure by which it had been modeled in the set M_{pp} was different from that which represented it in the set M'_{pp}, and there was no route from the latter to the former.

Nothing nearly so unequivocal can be said about the relation between successive theories of mathematical physics, the case to which Sneed and Stegmüller restrict their attention. Given a relativistic kinematic description of a moving rod, one can always compute the length and position functions that would be attributed to that rod in Newtonian physics.[24] It is, however, a special virtue of the Sneed formalism that it highlights the essential difference between that computation from relativity theory and the direct computation within Newtonian theory. In the latter case, one starts with a Newtonian core and computes values directly, moving from application to application with the aid of specified constraints. In the former, one starts with a relativistic core and moves through differently specified applications with the aid of constraints (on the length and time functions) that may also be different from the

23. It would be wrong to dismiss this loss of explanatory power by suggesting that the success of the phlogiston theory was only an accident which reflected no characteristic of nature. The metals do have common characteristics, and these can now be explained in terms of the similar arrangements of their valence electrons. Their compounds have less in common because combination with other atoms leads to a great variety in the arrangements of the electrons loosely bound to the resulting molecules. If the phlogiston theory missed the structure of the modern explanation, it was primarily by supposing that a source of similarity was added to dissimilar ores to create metals rather than that sources of difference were subtracted from them.

24. In Sneed's reconstruction, the field of particle kinematics is a low-level theory which supplies the M_{pp}'s required to formalize all varieties of particle mechanics (the latter being determined by the various possible ways of adding force and mass functions to the M_{pp}'s). Classical particle mechanics emerges only with specialization to the subset M (of the M_p's) which satisfy Newton's second law. But that mode of division will not serve, I think, when Newtonian mechanics is to be compared with relativistic, for the two must be built up from different space-time systems and thus from different kinematics of differently structured M_{pp}'s. Lacking a developed formalism for special relativity, I shall therefore continue to treat a kinematics loosely, as part of the mechanics by which it is presupposed.

Newtonian. Only in a last step does one, by setting $(v/c)^2 \ll 1$, obtain numerical values that agree with the earlier computations.

Sneed underscores this difference in the penultimate paragraph of his book:

> [T]he functions in the new theory appear in a different mathematical structure—they stand in different mathematical relations to each other; they admit of different possibilities of determining their values—than the corresponding functions in the old theory. . . . Of course, it is an interesting fact that classical particle mechanics stands in a reduction relation to special relativity and that the mass functions in the theories correspond in this reduction relation. But this should not obscure the fact that these functions have different formal properties and, in this sense, they are associated with different concepts. (pp. 305 f.)

These remarks seem to me precisely right,[25] and they suggest the following questions. Does not the reduction relation ρ between partial potential models demand an ability to relate the concepts, or formal properties, or mathematical structures underlying the M'_{pp}'s and the M_{pp}'s prior to a computation of the concrete numerical values which those structures in part determine? Is it merely the fact that those computations can be done that has made the existence of the relation ρ between partial potential models seem so little problematic?

To this point I have dealt exclusively with the difficulties presented by the reducibility relation between cores. In the Sneed formalism, however, the specification of a theory demands specification not only of a core but also of a set of intended applications, I. The reduction of a theory T by a theory T' must therefore require some restrictions on the permissible relations between members of the sets I and I'. In particular, if T' is to solve all the problems resolved by T and more besides, then I' must contain I. In the general case of qualitative theories, it is doubtful that this containment relation can be satisfied. (Certainly, as the preceding remarks about chemistry indicate, T' does not always solve all the problems resolved by T.) But in the absence of even a crude formalization for such theories the issue is difficult to analyze, and I shall therefore restrict myself here to the intended applications of Newtonian and relativistic mechanics, a case in which intuitions, at least, are more highly developed. Its consideration will rapidly direct

25. Kuhn, *Structure*, 2d ed., pp. 100–102.

attention to what is, for me, the single most striking aspect of the Sneed formalism and also the one which most requires further development, not necessarily formal.

If relativistic mechanics is to reduce Newtonian mechanics, then the intended applications of the latter (i.e., the structures to which Newtonian theory is expected to apply) must be restricted to velocities small compared to the velocity of light. There is not, to my knowledge, a bit of evidence that any restriction of that sort entered the mind of a single physicist before the end of the nineteenth century. The velocities to be found in applications of Newtonian mechanics were restricted only de facto, by the nature of the phenomena that physicists actually studied. It follows that the historical class I, made up of intended, not simply of actual, applications, included situations in which velocity might be appreciable compared with that of light. To apply the reduction relations, those members of I must be barred, creating a new and smaller constructed set of intended applications which I shall label I_c.

For traditional formalisms this restriction on the intended applications of Newtonian mechanics is of no evident importance, and it has uniformly been disregarded. The reduced theory was constituted by the *equations* of Newtonian mechanics, and they remained the same, whether posited directly or derived from the relativistic equations in the limit. But in the Sneed formalism the reduced theory is the ordered pair $\langle K, I_c \rangle$, and it differs from the original $\langle K, I \rangle$ because I_c differs systematically from I. If the difference were only in the membership of the two sets, it might be unimportant, because the excluded applications would uniformly be false. A close look at the way in which membership in I_c and I is determined suggests, however, that something far more essential is involved.

Giving reasons for that evaluation requires a short digression concerning one last, especially striking parallel between my views and Sneed's. His book emphasizes that membership in the class of intended applications I cannot be given extensionally, by a list, because theoretical functions would then be eliminable, and theories could not grow by acquiring new applications. In addition, he expresses doubts that membership in I is governed by anything quite like a set of necessary and sufficient conditions. Asking how it is determined, he refers cryptically to the Wittgensteinian predicate 'is a game', and he suggests that basketball, baseball, poker, etc. "might be 'paradigm examples' of games" (pp. 266–88, esp. p. 269). Stegmüller's section, "*Was ist ein Paradigma?*" considerably extends these points and calls explicitly upon

similarity relations (*Ähnlichkeitsbeziehungen*) to explain how membership in *I* is determined. Many of you will know that learned similarity
relations, acquired in the course of professional training, have also figured large in my own recent research.[26] I shall now very briefly extend
and apply what I have previously said about them.

In my view, one of the things (perhaps sometimes the only thing)
that changes in every scientific revolution is some part of the network
of similarity relations that determines and simultaneously gives structure
to the class of intended applications. Again, the very clearest examples
invoke qualitative scientific theories. I have elsewhere, for example,
pointed out that, before Dalton, solutions, alloys, and the compound
atmosphere were usually taken to be *like,* say, metallic oxides or sul-
phates, and *unlike* such physical mixtures as sulphur and iron filings.[27]
After Dalton, the pattern of similarities switched, so that solutions,
alloys, and the atmosphere were transferred from the class of chemical to
the class of physical applications (from chemical compounds to physical
mixtures).

Lack of even a sketchy formalism for chemistry prevents my pursuing that example, but a change of much the same sort is visible in the
transition from Newtonian to relativistic mechanics. In the former, neither the velocity of a moving body nor the velocity of light played any
role in determining the likeness between a candidate for membership
in *I* and other previously accepted members of that set; in relativistic
mechanics, on the other hand, both of these velocities enter into the
similarity relation which determines membership in the different class
I′. It is from the latter set, however, that the members of the constructed
class I_c are selected, and it is that set, not the historical set *I*, that is
used to specify the theory which can be reduced by relativistic mechanics. The important difference between them is not, therefore, that *I*
includes members excluded from I_c, but that even the members common
to the two sets are determined by quite different techniques and thus
have different structures and correspond to different concepts. The

26. Kuhn, *Structure,* 2d ed., pp. 187–91, 200 f.; Kuhn, "Second Thoughts." Note that neither
Dr. Sneed nor Professor Stegmüller had read these passages when their very similar views were
developed.

27. Kuhn, *Structure,* 2d ed., pp. 130–35. Note that what I have here been calling a similarity
relation depends not only on likeness to other members of the same class but also on difference
from the members of other classes (compare Kuhn, "Second Thoughts"). Failure to notice that
the similarity relation appropriate to determination of membership in natural families must be
triadic rather than diadic has, I believe, created some unnecessary philosophical problems which
I hope to discuss at a later date.

structural or conceptual shift required to make the transition from Newtonian to relativistic mechanics is thus also required by the transition
from the historical (and, in any usual sense, irreducible) theory $\langle K, I \rangle$
to the theory $\langle K, I_c \rangle$, constructed to satisfy Dr. Sneed's reduction relation. If that result reintroduces a rationality gap, it may be our notion
of rationality that is at fault.

These closing remarks should supply a fuller sense of the depth of
my pleasure in Dr. Sneed's formalism and in Professor Stegmüller's use
of it. Even where we disagree, interaction results in significant clarification and extension of at least my own viewpoint. It is not, after all, a
large step from Sneed's talk of 'different mathematical structure' or of
'different concepts' to my talk of 'seeing things differently' or of the
gestalt switches that separate the two ways of seeing. Sneed's vocabulary
gives promise of a precision and articulation impossible with mine, and
I welcome the prospect it affords. But, with respect to the comparison
of incompatible theories, it is entirely a prospect of things to come.
Having insisted, in the first paragraph of this paper, that Sneed's new
formalism makes important new territory accessible to analytic philosophy of science, I hope, in this closing section, to have indicated the
part of that territory which most urgently requires exploration. Until
it occurs, the Sneed formalism will have contributed little to the understanding of scientific revolutions, something I fully expect it will be able
to do.

Chapter Eight

Metaphor in Science

"Metaphor in Science" was one of two commentaries on "Metaphor and Theory Change: What is 'Metaphor' a Metaphor For?" by Richard Boyd, presented at a conference entitled "Metaphor and Thought" at the University of Illinois at Urbana-Champaign in September 1977. (The other comment was by Zenon Pylyshyn.) The proceedings of the entire conference were published as Metaphor and Thought, *edited by Andrew Ortony (Cambridge: Cambridge University Press, 1979). Reprinted with the permission of Cambridge University Press.*

———◦———

IF I HAD BEEN PREPARING the main paper on the role of metaphor in science, my point of departure would have been precisely the works chosen by Boyd: Max Black's well-known paper on metaphor, together with recent essays by Kripke and Putnam on the causal theory of reference.[1] My reasons for those choices would, furthermore, have been very nearly the same as his, for we share numerous concerns and convictions. But, as I moved away from the starting point that body of literature provides, I would quite early have turned in a direction different from Boyd's, following a path that would have brought me quickly to a central metaphorlike process in science, one which he passes by. That path I shall have to sketch if sense is to be made of my reactions to Boyd's proposals, and my remarks will therefore take the form of an excessively condensed epitome of parts of a position of my own, comments on

---

1. M. Black, "Metaphor," in *Models and Metaphors* (Ithaca, NY: Cornell University Press, 1962); S. A. Kripke, "Naming and Necessity," in *The Semantics of Natural Language*, ed. D. Davidson and G. Harman (Dordrecht: D. Reidel, 1972); H. Putnam, "The Meaning of Meaning" and "Explanation and Reference," in *Mind, Language, and Reality* (Cambridge: Cambridge University Press, 1975).

Boyd's paper emerging along the way. That format seems all the more essential inasmuch as detailed analysis of individual points presented by Boyd is not likely to make sense to an audience largely ignorant of the causal theory of reference.

Boyd begins by accepting Black's "interaction" view of metaphor. However metaphor functions, it neither presupposes nor supplies a list of the respects in which the subjects juxtaposed by metaphor are similar. On the contrary, as both Black and Boyd suggest, it is sometimes (perhaps always) revealing to view metaphor as creating or calling forth the similarities upon which its function depends. With that position I very much agree and, lacking time, I shall supply no arguments for it. In addition, and presently more significant, I agree entirely with Boyd's assertion that the open-endedness or inexplicitness of metaphor has an important (and I think precise) parallel in the process by which scientific terms are introduced and thereafter deployed. However scientists apply terms like 'mass', 'electricity', 'heat', 'mixture', or 'compound' to nature, it is not ordinarily by acquiring a list of criteria necessary and sufficient to determine the referents of the corresponding terms.

With respect to reference, however, I would go one step further than Boyd. In his chapter, the claims for a parallel to metaphor are usually restricted to the theoretical terms of science. I suppose that they often hold equally for what used to be called observation terms, for example 'distance', 'time', 'sulphur', 'bird', or 'fish'. The fact that the last of these terms figures large in Boyd's examples suggests that he is unlikely to disagree. He knows as well as I that recent developments in philosophy of science have deprived the theoretical/observational distinction of anything resembling its traditional cash value. Perhaps it can be preserved as a distinction between antecedently available terms and new ones introduced at particular times in response to new scientific discoveries or inventions. But, if so, the parallel to metaphor will hold for both. Boyd makes less than he might of the ambiguity of the word 'introduced'. Something with the properties of metaphor is often called upon when a new term is *introduced into* the vocabulary of science. But it is also called upon when such terms—by now established in the common parlance of the profession—are *introduced to* a new scientific generation by a generation that has already learned their use. Just as reference must be established for each new element in the vocabulary of science, so accepted patterns of reference must be reestablished for each new cohort of recruits to the sciences. The techniques involved in both modes of introduction are much the same, and they therefore apply on both sides of the divide between what used to be called "observational" and "theoretical" terms.

To establish and explore the parallels between metaphor and reference fixing, Boyd resorts both to the Wittgensteinian notion of natural families or kinds and to the causal theory of reference. I would do the same, but in a significantly different way. It is at this point that our paths begin to diverge. To see how they do so, look first at the causal theory of reference itself. As Boyd notes, that theory originated and still functions best in application to proper names like 'Sir Walter Scott'. Traditional empiricism suggested that proper names refer by virtue of an associated definite description chosen to provide a sort of definition of the name: for example, "Scott is the author of *Waverley*." Difficulties immediately arose, because the choice of the defining description seemed arbitrary. Why should being the author of the novel *Waverley* be a criterion governing the applicability of the name 'Walter Scott' rather than a historical fact about the individual to whom the name, by whatever techniques, does refer? Why should having written *Waverley* be a necessary characteristic of Sir Walter Scott but having written *Ivanhoe* a contingent one? Attempts to remove these difficulties by using more elaborate definite descriptions, or by restricting the characteristics on which definite descriptions may call, have uniformly failed. The causal theory of reference cuts the Gordian knot by denying that proper names have definitions or are associated with definite descriptions at all.

Instead, a name like 'Walter Scott' is a tag or label. That it attaches to one individual rather than to another or to no one at all is a product of history. At some particular point in time a particular infant was baptized or dubbed with the name 'Walter Scott' which he bore thereafter through whatever events he happened to experience or bring about (for example, writing *Waverley*). To find the referent of a name like 'Sir Walter Scott' or 'Professor Max Black', we ask someone who knows the individual about whom we inquire to point him out to us. Or else we use some contingent fact about him, like his authorship of *Waverley* or of the paper on metaphor, to locate the career line of the individual who happened to write that work. If, for some reason, we doubt that we have correctly identified the person to whom the name applies, we simply trace his life history or lifeline backward in time to see whether it includes the appropriate act of baptism or dubbing.

Like Boyd, I take this analysis of reference to be a great advance, and I also share the intuition of its authors that a similar analysis should apply to the naming of natural kinds: Wittgenstein's games, birds (or sparrows), metals (or copper), heat, and electricity. There is something right about Putnam's claim that the referent of 'electric charge' is fixed by pointing to the needle of a galvanometer and saying that 'electric charge' is the name of the physical magnitude responsible for its deflection. But, despite the amount that Putnam and Kripke have written on the subject, it is by no means clear just what is right about their intuition. My pointing to an individual, Sir Walter Scott, can tell you how to use the corresponding name correctly. But pointing to a galvanometer needle while supplying the name of the cause of its deflection attaches the name only to the cause of that particular deflection (or perhaps to an unspecified subset of galvanometer deflections). It supplies no information at all about the many other sorts of events to which the name 'electric charge' also unambiguously refers. When one makes the transition from proper names to the names of natural kinds, one loses access to the career line or lifeline which, in the case of proper names, enables one to check the correctness of different applications of the same term. The individuals which constitute natural families do have lifelines, but the natural family itself does not.

It is in dealing with difficulties like this one that Boyd makes what I take to be an unfortunate move. To get around them he introduces the notion of "epistemic access," explicitly abandoning in the process all use of 'dubbing' or 'baptism' and implicitly, so far as I can see, giving up recourse to ostension as well. Using the concept of epistemic access, Boyd has a number of cogent things to say both about what justifies the use of a particular scientific language and about the relation of a later scientific language to the earlier one from which it has evolved. To some of his points in this area I shall be returning. But despite these virtues, something essential is lost, I think, in the transition from "dubbing" to "epistemic access." However imperfectly developed, "dubbing" was introduced in an attempt to understand how, in the absence of definitions, the referents of individual terms could be established at all. When dubbing is abandoned or shoved aside, the link it provided between language and the world disappears as well. If I understand Boyd's chapter correctly—something I do not take for granted—the problems to which it is directed change abruptly when the notion of epistemic access is introduced. Thereafter, Boyd seems simply to assume that the adherents of a given theory somehow or other know to what their terms refer. How they can do so ceases to concern him. Rather than extending the causal theory of reference, he seems to have given it up.

Let me therefore attempt a different approach. Though ostension is basic in establishing referents both for proper names and for natural-kind terms, the two differ not only in complexity but also in nature. In the case of proper names, a single act of ostension suffices to fix reference. Those of you who have seen Richard Boyd once will, if your memories are good, be able to recognize him for some years. But, if I were to exhibit to you the deflected needle of a galvanometer, telling you that the cause of the deflection was called 'electric charge', you would need more than good memory to apply the term correctly in a thunderstorm or to the cause of the heating of your electric blanket. Where natural-kind terms are at issue, a number of acts of ostension are required.

For terms like 'electric charge', the role of multiple ostensions is difficult to make out, for laws and theories also enter into the establishment of reference. But my point does emerge clearly in the case of terms that are ordinarily applied by direct inspection. Wittgenstein's example, games, will do as well as another. A person who has watched chess, bridge, darts, tennis, and football, and who has also been told that each of them is a game, will have no trouble in recognizing that both backgammon and soccer are games as well. To establish reference in more puzzling cases—prizefights or fencing matches, for example— exposure is required also to members of neighboring families. Wars and gang rumbles, for example, share prominent characteristics with many games (in particular, they have sides and, potentially, a winner), but the term 'game' does not apply to them. Elsewhere I have suggested that exposure to swans and geese plays an essential role in learning to recognize ducks.[2] Galvanometer needles may be deflected by gravity or a bar magnet as well as by electric charge. In all these areas, establishing the referent of a natural-kind term requires exposure not only to varied members of that kind but also to members of others—to individuals, that is, to which the term might otherwise have been mistakenly applied. Only through a multiplicity of such exposures can the student acquire what other authors in this book (for example, Cohen and Or-

2. T. S. Kuhn, "Second Thoughts on Paradigms," in *The Structure of Scientific Theories*, ed. F. Suppe (Urbana: University of Illinois Press, 1974), pp. 459–82; reprinted in *The Essential Tension: Selected Studies in Scientific Tradition and Change* (Chicago: University of Chicago Press, 1977), pp. 293–319.

tony)[3] refer to as the *feature space* and the knowledge of *salience* required to link language to the world.

If that much seems plausible (I cannot, in a presentation so brief, hope to make it more so), then the parallel to metaphor at which I have been aiming may be apparent as well. Exposed to tennis and football as paradigms for the term 'game', the language learner is invited to examine the two (and soon, others as well) in an effort to discover the characteristics with respect to which they are alike, the features that render them similar, and which are therefore relevant to the determination of reference. As in the case of Black's interactive metaphors, the juxtaposition of examples calls forth the similarities upon which the function of metaphor or the determination of reference depend. As with metaphor, also, the end product of the interaction between examples is nothing like a definition, a list of characteristics shared by games and only games, or of the features common to both men and wolves and to them alone. No lists of that sort exist (not all games have either sides or a winner), but no loss of functional precision results. Both natural-kind terms and metaphors do just what they should without satisfying the criteria that a traditional empiricist would have required to declare them meaningful.

My talk of natural-kind terms has not yet, of course, quite brought me to metaphor. Juxtaposing a tennis match with a chess game may be part of what is required to establish the referents of 'game', but the two are not, in any usual sense, metaphorically related. More to the point, until the referents of 'game' and of other terms which might be juxtaposed with it in metaphor have been established, metaphor itself cannot begin. The person who has not yet learned to apply the terms 'game' and 'war' correctly can only be misled by the metaphor "War is a game," or "Professional football is war." Nevertheless, I take metaphor to be essentially a higher-level version of the process by which ostension enters into the establishment of reference for natural-kind terms. The actual juxtaposition of a series of exemplary games highlights features which permit the term 'game' to be applied to nature. The metaphorical juxtaposition of the terms 'game' and 'war' highlights other features, ones whose salience had to be reached in order that actual games and wars could constitute separate natural families. If Boyd is right that

3. L. J. Cohen, "The Semantics of Metaphor," in *Metaphor and Thought*, ed. A. Ortony (Cambridge: Cambridge University Press, 1979), pp. 64–77; A. Ortony, "The Role of Similarity in Similes and Metaphors," in *Metaphor and Thought*, pp. 186–201.

nature has "joints" which natural-kind terms aim to locate, then metaphor reminds us that another language might have located different
joints, cut up the world in another way.

Those last two sentences raise problems about the very notion of
joints in nature, and I shall return to them briefly in my concluding
remarks about Boyd's view of theory change. But one last point needs
first to be made about metaphor in science. Because I take it to be
both less obvious and more fundamental than metaphor, I have so far
emphasized the metaphorlike process which plays an important role in
fixing the referents of scientific terms. But, as Boyd quite rightly insists,
genuine metaphors (or, more properly, analogies) are also fundamental
to science, providing on occasions "an irreplacable part of the linguistic
machinery of a scientific theory," playing a role that is *constitutive* of
the theories they express, rather than merely exegetical." Those words
are Boyd's, and the examples which accompany them are good ones.
I particularly admire his discussion of the role of the metaphors which
relate cognitive psychology to computer science, information theory,
and related disciplines. In this area, I can add nothing useful to what
he has said.

Before changing the subject, however, I would suggest that what
Boyd does say about these "constitutive" metaphors may well have a
bearing wider than he sees. He discusses not only "constitutive" but
also what he calls "exegetical or pedagogical" metaphors, for example
those which describe atoms as "miniature solar systems." These, he
suggests, are useful in teaching or explaining theories, but their use is
only heuristic, for they can be replaced by nonmetaphorical techniques.
"One can say," he points out, "*exactly* in what respects Bohr thought
atoms were like solar systems without employing any metaphorical devices, and this was true when Bohr's theory was proposed."

Once again, I agree with Boyd but would nevertheless draw attention
to the way in which metaphors like that relating atoms and solar systems
are replaced. Bohr and his contemporaries supplied a model in which
electrons and nucleus were represented by tiny bits of charged matter
interacting under the laws of mechanics and electromagnetic theory.
That model replaced the solar system metaphor but not, by doing so,
a metaphorlike process. Bohr's atom model was intended to be taken
only more or less literally; electrons and nuclei were not thought to be
exactly like small billiard or Ping-Pong balls; only some of the laws of
mechanics and electromagnetic theory were thought to apply to them;
finding out which ones did apply and where the similarities to billiard

balls lay was a central task in the development of the quantum theory. Furthermore, even when that process of exploring potential similarities had gone as far as it could (it has never been completed), the model remained essential to the theory. Without its aid, one cannot even today write down the Schrödinger equation for a complex atom or molecule, for it is to the model, not directly to nature, that the various terms in that equation refer. Though not prepared here and now to argue the point, I would hazard the guess that the same interactive, similarity-creating process which Black has isolated in the functioning of metaphor is vital also to the function of models in science. Models are not, however, merely pedagogic or heuristic. They have been too much neglected in recent philosophy of science.

I come now to the large part of Boyd's chapter that deals with theory choice, and I shall have to devote disproportionately little time to my discussion of it. That may, however, be less of a drawback than it seems, for attention to theory choice will add nothing to our central topic, metaphor. In any case, with respect to the problem of theory change, there is a great deal about which Boyd and I agree. And in the remaining area, where we clearly differ, I have great difficulty articulating just what we disagree about. Both of us are unregenerate realists. Our differences have to do with the commitments that adherence to a realist's position implies. But neither of us has yet developed an account of those commitments. Boyd's are embodied in metaphors which seem to me misleading. When it comes to replacing them, however, I simply waffle. Under these circumstances, I shall attempt only a rough sketch of the areas in which our views coincide and in which they appear to diverge. For the sake of brevity in that attempt, furthermore, I shall henceforth drop the distinction on which I have previously insisted between metaphor itself and metaphorlike processes. In these concluding remarks, 'metaphor' refers to all those processes in which the juxtaposition either of terms or of concrete examples calls forth a network of similarities which help to determine the way in which language attaches to the world.

Presupposing what has already been said, let me summarize those portions of my own position with which I believe Boyd largely agrees. Metaphor plays an essential role in establishing links between scientific language and the world. Those links are not, however, given once and for all. Theory change, in particular, is accompanied by a change in some of the relevant metaphors and in the corresponding parts of the network of similarities through which terms attach to nature. The earth was like Mars (and was thus a planet) after Copernicus, but the two were in different natural families before. Salt-in-water belonged to the family of chemical compounds before Dalton, to that of physical mixtures afterward. And so on. I believe, too, though Boyd may not, that changes like these in the similarity network sometimes occur also in response to new discoveries, without any change in what would ordinarily be referred to as a scientific theory. Finally, these alterations in the way scientific terms attach to nature are not—logical empiricism to the contrary—purely formal or purely linguistic. On the contrary, they come about in response to pressures generated by observation or experiment, and they result in more effective ways of dealing with some aspects of some natural phenomena. They are thus substantive or cognitive.

These aspects of Boyd's and my agreement should occasion no surprise. Another one may, though it ought not. Boyd repeatedly emphasizes that the causal theory of reference or the concept of epistemic access makes it possible to compare successive scientific theories with each other. The opposing view, that scientific theories are incomparable, has repeatedly been attributed to me, and Boyd himself may believe I hold it. But the book on which this interpretation is imposed includes many explicit examples of comparisons between successive theories. I have never doubted either that they were possible or that they were essential at times of theory choice. Instead, I have tried to make two rather different points. First, comparisons of successive theories with each other and with the world are never sufficient to dictate theory choice. During the period when actual choices are made, two people fully committed to the values and methods of science, and sharing also what both concede to be data, may nevertheless legitimately differ in their choice of theory. Second, successive theories are incommensurable (which is not the same as incomparable) in the sense that the referents of some of the terms which occur in both are a function of the theory within which those terms appear. There is no neutral language into which both of the theories as well as the relevant data may be translated for purposes of comparison.

With all of this I believe, perhaps mistakenly, that Boyd agrees. If so, then our agreement extends one step further still. Both of us see in the causal theory of reference a significant technique for tracing the continuities between successive theories and, simultaneously, for revealing the nature of the differences between them. Let me provide an excessively cryptic and simplistic example of what I, at least, have in mind.

The techniques of *dubbing* and of *tracing lifelines* permit astronomical individuals—say, the earth and moon, Mars and Venus—to be traced through episodes of theory change, in this case the one due to Copernicus. The lifelines of these four individuals were continuous during the passage from heliocentric to geocentric theory, but the four were differently distributed among natural families as a result of that change. The moon belonged to the family of planets before Copernicus, not afterward; the earth to the family of planets afterward, but not before. Eliminating the moon and adding the earth to the list of individuals that could be juxtaposed as paradigms for the term 'planet' changed the list of features salient to determining the referents of that term. Removing the moon to a contrasting family increased the effect. That sort of redistribution of individuals among natural families or kinds, with its consequent alteration of the features salient to reference, is, I now feel, a central (perhaps *the* central) feature of the episodes I have previously labeled scientific revolutions.

Finally, I shall turn very briefly to the area in which Boyd's metaphors suggest that our paths diverge. One of those metaphors, reiterated throughout his chapter, is that scientific terms "cut [or can cut] nature at its joints." That metaphor and Field's notion of quasi-reference figure large in Boyd's discussion of the development of scientific terminology over time. Older languages succeeded, he believes, in cutting the world at, or close to, some of its joints. But they also often committed what he calls "real errors in classification of natural phenomena," many of which have since been corrected by "more sophisticated accounts of those joints." The older language may, for example, "have classified together certain things which have no important similarity, or [may] have failed to classify together, things which are, in fact, *fundamentally* similar" (italics added). This way of talking is, however, only a rephrased version of the classical empiricists' position that successive scientific theories provide successively closer approximations to nature. Boyd's whole chapter presupposes that nature has one and only one set of joints to which the evolving terminology of science comes closer and closer with time. At least, I can see no other way to make sense of what he says in the absence of some theory-independent way of distinguishing *fundamental* or *important* similarities from those that are *superficial* or *unimportant*.[4]

4. In revising the manuscript to which this paragraph and those following are addressed, Boyd has pointed out that both natural kinds and nature's joints may be context- or discipline- or interest-relative. But, as note 2 to his paper will indicate, that concession does not presently

To describe the successive-approximation view of theory change as a presupposition does not, of course, make it wrong, but it does point to the need for arguments missing from Boyd's paper. One form such arguments might take is the empirical examination of a succession of scientific theories. No pair of theories will do, for the more recent could, by definition, be declared the better approximation. But, given a succession of three or more theories directed to more or less the same aspects of nature, it should be possible, if Boyd is right, to display some process of bracketing and zeroing in on nature's real joints. The arguments which would be required are both complex and subtle. I am content to leave open the question to which they are directed. But my strong impression is that they will not succeed. Conceived as a set of instruments for solving technical puzzles in selected areas, science clearly gains in precision and scope with the passage of time. As an instrument, science undoubtedly does progress. But Boyd's claims are not about the instrumental effectiveness of science but rather about its ontology, about what really exists in nature, about the world's real joints. And in this area I see no historical evidence for a process of zeroing in. As I have suggested elsewhere, the ontology of relativistic physics is, in significant respects, more like that of Aristotelian than that of Newtonian physics. That example must here stand for many.

Boyd's metaphor of nature's joints relates closely to another, the last I shall attempt to discuss. Again and again, he speaks of the process of theory change as one which involves "the accommodation of language to the world." As before, the thrust of his metaphor is ontological; the world to which Boyd refers is the one real world, still unknown but toward which science proceeds by successive approximation. Reasons for being uneasy with that point of view have already been described, but this way of expressing the viewpoint enables me to phrase my reservations in a different way. What is the world, I ask, if it does not include most of the sorts of things to which the *actual* language spoken at a given time refers? Was the earth really a planet in the world of pre-Copernican astronomers who spoke a language in which the features

---

bring our positions closer together. It may do so in the future, however, for the same footnote undermines the position it defends. Boyd concedes (mistakenly, I think) that a kind is "un-'objective'" to the extent that it is context- or discipline-dependent. But that construal of "objective" requires that context-independent bounds be specified for context dependence. If any two objects could, in principle, be rendered similar by choice of an appropriate context, then objectivity, in Boyd's sense, would not exist. The problem is the same as the one suggested by the sentence to which this footnote is attached.

salient to the referent of the term 'planet' excluded its attachment to the earth? Does it obviously make better sense to speak of accommodating language to the world than of accommodating the world to language? Or is the way of talking which creates that distinction itself illusory? Is what we refer to as 'the world' perhaps a product of a mutual accommodation between experience and language?

I shall close with a metaphor of my own. Boyd's world with its joints seems to me, like Kant's "things in themselves," in principle unknowable. The view toward which I grope would also be Kantian, but without "things in themselves" and with categories of the mind which could change with time as the accommodation of language and experience proceeded. A view of that sort need not, I think, make the world less real.

Chapter Nine

Rationality and Theory Choice

"Rationality and Theory Choice" was presented to the American Philosophical Association at a symposium on the philosophy of Carl G. Hempel in December 1983. The proceedings of the symposium were published in The Journal of Philosophy *80 (1983). Reprinted with the permission of* The Journal of Philosophy.

———◦———

THE REMARKS THAT FOLLOW are a much compressed status report on one product of my continuing interaction with C. G. Hempel. That interaction began twenty years ago with my arrival at his University and my middle years. If new masters can be acquired at that age, then Hempel became mine. From him I learned to recognize philosophical distinctions centrally relevant to my enterprise. In him I learned to recognize the stance of a man who intends philosophical distinctions to advance truth rather than to win debates. Participation in a symposium that honors him gives me much pleasure.

Among the topics that have prompted frequent and lively exchanges between us is the evaluation of and choice between scientific theories. More than other philosophers of his persuasion, Hempel has examined my views in this area with care and sympathy: he is not one of those who suppose that I proclaim the irrationality of theory choice. But he sees why others have supposed so. Both in writing and in conversation, he has underscored the lack of argument or of apparent concern with which I switch from descriptive to normative generalizations, and he has repeatedly wondered whether I quite see the difference between

The final revisions of this paper owe much to the critical intervention of Ned Block.

explaining behavior, on the one hand, and justifying it, on the other.[1] It is to our continuing discussion of these questions that I now return. Under what circumstances may one properly claim that certain criteria which scientists are *observed* to use when evaluating theories are, in fact, also rational bases for their judgments?

I begin with a suggestion that I originally developed when commenting on a paper of Hempel's at Chapel Hill in 1976. Both he and I premise that the evaluation of criteria for theory choice requires the prior specification of the goals to be achieved by that choice. Now suppose— a simplistic assumption which will later prove dispensable—that the scientist's aim in selecting theories is to maximize efficiency in what I have elsewhere called "puzzle solving." Theories are, on this view, to be evaluated in terms of such considerations as their effectiveness in matching predictions with the results of experiment and observation. Both the number of matches and the closeness of fit then count in favor of any theory under scrutiny.

Clearly, a scientist who subscribed to this goal would be behaving irrationally if he sincerely said, "Replacing traditional theory X with new theory Y reduces the accuracy of puzzle solutions but has no effect with respect to the other criteria by which I judge theories; nevertheless, I shall select theory Y, setting X aside." Given the goal and the evaluation, that choice is obviously self-defeating. Similar considerations apply to a choice of theory whose *sole* effect with respect to criterial measures is to reduce the number of puzzle solutions, to decrease their simplicity (thus making them harder to achieve), or to increase the number of distinct theories (and thus the complexity of apparatus) required to maintain the puzzle-solving capacities of a scientific field. Each of these choices would be in prima facie conflict with the professed goal of the scientist who made it. There is no clearer sign of irrationality. Arguments of the same sort can be developed for other standard desiderata invoked when evaluating theories. If science can justifiably be described as a puzzle-solving enterprise, such arguments suffice to prove the rationality of the observed norms.

Since our encounter at Chapel Hill, Hempel has on occasion suggested what I take to be a deeper version of the same point. In the penultimate paragraph of a paper published in 1981, he points out that some of the difficulties with my published accounts of theory choice

---

1. See, for example, his "Scientific Rationality: Analytic vs. Pragmatic Perspectives," in *Rationality Today*, ed. Theodore F. Geraets (Ottawa: University of Ottawa Press, 1979), pp. 46–58.

would be avoided if desiderata like accuracy and scope, invoked when evaluating theories, were viewed not as means to an independently specified end, like puzzle solving, but as themselves goals at which scientific inquiry aims.[2] More recently still, he has written:

> Science is widely conceived as seeking to formulate an increasingly comprehensive, systematically organized, world view that is explanatory and predictive. It seems to me that the desiderata [which determine the goodness of a theory] may best be viewed as attempts to articulate this conception somewhat more fully and explicitly. And if the goals of pure scientific research are indicated by the desiderata, then it is obviously rational, in choosing between two competing theories, to opt for the one which satisfies the desiderata better. . . . [These considerations] might be viewed as *justifying* in a near-trivial way the choosing of theories in accordance with whatever constraints are imposed by the desiderata.[3]

Because it loosens the commitment to any particular prespecified goal like puzzle solving, Hempel's formulation is an improvement on mine; our points are otherwise the same. But if I read him correctly, Hempel is less satisfied than I with this approach to the problem of the rationality of theory choice. He refers to it as "near-trivial" in the passage just quoted, apparently because it rests on something very like tautology, and he finds it correspondingly lacking in the philosophical bite one expects from a satisfactory justification of the norms for rational theory choice. In particular, he underscores two respects in which near-trivial justification seems to fail. "The problem of formulating norms for the critical appraisal of theories may," he points out, "be regarded as a modern outgrowth of the classical problem of induction," a problem that the near-trivial justification "does not address at all" (92). Elsewhere he emphasizes that, if norms are to be derived from a description of the essential aspects of science (my "puzzle-solving enterprise" or his "increasingly comprehensive, systematically organized, world view"), then the choice of the description that serves as premise for the near-trivial

2. "Turns in the Evolution of the Problem of Induction," *Synthese* 46 (1981): 389–404. This position is foreshadowed on p. 42 of the paper cited above, where Hempel notes the difficulties in deciding whether a particular desideratum, e.g., simplicity, should be viewed as a goal or as a means to its attainment.

3. "Valuation and Objectivity in Science," in *Physics, Philosophy and Psychoanalysis: Essays in Honor of Adolf Grünbaum*, ed. R. S. Cohen and L. Laudan (Boston: Reidel, 1983), pp. 73–100; quotation from pp. 91 f. Subsequent references to this paper will be by parenthetical page number in the text.

approach itself requires justification which neither of us appears to provide (86 f., 93). The activities observed by a science watcher can be described in countless different ways, each the source of different desiderata. What justifies the choice of one of these, the rejection of another?

These examples of the shortcomings of the near-trivial approach are well chosen, and I shall shortly return to them. At that point I shall sketch an argument suggesting that a particular sort of descriptive premise requires no further justification and that the near-trivial approach itself is therefore deeper and more fundamental than Hempel supposes. In doing so, however, I shall be venturing into what is for me new territory, and I want first to clarify the argument by indicating its relation to positions that, for another territory, I have developed in some detail before. If I am right, the descriptive premise of the near-trivial approach exhibits, within the language used to describe human actions, two closely related characteristics that I have previously insisted are essential features also of the language used to describe natural phenomena.[4] Before returning to the problem of rational justification, let me briefly describe the manifestations of those characteristics in the area where I have previously encountered them.

The first characteristic is one I have recently been calling "local holism." Many of the referring terms of at least scientific languages cannot be acquired or defined one at a time but must instead be learned in clusters. In the learning process, furthermore, an essential role is played by explicit or implicit generalizations about the members of the taxonomic categories into which those terms divide the world. The Newtonian terms 'force' and 'mass' provide the simplest sort of example. One cannot learn how to use either one without simultaneously learning how to use the other. Nor can this part of the language acquisition process go forward without resort to Newton's second law of motion. Only with its aid can one learn how to pick out Newtonian forces and masses, how to attach the corresponding terms to nature.

From this holistic acquisition procedure a second characteristic of

---

4. The most explicit and developed formulations are recent: T. S. Kuhn, "What Are Scientific Revolutions?" Occasional Paper 18, Center for Cognitive Science (Cambridge, MA: Massachusetts Institute of Technology, 1981), reprinted in *The Probabilistic Revolution*, vol. 1, *Ideas in History*, ed. L. Krüger, L. J. Daston, and M. Heidelberger (Cambridge, MA: MIT Press, 1987), pp. 7–22; also reprinted in this volume as essay 1; "Commensurability, Comparability, Communicability," in *PSA 1982: Proceedings of the 1982 Biennial Meeting of the Philosophy of Science Association*, vol. 2, ed. P. D. Asquith and T. Nickles (East Lansing, MI: Philosophy of Science Association, 1983), pp. 669–88; reprinted in this volume as essay 2. For what I now take to be an implicit, but perhaps more sophisticated, version of the same themes, see my far older paper,

scientific languages follows. Once acquired, the member terms of an interrelated set can be used to formulate infinitely many new generalizations, all of them contingent. But some of the original generalizations or others compounded from them prove to be necessary. Look again at Newtonian force and mass. The force of gravity might have been inverse cube rather than inverse square; Hooke might have discovered that the restoring force of elasticity was proportional to the square of the displacement. These laws were fully contingent. But no imaginable experiment could change merely the form of Newton's second law. If the second law failed, replacing it with another would result also in a local alteration of the language in which Newton's laws had previously been stated. Conversely, the Newtonian terms 'force' and 'mass' can function successfully only in a world in which Newton's second law holds.

I have called the second law necessary, but the sense in which that is so needs further specification. In two respects, the law is not a tautology. First, neither 'force' nor 'mass' is independently available for use in a definition of the other. In any case, the second law, unlike a tautology, can be tested. One can, that is, measure Newtonian force and mass, insert the result in the second law, and discover that the law fails. Nevertheless, I take the second law to be necessary in the following language-relative sense: if the law fails, the Newtonian terms in its statement are shown not to refer. No substitute for the second law is compatible with Newtonian language. One can use the relevant parts of the language unproblematically only so long as one is committed to the law. For this situation, the term 'necessary' is perhaps inappropriate, but I have no better. 'Analytic' clearly will not do.

Return now to the near-trivial justification of the norms or desiderata for theory choice, and begin by asking about the people who embody those norms. What is it to be a scientist? What does the term 'scientist' mean? The word itself was coined around 1840 by William Whewell. What evoked it was the emergence, beginning at the end of the previous century, of the modern use of the term 'science' to label a still-forming set of disciplines that were to be set beside and contrasted with such other disciplinary clusters as those labeled 'fine arts', 'medicine', 'law', 'engineering', 'philosophy', and 'theology'.

Few or none of these disciplinary clusters can be characterized by a

"A Function for Thought Experiments," reprinted in *The Essential Tension: Selected Studies in Scientific Tradition and Change* (Chicago: University of Chicago Press, 1977), pp. 240–65.

set of necessary and sufficient conditions for membership. Instead, one recognizes a group's activity as scientific (or artistic, or medical) in part by its resemblance to other fields in the same cluster and in part by its difference from the activities belonging to other disciplinary clusters. To learn to use the term 'science', one must therefore learn also to use some other disciplinary terms like 'art', 'engineering', 'medicine', 'philosophy', and perhaps 'theology'. And what thereafter makes possible the identification of a given activity as science (or art or medicine, etc.) is its position within the acquired semantic field that also contains these other disciplines. To know that position among the disciplines is to know what the term 'science' means or, equivalently, what a science is.

The names of disciplines thus label taxonomic categories, several of which must, like the terms 'mass' and 'force', be learned together. That local linguistic holism was the first of the characteristics isolated above, and, again, a second characteristic goes with it. The terms that name the disciplines function effectively only in a world that possesses disciplines quite like our own. To say, for example, that in Hellenic antiquity science and philosophy were one is to say also, and paradoxically, that in Greece before the death of Aristotle there was no enterprise quite classifiable as philosophy or as science. The modern disciplines have, of course, evolved from ancient ones, but not one for one, not each from an ancient progenitor appropriately viewed as a (perhaps more primitive) form of the same thing. The actual progenitors require description in their own terms, not in ours, and that task calls for a vocabulary that divides up, categorizes, intellectual activities in a way different from our own. Finding and disseminating a vocabulary that permits description and understanding of older times or of other cultures is central to what historians and anthropologists do.[5] Anthropologists who refuse the challenge are called "ethnocentric"; historians who refuse it are called "Whig."

This thesis—the need for other languages to describe other times and cultures—again has a converse. While we speak our own language,

---

5. The force of this point depends critically on the claim, developed and defended in "Commensurability, Comparability, Communicability," that the language required to describe some aspects of the past (or another culture) is not translatable into the native language of the person who provides the description. I have provided an extended example of the difficulties created by forcing a modern disciplinary taxonomy on the past in my "Mathematical versus Experimental Traditions in the Development of Physical Science," reprinted in *The Essential Tension*, pp. 31–65.

any activity that we label 'science', or 'philosophy' or 'art', and so on, must necessarily display pretty much the same characteristics as the activities to which we customarily apply those terms. Just as access to Newton's second law is required in order to pick out Newtonian forces and masses, so picking out the referents of the modern vocabulary of disciplines requires access to a semantic field that clusters activities with respect to such dimensions as accuracy, beauty, predictive power, normativeness, generality, and so on. Though a given sample of activity can be referred to under many descriptions, only those cast in this vocabulary of disciplinary characteristics permit its identification as, say, science; for that vocabulary alone can locate the activity close to other scientific disciplines and at a distance from disciplines other than science. That position, in turn, is a necessary property of all referents of the modern term 'science'.

Of course a science need not possess all the characteristics (positive or negative) that prove useful in identifying disciplines as sciences: not all sciences are predictive; not all are experimental. Nor need it always be possible, using these characteristics, to decide whether a given activity is science or not: that question need not have an answer. But a speaker of the relevant disciplinary language may not, on pain of self-contradiction, utter statements like the following: "The science X is *less* accurate than the non-science $Y;$ otherwise the two occupy the same position with respect to all disciplinary characteristics." Statements of that sort place the person who makes them outside of his or her language community. Persistence in them results in communication breakdown and, if elaborated, often in charges of irrationality as well. One can no more decide for oneself what 'science' means than what science is.

Now, of course, I am back where I began. The person who named X a science, Y not, was doing the same thing as the person who, earlier in this paper, preferred X to Y when both were scientific theories. Both violated some of the semantic rules that enable language to describe the world. An interlocutor who supposed their usage normal would find them guilty of self-contradiction. An interlocutor who recognized their usage as aberrant would be hard put to imagine what they could be trying to say. It is not, however, merely language that these statements violate. The rules involved are not conventions, and the contradiction that results from their abrogation is not the negation of a tautology. Rather, what is being set aside is the empirically derived taxonomy of disciplines, one that is embodied in the vocabulary of disciplines and applied by virtue of the associated field of disciplinary characteristics.

That vocabulary can fail to describe, but not, I have argued, merely term by term. Instead, failure must be met by the simultaneous adjustment of large parts of the disciplinary vocabulary. And until the adjustment has occurred, the person who preferred X to Y is simply opting out of the scientific language game. That, I believe, is where the near-trivial approach to justifying norms for theory choice gets its bite.

That bite is, of course, limited. Hempel is right to point out that the near-trivial approach provides no solution to the problem of induction. But the two now do make contact. Like 'mass' and 'force', or 'science' and 'art', 'rationality' and 'justification' are interdefined terms. One requisite for either is conforming to the constraints of logic, and I have made use of it to show that the usual norms for theory choice are justified ("rationally justified" was redundant). Another requisite is conforming to the constraints of experience in the absence of good reasons to the contrary. Both display part of what it is to be rational. One does not know what a person who denies the rationality of learning from experience (or denies that conclusions based upon it are justified) is trying to say. But all that simply provides background for the problem of induction, which, viewed from the perspective developed here, acknowledges that we have no rational alternative to learning from experience, and asks why that should be the case. It asks, that is, not for a justification of learning from experience, but for an explanation of the viability of the whole language game that involves 'induction' and underpins the form of life we live.

To that question I attempt no answer, but I would like one. Together with most of you, I share Hume's itch. Preparing this paper has made me realize that the itch may be intrinsic to the game, but I am not ready for that conclusion.

Chapter Ten

The Natural and the Human Sciences

"The Natural and the Human Sciences" was a prepared contribution to a panel discussion at LaSalle University, sponsored by the Greater Philadelphia Philosophy Consortium, on February 11, 1989. (Charles Taylor was to have been on the panel as well, but had to withdraw at the last minute.) It was published in The Interpretive Turn: Philosophy, Science, Culture, *edited by David R. Hiley, James F. Bohman, and Richard Shusterman (Ithaca: Cornell University Press, 1991). Used by permission of Cornell University Press.*

LET ME BEGIN WITH a fragment of autobiography. Forty years ago, when I first began to develop heterodox ideas about the nature of natural science, especially physical science, I came upon a few pieces of the Continental literature on the methodology of social science. In particular, if memory serves, I read a couple of Max Weber's methodological essays, then recently translated by Talcott Parsons and Edward Shils, as well as some relevant chapters from Ernst Cassirer's *Essay on Man*. What I found in them thrilled and encouraged me. These eminent authors were describing the social sciences in ways that closely paralleled the sort of description I hoped to provide for the physical sciences. Perhaps I really was onto something worthwhile.

My euphoria was, however, regularly damped by the closing paragraphs of these discussions, which reminded readers that their analyses applied only to the *Geisteswissenschaften,* the social sciences. *"Die Naturwissenschaften,"* their authors loudly proclaimed, *"sind ganz anders"* ("The natural sciences are entirely different"). What then followed was a relatively standard, quasi-positivist, empiricist account of natural science, just the image that I hoped to set aside.

Under those circumstances, I promptly returned to my own knitting, the materials for which were the physical sciences in which I had taken my Ph.D. Then and now, my acquaintance with the social sciences was extremely limited. My present topic—the relation of the natural and human sciences—is not one I have thought a great deal about, nor do I have the background to do so. Nevertheless, though maintaining my distance from the social sciences, I've from time to time encountered other papers to which I reacted as I had to Weber's and Cassirer's. Brilliant, penetrating essays on the social or human sciences, they seemed to me, but papers that apparently needed to define their position by using as foil an image of the natural sciences to which I remain deeply opposed. One such essay supplies the reason for my presence here.

That paper is Charles Taylor's "Interpretation and the Sciences of Man."[1] For me it's a special favorite: I've read it often, learned a great deal from it, and used it regularly in my teaching. As a result, I took particular pleasure in the opportunity to participate with its author in an NEH Summer Institute on Interpretation held during the summer of 1988. The two of us had not had the opportunity to talk together before, but we quickly started a spirited dialogue, and we undertook to continue it before this panel. As I planned my introductory contribution, I was confident of a lively and fruitful exchange to follow. Professor Taylor's forced withdrawal has been correspondingly disappointing, but by the time it occurred, it was too late for a radical change of plans. Though I'm reluctant to talk about Professor Taylor behind his back, I've had no alternative but to play a role close to the one for which I was originally cast.

———◇———

To avoid confusion, I shall start by locating what Taylor and I, in our discussions at the 1988 institute, primarily differed about. It was not the question whether the human and natural sciences were of the same kind. He insisted they were not, and I, though a bit of an agnostic, was inclined to agree. But we did differ, often sharply, about how the line

1. C. Taylor, "Interpretation and the Sciences of Man," in *Philosophy and the Human Sciences* (Cambridge: Cambridge University Press, 1985).

between the two enterprises might be drawn. I did not think his way would do at all. But my notions of how to replace it—about which I shall later have just a bit to say—remained extremely vague and uncertain.

To make our difference more concrete, let me start from a too simple version of what most of you know. For Taylor, human actions constitute a text written in behavioral characters. To understand the actions, recover the meaning of the behavior, requires hermeneutic interpretation, and the interpretation appropriate to a particular piece of behavior will, Taylor emphasizes, differ systematically from culture to culture, sometimes even from individual to individual. It is this characteristic— the intentionality of behavior—that, in Taylor's view, distinguishes the study of human actions from that of natural phenomena. Early in the classic paper to which I previously referred, he says, for example, that even objects like rock patterns and snow crystals, though they have a coherent pattern, have no meaning, nothing that they express. And later in the same essay he insists that the heavens are the same for all cultures, say, for the Japanese and for us. Nothing like hermeneutic interpretation, Taylor insists, is required to study objects like these. If they can properly be said to have meaning, those meanings are the same for all. They are, as he has more recently put it, absolute, independent of interpretation by human subjects.

That viewpoint seems to me mistaken. To suggest why, I shall also use the example of the heavens, which, as it happens, I had used also in the set of manuscript lectures that provided my primary text at the 1988 institute. It is not, perhaps, the most conclusive example, but it is surely the least complex and thus the most suitable for brief presentation. I did not and cannot compare our heavens with those of the Japanese, but I did and will here insist that ours are different from the ancient Greeks'. More particularly, I want to emphasize that we and the Greeks divided the population of the heavens into different kinds, different categories of things. Our celestial taxonomies are systematically distinct. For the Greeks, heavenly objects divided into three categories: stars, planets, and meteors. We have categories with those names, but what the Greeks put into theirs was very different from what we put into ours. The sun and moon went into the same category as Jupiter, Mars, Mercury, Saturn, and Venus. For them these bodies were like each other, and unlike members of the categories 'star' and 'meteor'. On the other hand, they placed the Milky Way, which for us is populated by

stars, in the same category as the rainbow, rings round the moon, shoot-
ing stars and other meteors. There are other similar classificatory differences. Things like each other in one system were unlike in the other.
Since Greek antiquity, the taxonomy of the heavens, the patterns of
celestial similarity and difference, have systematically changed.

Many of you will, I know, wish to join Charles Taylor in telling me
that these are merely differences in beliefs about objects that themselves
remained the same for the Greeks as for us—something that could be
shown, for example, by getting observers to point at them or to describe
their relative positions. This is not the place for me to try very seriously
to talk you out of that plausible position. But given more time, I would
certainly make the attempt, and I want here to indicate what the structure of my argument would be.

It would begin with some points about which Charles Taylor and
I agree. Concepts—whether of the natural or social world—are the
possession of communities (cultures or subcultures). At any given time
they are largely shared by members of the community, and their transmission from generation to generation (sometimes with changes) plays
a key role in the process by which the community accredits new members. What I take "sharing a concept" to be must here remain mysteri-
ous, but I am at one with Taylor in vehemently rejecting a long-stan-
dard view. To have grasped a concept—of planets or stars, on the one
hand, of equity or negotiation, on the other—is not to have internalized
a set of features that provide necessary and sufficient conditions for the
concept's application. Though anyone who understands a concept must
know *some* salient features of the objects or situations that fall under
it, those features may vary from individual to individual, and no one
of them need be shared to permit the concept's proper application. Two
people could, that is, share a concept without sharing a single belief
about the feature or features of the objects or situations to which it
applied. I don't suppose that often occurs, but in principle it could.

This much is largely common ground for Taylor and me. We part
company, however, when he insists that, though social concepts shape
the world to which they are applied, concepts of the natural world do
not. For him but not for me, the heavens are culture-independent. To
make that point, he would, I believe, emphasize that an American or
European can, for example, point out planets or stars to a Japanese but
cannot do the same for equity or negotiation. I would counter that one
can point only to individual exemplifications of a concept—to this star

or that planet, this episode of negotiation or that of equity—and that the difficulties involved in doing so are of the same nature in the natural and social worlds.

For the social world Taylor has himself supplied the arguments. For the natural world the basic arguments are supplied by David Wiggins in, among other places, *Sameness and Substance*.[2] To point usefully, informatively, to a particular planet or star, one must be able to point to it more than once, to pick out the same individual object again. And this one cannot do unless one has already grasped the sortal concept under which the individual falls. Hesperus and Phosphorus are the same *planet*, but it is only under that description, only as planets, that they can be recognized as one and the same. Until identity can be made out, there is nothing to be learned (or taught) by pointing. As in the case of equity or negotiation, neither the presentation nor the study of examples can begin until the concept of the object to be exemplified or studied is available. And what makes it available, whether in the natural or the social sciences, is a culture, within which it is transmitted by exemplification, sometimes in altered form, from one generation to the next.

I do, in short, really believe some—though by no means all—of the nonsense attributed to me. The heavens of the Greeks were irreducibly different from ours. The nature of the difference is the same as that Taylor so brilliantly describes between the social practices of different cultures. In both cases the difference is rooted in conceptual vocabulary. In neither can it be bridged by description in a brute data, behavioral vocabulary. And in the absence of a brute data vocabulary, any attempt to describe one set of practices in the conceptual vocabulary, the meaning system, used to express the other, can only do violence. That does not mean that one cannot, with sufficient patience and effort, discover the categories of another culture or of an earlier stage of one's own. But it does indicate that discovery is required and that hermeneutic interpretation—whether by the anthropologist or the historian—is how such discovering is done. No more in the natural than in the human sciences is there some neutral, culture-independent, set of categories within which the population—whether of objects or of actions—can be described.

Most of you will long since have recognized these theses as redevelopments of themes to be found in my *Structure of Scientific Revolutions* and related writings. Letting a single example serve for all, the gap that

2. D. Wiggins, *Sameness and Substance* (Cambridge, MA: Harvard University Press, 1980).

I have here described as separating the Greek heavens from our own is the sort that could only have resulted from what I earlier called a scientific revolution. The violence and misrepresentation consequent on describing their heavens in the conceptual vocabulary required to describe our own is an example of what I then called incommensurability. And the shock generated by substituting their conceptual spectacles for our own is the one I ascribed, however inadequately, to their living in a different world. Where the social world of another culture is at issue, we have learned, against our own deep-seated ethnocentric resistance, to take shock for granted. We can, and in my view must, learn to do the same for their natural worlds.

<hr />

What does all of this, supposing it cogent, have to tell us about the natural and human sciences? Does it indicate that they are alike except perhaps in their degree of maturity? Certainly it reopens that possibility, but it need not force that conclusion. My disagreement with Taylor was not, I remind you, about the existence of a line between natural and human sciences, but rather about the way in which that line may be drawn. Though the classic way to draw it is unavailable to those who take the viewpoint developed here, another way to draw the line emerges clearly. What I'm uncertain about is not whether differences exist, but whether they are principled or merely a consequence of the relative states of development of the two sets of fields.

Let me therefore conclude these reflections with a few tentative remarks about this alternate way of line-drawing. My argument has so far been that the natural sciences of any period are grounded in a set of concepts that the current generation of practitioners inherit from their immediate predecessors. That set of concepts is a historical product, embedded in the culture to which current practitioners are initiated by training, and it is accessible to nonmembers only through the hermeneutic techniques by which historians and anthropologists come to understand other modes of thought. Sometimes I have spoken of it as the hermeneutic basis for the science of a particular period, and you may note that it bears a considerable resemblance to one of the senses of what I once called a paradigm. Though I seldom use that term these days, having totally lost control of it, I shall for brevity sometimes use it here.

If one adopts the viewpoint I've been describing toward the natural sciences, it is striking that what their practitioners mostly do, given a paradigm or hermeneutic basis, is not ordinarily hermeneutic. Rather, they put to use the paradigm received from their teachers in an endeavor I've spoken of as normal science, an enterprise that attempts to solve puzzles like those of improving and extending the match between theory and experiment at the advancing forefront of the field. The social sciences, on the other hand—at least for scholars like Taylor, for whose view I have the deepest respect—appear to be hermeneutic, interpretive, through and through. Very little of what goes on in them at all resembles the normal puzzle-solving research of the natural sciences. Their aim is, or should be in Taylor's view, to understand behavior, not to discover the laws, if any, that govern it. That difference has a converse that seems to me equally striking. In the natural sciences the practice of research does occasionally produce new paradigms, new ways of understanding nature, of reading its texts. But the people responsible for those changes were not looking for them. The reinterpretation that resulted from their work was involuntary, often the work of the next generation. The people responsible typically failed to recognize the nature of what they had done. Contrast that pattern with the one normal to Taylor's social sciences. In the latter, new and deeper interpretations are the recognized object of the game.

The natural sciences, therefore, though they may require what I have called a hermeneutic base, are not themselves hermeneutic enterprises. The human sciences, on the other hand, often are, and they may have no alternative. Even if that's right, however, one may still reasonably ask whether they are restricted to the hermeneutic, to interpretation. Isn't it possible that here and there, over time, an increasing number of specialties will find paradigms that can support normal, puzzle-solving research?

About the answer to that question, I am totally uncertain. But I shall venture two remarks, pointing in opposite directions. First, I'm aware of no principle that bars the possibility that one or another part of some human science might find a paradigm capable of supporting normal, puzzle-solving research. And the likelihood of that transition's occurring is for me increased by a strong sense of déjà vu. Much of what is ordinarily said to argue the impossibility of puzzle-solving research in the human sciences was said two centuries ago to bar the possibility of a science of chemistry and was repeated a century later to show the impossibility of a science of living things. Very probably the transition I'm suggesting is already under way in some current specialties within the human sciences. My impression is that in parts of economics and psychology, the case might already be made.

On the other hand, in some major parts of the human sciences there is a strong and well-known argument against the possibility of anything quite like normal, puzzle-solving research. I earlier insisted that the Greek heavens were different from ours. I should now also insist that the transition between them was relatively sudden, that it resulted from research done on the prior version of the heavens, and that the heavens remained the same while that research was under way. Without that stability, the research responsible for the change could not have occurred. But stability of that sort cannot be expected when the unit under study is a social or political system. No lasting base for normal, puzzle-solving science need be available to those who investigate them; hermeneutic reinterpretation may constantly be required. Where that is the case, the line that Charles Taylor seeks between the human and the natural sciences may be firmly in place. I expect that in some areas it may forever remain there.

Chapter Eleven

Afterwords

"Afterwords" is Kuhn's reply to nine papers—all inspired by or about his work—by John Earman, Michael Friedman, Ernan McMullin, J. L. Heilbron, N. M. Swerdlow, Jed Z. Buchwald, M. Norton Wise, Nancy Cartwright, and Ian Hacking. The initial versions of these papers and the reply were all presented at a two-day conference held in honor of Kuhn at MIT in May 1990. The revised proceedings of that conference were published as World Changes: Thomas Kuhn and the Nature of Science, *edited by Paul Horwich (Cambridge, MA: Bradford/MIT Press, 1993). When Kuhn discusses the views of the authors listed above, unless otherwise specified, he is referring to their essays in that volume.*

—◦—

REREADING THE PAPERS that make up this volume has recalled the feelings with which, almost two years ago, I rose to present my original response to them. C. G. Hempel, who for more than two decades has been a beloved mentor, had just delivered the remarks with which this volume now opens. They were the penultimate event of an intense day-and-a-half conference characterized by splendid papers and warm constructive discussion. Only a few personal occasions—deaths, births, and other salient comings together or apart—had moved me as deeply. When I reached the podium I was not sure I would be able to speak, and it took a few moments to find out. After the conference my wife said to me that I'd never be the same again, and time is proving her right. On this occasion, as on that, I begin with heartfelt thanks to those who made the occasion possible: its inventors, organizers, contributors, and participants.[1] They have presented me with a gift I did not know existed.

In accepting that gift, I begin by returning to the remarks made by Professor Hempel. I too remember our first meeting: I was at Berkeley but considering an attractive invitation from Princeton; he was in residence across the Bay at the Center for Advanced Study in the Behavioral Sciences. I called on him there to ask for clues to what life and work at Princeton might be like. If that visit had gone badly, I might not have accepted Princeton's offer. But it did not, and I did. Our meeting in Palo Alto was only the first of a still continuing series of warm and fruitful interactions. As Professor Hempel (for me he long ago became Peter) has said, our views at the start were very different, far more so than they have become through our interactions. But they were perhaps not quite so different as we both then thought, for I had begun to learn from him almost fifteen years before.

By the late 1940s I was deeply convinced that the received view of meaning, including its various positivist formulations, would not do: scientists did not, it seemed to me, understand the terms they used in the way described by the various versions of the tradition, and there was no evidence that they needed to do so. That was my state of mind when I first encountered Peter's old monograph on concept formation. Though it was many years before I saw its full relevance to my emerging position, it fascinated me from the start, and its role in my intellectual development must have been considerable. Four essential elements of my developed position are, in any case, to be found there: scientific terms are regularly learned in use; that use involves the description of one or another paradigmatic example of nature's behavior; a number of such examples are required for the process to work; and, finally, when the process is complete, the language or concept learner has acquired not only meanings but also, inseparably, generalizations about nature.[2]

1. My special thanks go to Judy Thomson, who conceived the voyage; to Paul Horwich, who captained the ship; and to my secretary, Carolyn Farrow, who has been his able first mate.
2. C. G. Hempel, *Fundamentals of Concept Formation in Empirical Science*, International Encyclopedia of Unified Science, vol. 2, no. 7 (Chicago: University of Chicago Press, 1952). I could have found similar elements in the discussion of Ramsey sentences in R. B. Braithwaite's *Scientific Explanation* (Cambridge: Cambridge University Press, 1953), but my encounter with it came later.

A more general, broader, and deeper version of these views appeared a few years later in the classic paper that Peter significantly entitled "The Theoretician's Dilemma."[3] The dilemma was how to preserve a principled distinction between what he then still called "observational" and "theoretical terms." When, a few years later still, he began instead to describe the distinction as one between "antecedently available terms" and those learned together with a new theory, I could see him as having implicitly adopted a developmental or historical stance. I am uncertain whether that change of vocabulary occurred before or after we first met, but the basis for our convergence was by then clearly in place and our rapprochement perhaps already under way.

After I got to Princeton, Peter and I talked regularly, and occasionally we also taught together. When I later briefly took over the course in which I had assisted him, I opened by telling the class that my object was to show them the extra benefits of putting to work, within the historical or developmental approach to philosophy of science, some of the splendid analytic tools developed within the more static logical-empiricist tradition. I continue to think of my philosophical work as pursuing that goal. There are still other products of my interactions with Peter, and I shall turn to a significant one below. But what I primarily owe him is not from the realm of ideas. Rather it is the experience of working with a philosopher who cares more about arriving at truth than about winning arguments. I love him most, that is, for the noble uses to which he puts a distinguished mind. How could I not have been deeply moved when I followed him once more to the podium?

These remarks should suggest that I have known from the start of my meddling with philosophy that the historical approach I joined in developing owed as much to difficulties encountered by the logical-empiricist tradition as it did to history of science. Quine's "Two Dog-

3. Originally published in 1958, the paper is most conveniently available as chapter 8 in C. G. Hempel, *Aspects of Scientific Explanation and Other Essays in the Philosophy of Science* (New York: Free Press, 1965). That formulation I still regularly use in my teaching. A more fully articulated version of a similar position is explicitly applied to my views in Wolfgang Stegmüller's *The Structure and Dynamics of Theories*, trans. W. Wohlhueter (New York: Springer-Verlag, 1976), a book whose influence is also reflected in some of my more recent work; in particular, see "Possible Worlds in History of Science," in *Possible Worlds in Humanities, Arts and Sciences: Proceedings of Nobel Symposium 65*, ed. Sture Allén, Research in Text Theory, vol. 14 (Berlin: Walter de Gruyter, 1989), pp. 9–32; reprinted in this volume as essay 6. A slightly abridged version of this paper is available in *Scientific Theories*, ed. C. W. Savage, Minnesota Studies in the Philosophy of Science, vol. 14 (Minneapolis: University of Minnesota Press, 1990), pp. 298–318.

mas" provides a second, for me formative, example of what I took those difficulties to be.[4] About all of this Michael Friedman's elegant sketch is quite right, and I look forward to the fuller version he promises. In his original conference paper he added another telling remark, one that is here elaborated by John Earman in appropriate, but for me excruciating, detail. Whatever role the problems encountered by positivism may have played in the background for *The Structure of Scientific Revolutions*, my knowledge of the literature that attempted to deal with those problems was decidedly sketchy when the book was written. In particular, I was almost totally innocent of the post-*Aufbau* Carnap, and discovering him has distressed me acutely. Part of my embarrassment results from my sense that responsibility required that I know my target better, but there is more. When I received the kind letter in which Carnap told me of his pleasure in my manuscript, I interpreted it as mere politeness, not as an indication that he and I might usefully talk. That reaction I repeated to my loss on a later occasion.

Nevertheless, the passages which John quotes to show the deep parallels between Carnap's position and mine also show, when read in the context of his paper, a correspondingly deep difference. Carnap emphasized untranslatability as I do. But, if I understand Carnap's position correctly, the cognitive importance of language change was for him merely pragmatic. One language might permit statements that could not be translated into another, but anything properly classified as scientific knowledge could be both stated and scrutinized in either language, using the same method and gaining the same result. The factors responsible for the use of one language rather than another were irrelevant both to the results achieved and, more especially, to their cognitive status.

This aspect of Carnap's position has never been available to me. Concerned from the start with the *development* of knowledge, I have seen each stage in the evolution of a given field as built—not quite squarely—upon its predecessor, the earlier stage providing the problems, the data, and most of the concepts prerequisite to the emergence of the stage that followed. In addition, I have insisted that some changes in conceptual vocabulary are required for the assimilation and development of the observations, laws, and theories deployed in the later stage (whence the phrase "not quite squarely" above). Given those beliefs, the process of transition from old state to new becomes an integral part

4. W. V. O. Quine, "Two Dogmas of Empiricism," in *From a Logical Point of View*, 2d ed. (Cambridge, MA: Harvard University Press, 1961).

of science, a process that must be understood by the methodologist concerned to analyze the cognitive basis for scientific beliefs. Language change is *cognitively* significant for me as it was not for Carnap.

To my dismay, what John not unfairly labels my "purple passages" led many readers of *Structure* to suppose that I was attempting to undermine the cognitive authority of science rather than to suggest a different view of its nature. And even for those who understood my intent, the book had little constructive to say about how the transition between stages comes about or what its cognitive significance can be. I can do better on these and related subjects now, and the book on which I am presently at work will have much to say about them. Obviously, I cannot even sketch the book's content here, but I shall use my license as commentator to suggest as best I can what my position has become in the years since *Structure*. I will, that is, use the papers in this volume as grist for my current mill. To my great pleasure, all of them contribute to my purpose, though the treatment that results is inevitably incompletely balanced.

I begin with some anticipatory remarks on the topic that dominates my project: incommensurability and the nature of the conceptual divide between the developmental stages separated by what I once called "scientific revolutions." My own encounter with incommensurability was the first step on the road to *Structure*, and the notion still seems to me the central innovation introduced by the book. Even before *Structure* appeared, however, I knew that my attempts to describe its central conception were extremely crude. Efforts to understand and refine it have been my primary and increasingly obsessive concern for thirty years, during the last five of which I've made what I take to be a rapid series of significant breakthroughs.[5] The earliest of these first surfaced in a series of three unpublished Shearman lectures delivered in 1987 at University College, London. A manuscript of those lectures is, as Ian Hacking says, the primary source for the taxonomic solution to what he calls the new-world problem. Though the solution he describes was never quite my own and though my own has developed substantially since the manuscript he cites was written, I take immense pleasure in his paper. I will presuppose acquaintance with it in this attempt to suggest what my position has become.

---

5. An excellent account of the earlier stages of those attempts is included in P. Hoyningen-Huene, *Reconstructing Scientific Revolutions: Thomas S. Kuhn's Philosophy of Science*, trans. A. T. Levine (Chicago: University of Chicago Press, 1993).

First, though natural kinds provided me with a point of entry, they will not—for reasons Ian cites—resolve the full range of problems that incommensurability poses. The kind concepts I require range far beyond anything to which the phrase 'natural kinds' has ordinarily referred. But, for the same reason, Ian's "scientific kinds" will not do either: what is required is a characteristic of kinds and kind terms in general. In the book I will suggest that this characteristic can be traced to, and on from, the evolution of neural mechanisms for reidentifying what Aristotle called "substances": things that, between their origin and demise, trace a lifeline through space over time.[6] What emerges is a mental module that permits us to learn to recognize not only kinds of physical object (e.g., elements, fields, and forces), but also kinds of furniture, of government, of personality, and so on. In what follows I shall refer to it frequently as the lexicon, the module in which members of a speech community store the community's kind terms.

That required generality reinforces, though it does not cause, a second difference between my position and the one Ian presents. His nominalist version of my position—there are real individuals out there, and we divide them into kinds at will—does not quite face my problems. The reasons are numerous, and I mention only one here: how can the referents of terms like 'force' and 'wave front' (much less 'personality') be construed as individuals? I need a notion of 'kinds', including social kinds, that will populate the world as well as divide up a preexisting population. That need in turn introduces a last significant difference between me and Ian. He hopes to eliminate all residues of a theory of meaning from my position; I do not believe that that can be done. Though I no longer speak of anything so vague and general as "language change," I do talk of change in concepts and their names, in conceptual vocabulary, and in the structured conceptual lexicon that contains both kind concepts and their names. A schematic theory intended to provide a basis for talk of this kind is central to my projected book. With respect to kind terms, aspects of a theory of meaning remain at the heart of my position.

Here I can hope only to sketch what my position has become since the Shearman lectures, and the sketch must be both dogmatic and incomplete. Kind concepts need not have names, but in linguistically en-

---

6. As this sentence may suggest, a significant role in the recent development of my ideas has been played by David Wiggins's, *Sameness and Substance* (Cambridge, MA: Harvard University Press, 1980).

dowed populations they mostly do, and I will restrict my attention to them. Among English words, they can be identified by grammatical criteria: for example, most of them are nouns that take an indefinite article either by themselves or, in the case of mass nouns, when conjoined with a count noun, as in '*gold* ring'. Such terms share a number of important properties, the first set of which was enumerated in my earlier acknowledgement of my debt to Peter Hempel's work on concept formation. Kind terms are learned in use: someone already adept in their use provides the learner with examples of their proper application. Several such exposures are always required, and their outcome is the acquisition of more than one concept. By the time the learning process has been completed, the learner has acquired knowledge not only of the concepts but also of the properties of the world to which they apply.

Those characteristics introduce a second shared property of kind terms. They are projectible: to know any kind term at all is to know some generalizations satisfied by its referents and to be equipped to look for others. Some of these generalizations are normic, admit exceptions.[7] "Liquids expand when heated" is an example even though it sometimes fails, e.g., for water between o and 4 degrees centigrade. Other generalizations, though often only approximate, are nomic, exceptionless. In the sciences, where they mainly function, these generalizations are usually laws of nature: Boyle's law for gases or Kepler's laws for planetary motions are examples.

These differences in the nature of the generalizations acquired in learning kind terms correspond to a necessary difference in the way the terms are learned. Most kind terms must be learned as members of one or another contrast set. To learn the term 'liquid', for example, as it is used in contemporary nontechnical English, one must also master the terms 'solid' and 'gas'. The ability to pick out referents for any of these terms depends critically upon the characteristics that differentiate its referents from those of the other terms in the set, which is why the terms involved must be learned together and why they collectively constitute a contrast set. When terms are learned together in this way, each comes with attached normic generalizations about the properties likely to be shared by its referents. The other sort of kind term—'force', for exam-

---

7. On "normic generalizations," see Michael Scriven's too much neglected paper "Truisms as the Ground for Historical Explanations," in *Theories of History*, ed. Patrick Gardiner (New York: Free Press, 1959).

ple—stands alone. The terms with which it needs to be learned are closely related, but not by contrast. Like 'force' itself, they are not normally in any contrast set at all. Instead, 'force' must be learned with terms like 'mass' and 'weight'. And they are learned from situations in which they occur together, situations exemplifying laws of nature. I have elsewhere argued that one cannot learn 'force' (and thus acquire the corresponding concept) without recourse to Hooke's law and either Newton's three laws of motion or else his first and third laws together with the law of gravity.[8]

These two characteristics of kind terms necessitate a third, the one at which this exercise has been aimed. In a sense that I will not further explicate here, the expectations acquired in learning a kind term, though they may differ from individual to individual, supply the individuals who have acquired them with the meaning of the term.[9] Changes in expectations about a kind term's referents are therefore changes in its meaning, so that only a limited variety of expectations may be accommodated within a single speech community. So long as two community members have compatible expectations about the referents of a term they share, there will be no difficulty. One or both of them may know things about those referents that the other does not, but they will both pick out the same things, and they can learn more about those things from each other. But if the two have incompatible expectations, one will occasionally apply the term to a referent to which the other categorically denies that it applies. Communication is then jeopardized, and the jeopardy is especially severe because, like meaning differences in general, the difference between the two cannot be rationally adjudicated. One or both of the individuals involved may be failing to conform to standard social usage, but it is only with respect to social usage that either of them can be said to be right or wrong. What they differ about is, in that sense, convention rather than fact.

One way to describe this difficulty is as a case of polysemy: the two

8. On the problem of learning 'force', see my "Possible Worlds in History of Science." The paper also discusses, though in the context of concept development rather than of concept acquisition, the significance of the contrast set which contains 'liquid' to the determination of the referents of 'water'.

9. Explicating this sense of 'meaning' would require giving body to the claim that kind terms do not have meanings by themselves but only in their relations to other terms in an isolatable region of a structured lexicon. It is congruence of structure that makes meanings the same for those who have acquired different expectations from their learning experience.

individuals are applying the same name to different concepts. But that description, though correct as far as it goes, fails to catch the depth of the difficulty. Polysemy has a standard remedy, widely deployed in analytic philosophy: two names are introduced where there had been only one before. If the polysemous term is 'water', the difficulties are to be lifted by replacing it with a pair of terms, say 'water1' and 'water2,' one for each of the concepts that previously shared the name 'water'. Though the two new terms differ in meaning, most referents of 'water1' are referents of 'water2' and vice versa. But each term also refers to a few items to which the other does not, and it was about the applicability of 'water' in such cases that the two community members disagreed. Introducing two terms where there was one before appears to resolve the difficulty by enabling the disputants to see that their difference was simply semantic. They were disagreeing about words, not about things.

That way of resolving the disagreement is, however, linguistically unsupportable. Both 'water1' and 'water2' are kind terms: the expectations they embody are therefore projectible. Some of those expectations are different, however, which results in difficulties in the region where both apply. Calling an item in the overlap region 'water1' induces one set of expectations about it; calling the same item 'water2' induces another, partly incompatible, set. Both names cannot apply, and which to choose is no longer about linguistic conventions but rather about matters of evidence and fact. And if the matters of fact are taken seriously, then in the long run only one of the two terms can survive within any single language community. The difficulty is most obvious with terms like 'force' that bring with them nomic expectations. If a referent lay in the overlap region (say between Aristotelian and Newtonian usage), it would be subject to two incompatible natural laws. For normic expectations the prohibition must be slightly weakened: only terms which belong to the same contrast set are prohibited from overlapping in membership. 'Male' and 'horse' may overlap but not 'horse' and 'cow'.[10] Periods in which a speech community does deploy overlapping kind terms end in one of two ways: either one entirely displaces the other, or the community divides into two, a process not unlike speciation and

---

10. Reference to the contrast set containing 'male' and 'female' indicates both the difficulties and the importance of developing more refined versions of this "no-overlap" principle. I think no individual creature is both *a* male and *a* female, though it may exhibit both male and female characteristics. Perhaps good usage also permits describing an individual as both male and female, the terms being used adjectivally, but the locution seems to me strained.

one that I will later suggest is the reason for the ever-increasing specialization of the sciences.

What I have just been saying is, of course, my version of the solution
to what Ian has dubbed the new-world problem. Kind terms supply the
categories prerequisite to description of and generalization about the
world. If two communities differ in their conceptual vocabularies, their
members will describe the world differently and make different generalizations about it. Sometimes such differences can be resolved by importing the concepts of one into the conceptual vocabulary of the other.
But if the terms to be imported are kind terms that overlap kind terms
already in place, no importation is possible, at least no importation
which allows both terms to retain their meaning, their projectibility,
their status as kind terms. Some of the kinds that populate the worlds
of the two communities are then irreconcilably different, and the difference is no longer between descriptions but between the populations
described. Is it, in these circumstances, inappropriate to say that the
members of the two communities live in different worlds?

———◇———

I have so far been discussing what Ian calls scientific kinds, or at least
the kinds that nature exhibits to the members of a culture, and I will
be returning to them when discussing Jed Buchwald's paper in the next
section. But it will help to consider first an example of the significance
of the no-overlap principle for social kinds. John Heilbron's and Noel
Swerdlow's papers provide a central illustration.

John's "Mathematicians' Mutiny" is a splendid example of the historian's craft. It is also thoroughly relevant to the old paper of mine to
which he applies it. Though he fits to that paper an even narrower
straitjacket than the one I made myself, I've learned from and accept
in full the more complex and nuanced studies of the development and
interrelations of scientific fields that he here and elsewhere provides.
Taken together, his studies constitute a major and still advancing
achievement. But John's methodological remarks about the vocabulary
the historian requires in order to describe the phenomena he or she
studies seem to me mistaken in ways that have often damaged historical
understanding.

The fundamental product of historical research is narratives of development over time. Whatever its subject, the narrative must always begin

by setting the stage. If its subject is beliefs about nature, it must open with a description of what beliefs were accepted at the time and place from which the narrative begins. That description must include as well a specification of the vocabulary in which natural phenomena were described and beliefs about them stated. If, instead, the narrative deals with group activities or practices, it must open with a description of the various practices recognized at the time the narrative begins, and it must indicate what was expected of those practices both by practitioners and those around them. In addition, the stage setting must introduce names for those practices (preferably the names used by practitioners) and display contemporary expectations about them: how were they justified and how criticized in their own time?

To learn the nature and objects of these beliefs and expectations, the historian deploys the techniques that I once sketched under the rubric of translation but would now insist are directed to language learning, a distinction to which I will return below. To communicate the results to readers, the historian becomes a language teacher and shows readers how to use the terms, most or all of them kind terms, current when the narrative began but no longer accessible in the language shared by the historian and his or her readers. Some of those terms—'science' or 'physics', for example—still exist in the readers' language but with altered meanings, and these must be unlearned and replaced by their predecessors. When the process is complete, or sufficiently complete for the historian's purposes, the required stage setting has been provided, and the narrative can begin. It can furthermore be related entirely in the terms taught at the start or in their successors, the latter being introduced within the narrative. It is only in the initial teaching operation, in setting the stage, that the historian must, like other sorts of language teachers, make use of the language that readers bring with them. (See John's remark about my use of anachronistic terminology in the title of my paper "Mathematical versus Experimental Traditions in the Development of Physical Science.") It is, of course, always tempting and sometimes irresistibly convenient to use later, already familiar, terms or other terms that, like John's synchronic usages, depart from those deployed at the time. Circumlocution is avoided and the result is not invariably damaging. But the price of convenience is always great risk: exquisite sensitivity and great restraint are required to avoid damage. Experience suggests that few historians develop these qualities in sufficient measure; certainly, I have repeatedly fallen short myself.

The danger in using the names of contemporary scientific fields when discussing past scientific development is the same as that of applying modern scientific terminology when describing past belief. Like 'force' and 'element', 'physics' and 'astronomy' are kind terms, and they carry behavioral expectations with them. These and other names of individual sciences are acquired together in a contrast set, and the expectations that enable one to pick out examples of the practice of each are rich in characteristics that differentiate the examples of one practice from the examples of another, poor in characteristics that examples of a single practice share. That is why one has to learn the names of a number of sciences together before one can pick out examples of the practice of any one of them. Interjection of a name not current at the time thus usually results in a violation of the no-overlap principle and generates conflicting expectations about behavior. That I take to be the lesson to be learned from John's Lavoisier and Poisson examples. His three orthographically distinguished usages have the merit of showing why the debates arose, but they neither point the way to nor play a role in their resolution. To me, those examples suggest not the need for the three uses in historical descriptions but the trouble caused by failure to avoid two of them.

The most obvious danger results from what John calls diachronic usage—the resort to modern terminology—which he suggests might be distinguished by italics. Noel Swerdlow's paper is a strikingly successful and still badly needed attempt to undo one prominent example. When I entered history of science, it was customary, largely due to the influence of Pierre Duhem, to speak of 'medieval science', and I often used that highly questionable phrase myself. Many people, likely still including me, regularly spoke also of 'medieval physics' and sometimes of 'medieval chemistry' as well. Some experts spoke also of 'medieval dynamics' and 'kinematics', drawing a distinction for which I could find neither need nor basis in the texts. At its narrowest, this introduction of modern conceptual distinctions led to misreading, and some of these directly influenced the understanding of figures as recent as Galileo. At its broadest, represented by phrases like 'medieval science' and 'medieval physics', the use of a modern vocabulary led to debates about whether the Renaissance had played any role in the origin of modern science, a debate that, though never conclusive, regularly minimized the role of the Renaissance in scientific development. Though the situation has considerably improved in the forty years since I entered the field, important residues of that debate still linger, residues that Noel is insisting must be set aside. For all my pretense of a position beyond anachronism, I've learned important lessons from his paper and am sure others will do the same.

So far I have been discussing the use of the field names that John labels "diachronic," not the use he calls "synchronic" and employs to refer to "a science or sciences within a narrowly restricted period of time." That use presents subtler problems than the diachronic, but they are problems of the same general sort. In a letter to me John motivates the introduction of synchronic names for fields by pointing out that "contemporary usage is seldom uniform, even at a single historical moment and certainly not over a period long enough to interest the historian," and I know what he has in mind. But the historian has no need to introduce special terms that average out variations in usage with, say, time, place, and affiliation. As with the very similar differences between the idiolects of different individuals, the averaging process takes care of itself. If variations in usage, whether from individual to individual or group to group, did not, at the time under study, interfere with successful communication concerning problems relevant to the narrative, the historian may simply use the terms deployed by his or her subjects. If the variations did make a historical difference, the historian is required to discuss them. In neither case is averaging appropriate. The same is true of variation over time. If the variation is systematic and large enough to make it hard for members of a later generation to understand predecessors who matter to them, then the historian must show how and why those changes came about. If understanding is not affected by the passage of time, then there is no more reason to introduce a new term than to choose whether to use the older or the newer version. Indeed, in the latter case, it is hard to see in what sense there are then two versions to choose between.

I am not, let me be clear, suggesting that the historian is required to report every change of usage, whether from place to place, from group to group, or from time to time. Historical narratives are by their nature intensely selective. Historians are required to include in them only those aspects of the historical record that affect the accuracy and plausibility of their narrative. If they ignore such items—changes in usage included—they risk both criticism and correction. But omission of changes and acceptance of the consequent risk is one thing; introducing new terms is another. As with John's diachronic use, so with his synchronic: new terms can disguise problems that historians are required to face. License to alter the descriptive language of the times they describe should, I think, be denied.

---◂◉▸---

Jed Buchwald's rich and evocative paper returns the topic from social to scientific kinds, and Norton Wise's raises the issue of the relationship between them. The most obvious and direct ties between Jed's paper and the problematic I've been developing are his brief discussions of the difference between the concepts of light rays and of polarization as they are found in the wave and in the emission theory of light.[11] (For rays, geometric optics is also relevant.) Jed's discussions make no reference to kinds or to the no-overlap principle, and none is required. But these examples can easily be recast. 'Ray', for example, is used as a kind term by both the wave and emission theories: the overlap between its referents in the two cases (together with the overlap between the kinds of polarization appropriate to the two theories) cause the difficulties that Jed's paper discusses. In a brilliant paper that takes off from the one he presented to the conference, Jed has now systematically analyzed numerous aspects of the transition from the emission to the wave theory of light as the result of changes in kinds. His paper is likely, I think, to introduce a new stage in the historical analysis of episodes involving conceptual change.[12]

The second point of contact between Jed's paper and my remarks on kinds concerns translation. In *Structure* I spoke of meaning change as a characteristic feature of scientific revolutions; later, as I increasingly identified incommensurability with difference of meaning, I repeatedly referred to the difficulties of translation. But I was then torn, usually without quite realizing it, between my sense that translation between an old theory and a new one was possible and my competing sense that it was not. Jed quotes a long passage (from the Postscript added to the second edition of *Structure*) in which I took the first of these alternatives and described, under the rubric of translation, a process through which

11. A more unified and correspondingly clearer presentation of these concepts is to be found in the introduction of Jed's book *The Rise of the Wave Theory of Light* (Chicago: University of Chicago Press, 1989), pp. xiii–xx.

12. See his "Kinds and the Wave Theory of Light," *Studies in the History and Philosophy of Science* 23 (1992): 39–74. The main diagrams in that paper were originally intended for an appendix to the one presented at this conference.

"participants in a communication breakdown" could reestablish communication by studying each other's use of language and learning, finally, to understand each other's behavior. With what he says in discussing that passage, I fully agree; in particular, though the process described is vital to historians, scientists themselves seldom or never use it. But it is important also to recognize that I was wrong to speak of translation.[13] What I described, I now realize, was language learning, a process that need not, and ordinarily does not, make full translation possible.

Language learning and translation are, I have in recent years been emphasizing, very different processes: the outcome of the former is bilingualism, and bilinguals repeatedly report that there are things they can express in one language that they cannot express in the other. Such barriers to translation are taken for granted if the matter to be translated is literature, especially poetry. My remarks on kinds and kind terms were intended to suggest that the same difficulties in communication arise between members of different scientific communities, whether what separates them is the passage of time or the different training required for the practice of different specialties. For both literature and science, furthermore, the difficulties in translation arise from the same cause: the frequent failure of different languages to preserve the structural relations among words, or in the case of science, among kind terms. The associations and overtones so basic to literary expression obviously depend upon these relations. But so, I have been suggesting, do the criteria for determining the reference of scientific terms, criteria vital to the precision of scientific generalizations.

The third way in which Jed's paper intersects my remarks on kinds relates to Norton Wise's paper as well, and the relationship is in both cases more problematic and more speculative than those I've so far been discussing. Jed's paper speaks of an unarticulated core or substructure which he contrasts with an explicitly articulated superstructure. People who share substructure, he suggests, may disagree about appropriate articulations, but people who differ with respect to substructure will simply misunderstand each other's points, usually without realizing that anything different from disagreement is involved. These properties echo those of the mental module I called "the lexicon" when updating my solution of the new-world problem, the module in which each member

13. The same use of 'translation' is quoted by Ernan McMullin from another place. With what he has to say about the phenomena I refer to, I again fully agree, but again my reference should not have been to translation.

of a speech community stores the kind terms and kind concepts used by community members to describe and analyze the natural and social worlds. It would be too much to suggest that Jed and I are talking about the same thing, but we are certainly exploring the same terrain, and it is worth specifying that shared terrain more closely.

For brevity I restrict attention to the most populous part of the lexicon, the one that contains concepts learned in contrast sets and carrying normic expectations. What this part of their lexicons supplies to community members is a set of learned expectations about the similarities and differences between the objects and situations that populate their world. Presented with examples drawn from various kinds, any member of the community can tell which presentations belong to which kind, but the techniques by which they do so depend less on the characteristics shared by members of a given kind than on those which distinguish members of one kind from those of another. All competent community members will produce the same results, but they need not, as I've previously indicated, make use of the same set of expectations in doing so. Full communication between community members requires only that they refer to the same objects and situations, not that they have the same expectations about them. The ongoing process of communication which unanimity in identification permits allows individual community members to learn each others' expectations, making it likely that the congruence of their bodies of expectations will increase with time. But, though the expectations of individual community members need not be the same, success in communication requires that the differences between them be heavily constrained. Lacking time to develop the nature of the constraint, I will simply label it with a term of art. The lexicons of the various members of a speech community may vary in the expectations they induce, but they must all have the same *structure*. If they do not, then mutual incomprehension and an ultimate breakdown of communication will result.

To see how closely this position maps onto Jed's, read my lexically induced "expectations" as Jed's "articulations" of a core. People who share a core, like those who share a lexical structure, can understand each other, communicate about their differences, and so on. If, on the other hand, cores or lexical structures differ, then what appears to be disagreement about fact (which kind does a particular item belong to?) proves to be incomprehension (the two are using the same name for different kinds). The would-be communicants have encountered incommensurability, and communication breaks down in an especially frustrating way. But because what's involved is incommensurability, the missing prerequisite to communication—a "core" for Jed, a "lexical structure" for me—can only be exhibited, not articulated. What the participants in communication fail to share is not so much belief as a common culture.

Norton's paper is also concerned with the commonalities that make a shared scientific culture, and his mediating balances behave somewhat like Jed's core in that they isolate characteristics shared by items located at the various different nodes of his network. In this case, however, the likenesses are between items in the social world rather than the natural. They hold, that is, between practices in the various scientific fields as well as between them and the larger culture (note Norton's introduction of the figure of Republican France). Having been on record for many years as deeply skeptical about the extent to which grounded ties of this wide-ranging sort could be found, I feel bound to announce that I've been largely converted, mostly as a consequence of Norton's work, especially that on nineteenth-century Britain.[14] These ties between the practices Norton discusses cannot, I think, be coincidental or mere imaginative fabrications. They signify, I am convinced, something of great importance to an understanding of science. But at this early stage of the development of these ties, I'm deeply uncertain what they can signify: essential parts of the story his points require seem to me to be missing.

In the first place, I do not know what Norton's "rationalist scientific culture" is, how the practices between which his balances mediate are recognized or selected. My early readings of his paper suggested that they were simply the sciences as practiced in the national culture of late eighteenth-century France, but Norton has assured me that that is not at all his intent. Not all French scientific practices belong to his network, he insists, and some of the practices that do are located in other national cultures. But neither can his network be identified simply by the bridge with respect to which its nodes are similar: an arbitrarily selected set of practices will ordinarily be similar in some respect or other. I am not suggesting that a *definition* of rationalist scientific culture is required, but I do feel the need for some description of its salient characteristics, of characteristics that would collectively permit me to pick out some

14. See especially C. Smith and M. Norton Wise, *Energy and Empire: A Biographical Study of Lord Kelvin* (Cambridge: Cambridge University Press, 1989), and earlier articles of Norton's there cited.

practices as exemplifying the culture and others as not. The point is not that I want to check Norton's story. He recognizes the practices involved, and I've great confidence in his judgment. I'm sure he will, in the long run, provide an answer to my question. But until I know something about how Norton recognizes the practices which are the nodes of his network, I am quite literally not going to understand what he's trying to tell me.

That difficulty is aggravated by another, one that I'm far less sure can be resolved and to which I'm especially sensitive because it exemplifies a trap into which I fell again and again in *Structure*. Norton illustrates the bridges supplied by his balances by showing a small number of individuals interacting across them: Lavoisier with Laplace, Condillac with Lavoisier, Lavoisier with Condorcet. But he uses those illustrations to suggest that the bridges link not only people but also practices— chemistry, physical astronomy, electricity, political economy, and others—and that suggestion leads to three difficulties to which I'll point in order of increasing importance.

The first difficulty I mention only for one of its consequences. To generalize these bridges from individuals to the various scientific practices, one would need to show that they operated for a considerable number of practitioners and to illustrate the differences their existence made to the practices that they linked. I am not, however, the one to throw stones at people who overgeneralize, and the point I'm after is different. The speed of Norton's transition from individual to group obscures another possible explanation of the individual behavior he reports. Perhaps the mediating balances are not characteristics of the various scientific cultures but rather of the larger culture within which those practices take place. That could make bridges available to individuals without affecting the mode of group practice at all. Such an explanation may not be right, but room needs to be allowed for its consideration. And considering it would, among other desiderata, provide the room needed to ask what's right about the position of the diehards who insist, for example, that chemistry is chemistry, physics physics, and math math in whatever culture they occur.

The third difficulty is of another and more important sort. In my view, Norton's passage from individual to group involves a damaging category mistake, one of which I was repeatedly guilty in *Structure* and which is endemic also in the writings of historians, sociologists, social psychologists, and others. The mistake is to treat groups as individuals writ large or else individuals as groups writ small. It results, at its crudest, in talk of the group mind (or group interest) and, in its subtler forms, in attributing to the group a characteristic shared by all or most of its members. The most egregious example of this mistake in *Structure* is my repeated talk of gestalt switches as characteristic of the experience undergone by the group. In all these cases the error is grammatical. A group would not experience a gestalt switch even in the unlikely event that every one of its members did so. A group does not have a mind (or interests), though each of its members presumably does. By the same token, it does not make choices or decisions even if each of its members does so. The outcome of a vote, for example, may result from the thoughts, interests, and decisions of group members, but neither the vote nor its outcome is a decision. If, as has traditionally been taken for granted, a group were nothing but the aggregate of its individual atomic members, this grammatical error would be inconsequential. But it is increasingly recognized that a group is not just the sum of its parts and that an individual's identity in part consists in (not just: is determined by) the groups of which he or she is a member. We badly need to learn ways of understanding and describing groups that do not rely upon the concepts and terms we apply unproblematically to individuals.

I do not command the required ways of understanding, but in recent years I have made two steps toward doing so. The first, which I've already mentioned but still lack space to explain, is the distinction between a lexicon and a lexical structure. Each member of a community possesses a lexicon, the module that contains the community's kind concepts, and in each lexicon the kind concepts are clothed with expectations about the properties of their various referents. But though the kinds must be the same in the lexicons of all community members, the expectations need not be. Indeed, in principle, the expectations need not even overlap. What is required is only that they give the lexicons of all community members the same structure, and it is this structure, not the varied expectations through which different members express it, that characterizes the community as a whole.

My other step forward is the discovery of a tool that I've as yet scarcely learned to put to use. But I am currently being much instructed by the discovery that the puzzles about the relation of group members to group have a quite precise parallel in the field of evolutionary biology: the vexing relation between individual organisms and the species to which they belong. What characterizes the individual organism is a particular set of genes; what characterizes the species is the gene pool of the entire interbreeding population which, geographical isolation aside, constitutes the species. Understanding the process of evolution has in recent years seemed increasingly to require conceiving the gene pool, not as the mere aggregate of the genes of individual organisms, but as itself a sort of individual of which the members of the species are parts.[15] I am persuaded that this example contains important clues to the sense in which science is intrinsically a community activity. Methodological solipsism, the traditional view of science as, at least in principle, a one-person game, will prove, I am quite sure, to have been an especially harmful mistake.

That Norton's paper calls forth thoughts of this sort is an index of the seriousness with which I take it. He is, I'm quite certain, on his way to significant discoveries, and I look forward to them with considerable excitement. But those discoveries are still emerging. So far, I am finding them extraordinarily difficult to grasp.

———◦———

I come at last to relativism and realism, issues central to the papers of Ernan McMullin and Nancy Cartwright but implicit in several of the other papers as well. As in the past Ernan proves to be among my most discerning and sympathetic critics, and I will presuppose much that he has said in order to concentrate on points at which our views depart. Of these the most important to both of us involves what Ernan takes to be my antirealist stance and my corresponding lack of concern with epistemic (as against puzzle-solving) values. But that characterization does not quite catch the nature of my enterprise. My goal is double. On the one hand, I aim to justify claims that science is cognitive, that its product is knowledge of nature, and that the criteria it uses in evaluating beliefs are in that sense epistemic. But on the other, I aim to deny all meaning to claims that successive scientific beliefs become more and more probable or better and better approximations to the truth and simultaneously to suggest that the subject of truth claims cannot be a relation between beliefs and a putatively mind-independent or "external" world.

Postponing remarks on the nature of truth claims, I begin with the question of science's zeroing in on, getting closer and closer to, the truth. That claims to that effect are meaningless is a consequence of

15. See David Hull, "Are Species Really Individuals?" *Systematic Zoology* 25 (1976): 174–91.

incommensurability. This is not the place to elaborate the needed arguments, but their nature is suggested by my earlier remarks on kinds, the no-overlap principle, and the distinction between translation and language learning. There is, for example, no way, even in an enriched Newtonian vocabulary, to convey the Aristotelian propositions regularly misconstrued as asserting the proportionality of force and motion or the impossibility of a void. Using our conceptual lexicon, these Aristotelian propositions cannot be expressed—they are simply ineffable—and we are barred by the no-overlap principle from access to the concepts required to express them. It follows that no shared metric is available to compare our assertions about force and motion with Aristotle's and thus to provide a basis for a claim that ours (or, for that matter, his) are closer to the truth.[16] We may, of course, conclude that our lexicon permits a more powerful and precise way than his of dealing with what are *for us* the problems of dynamics, but these were not his problems, and lexicons are not, in any case, the sorts of things that can be true or false.

A lexicon or lexical structure is the long-term product of tribal experience in the natural and social worlds, but its logical status, like that of word meanings in general, is that of convention. Each lexicon makes possible a corresponding form of life within which the truth or falsity of propositions may be both claimed and rationally justified, but the justification of lexicons or of lexical change can only be pragmatic. With the Aristotelian lexicon in place it does make sense to speak of the truth or falsity of Aristotelian assertions in which terms like 'force' or 'void' play an essential role, but the truth values arrived at need have no bearing on the truth or falsity of apparently similar assertions made with the Newtonian lexicon. Whatever I may have believed when I wrote *The Copernican Revolution*, I would not now assume (*pace* Ernan) "that the simpler, the more beautiful [astronomical] models are more

---

16. Aristotle's discussions of force and motion did, of course, include statements that we can make in a Newtonian vocabulary and can then criticize. His explanations of the continued motion of a projectile after it leaves the mover's hand are especially well known examples. But the basis of our criticisms is observations that could in many cases have been made by Aristotle, that were made explicit by his successors, and that led to the development of the so-called impetus theory, a theory that avoided the difficulties that Aristotle's had faced but that did not directly affect the Aristotelian conceptions of force and motion. This example is developed in my "What Are Scientific Revolutions?" Occasional Paper #18, Center for Cognitive Science (Cambridge, MA: Massachusetts Institute of Technology, 1981); reprinted in *The Probabilistic Revolution*, vol. 1, *Ideas in History*, ed. L. Krüger, L. J. Daston, and M. Heidelberger (Cambridge, MA: MIT Press, 1987), pp. 7–22; also reprinted in this volume as essay 1.

likely to be true." Though simplicity and beauty provide important criteria of choice in the sciences (as does making causal sense of phenomena, which Ernan also cites), they are instrumental rather than epistemic where lexical change is involved. What they are instrumental for will be my closing topic below.

At least all this is the case if the sense of 'epistemic' is the one I take Ernan to have in mind, the sense in which the truth or falsity of a statement or theory is a function of its relation to a real world, independent of mind and culture. There is, however, another sense in which criteria like simplicity may be called epistemic, and it has already figured, implicitly or explicitly, in several of the papers presented at this conference. Its most suggestive appearance is also the briefest: Michael Friedman's description of Reichenbach's distinction between two meanings of the Kantian a priori, one which "involves unrevisability and . . . absolute fixity for all times" while the other means " 'constitutive of the concept of the object of knowledge'." Both meanings make the world in some sense mind-dependent, but the first disarms the apparent threat to objectivity by insisting on the absolute fixity of the categories, while the second relativizes the categories (and the experienced world with them) to time, place, and culture.

Though it is a more articulated source of constitutive categories, my structured lexicon resembles Kant's a priori when the latter is taken in its second, relativized sense. Both are constitutive of *possible experience* of the world, but neither dictates what that experience must be. Rather they are constitutive of the infinite range of possible experiences that might conceivably occur in the actual world to which they give access. Which of these conceivable experiences occurs in that actual world is something that must be learned, both from everyday experience and from the more systematic and refined experience that characterizes scientific practice. They are both stern teachers, firmly resisting the promulgation of beliefs unsuited to the form of life the lexicon permits. What results from respectful attention to them is knowledge of nature, and the criteria that serve to evaluate contributions to that knowledge are, correspondingly, epistemic. The fact that experience within another form of life—another time, place, or culture—might have constituted knowledge differently is irrelevant to its status as knowledge.

Norton Wise's closing pages I take to be making a very similar point. Through much of his paper, technology (for his culture, the various balances) is seen as providing a culturally based mediator between instruments and reality at one end of his cylinder (his figure 18) and between instruments and theories at the other.[17] Except that the technologies are conceived as situated in local cultures, no traditional philosopher of science would find fault with that model. Of course instruments, including the sense organs, are required to mediate between reality and theory. To this point in the argument, no references to anything like a constructed or mind-dependent reality are called for. But Norton then folds his cylinder back on itself to form a doughnut, and the picture changes decisively (his figure 19). A geometry that requires a figure with two ends is replaced by one that calls for three symmetrically placed slices. Technology continues to provide a two-way route between theories and reality, but reality provides the same sort of route between theory and technology, and theories provide a third route between reality and technology. Scientific practice requires all three of these sorts of mediation, and none of them has priority. Each of his three slices—technology, theory, and reality—is constitutive for the other two. And all three are required for the practice whose product is knowledge. When Norton closes by describing what he has been doing as "depicting cultural epistemology," I think he gets it just right. But I would add "cultural ontology" as well.

Nancy Cartwright's exciting paper indicates ways to move further in the same direction, but for my purposes, its opening remarks on the theory/observation distinction need first to be somewhat recast. I agree that the distinction is needed, but it cannot be just that between the "peculiarly recondite terms [of modern science and] those we are more used to in our day-to-day life." Rather, the concepts of theoretical terms must be relativized to one or another particular theory. Terms are theoretical with respect to a particular theory if they can only be acquired with that theory's aid. They are observation terms if they must first have been acquired elsewhere before the theory can be learned.[18] 'Force' is thus theoretical with respect to Newtonian dynamics but observational with respect to electromagnetic theory. This view is very close to the third of the interpretations that Nancy suggests for Peter Hempel's phrase 'antecedently available', and that interpretation has been

17. Norton would probably use the term 'ideology' rather than 'theory', but he regularly conflates the two, as in figure 18, which illustrates the present point. Reasons for the difference in our choice of terms will be apparent.

18. For this view, see particularly Stegmüller, *Structure and Dynamics of Theories*, pp. 40–57. Its origin is in J. D. Sneed, *The Logical Structure of Mathematical Physics* (Dordrecht: Reidel, 1971), but its presentation there is widely scattered.

instrumental, as my opening remarks will have suggested, in Peter's and my very considerable rapprochement.

Replacing the concept of observational terms with that of antecedently available terms has three special advantages. First, it ends the apparent equivalence of 'observational' and 'nontheoretical': many of the recondite terms of modern science are both theoretical and observational, though the observation of their referents requires recondite instruments. Second, unlike its predecessor, the distinction between theoretical and antecedently available terms is freed to become developmental, as I think it must: terms antecedently available, whether to an individual or a culture, are the base for the further extension of both vocabulary and knowledge. And third, viewing the distinction as developmental focuses attention on the process by which a conceptual vocabulary is transmitted from one generation to the next—first to young children being prepared (socialized) for the adult society of their culture, and second to young adults being prepared (again, socialized) to take their place among the practitioners of their discipline.

For present purposes the last point is the crucial one, for it will bring me rapidly back to realism. In the theory of the lexicon, to which I've already repeatedly referred, a key role is played by the process by which lexicons are transmitted from one generation to the next, whether from parents to children or from practitioners to apprentices. In that process the exhibition of concrete examples plays the central role, where the "exhibition" may be accomplished either by pointing to real-life examples in the everyday world or the laboratory or else by describing these potential examples in the vocabulary antecedently available to the student or inductee. What is acquired in this process is, of course, the kind concepts of a culture or subculture. But what comes with them, inseparably, is the world in which members of the culture live.

Nancy omits the developmental context, which I take to be central. But the process is one she twice illustrates: for scientific kinds, by the passage about Newton's second law that she resurrects from the second edition of *Structure*, and for social kinds, by her discussion of fables. The pendulum, the inclined plane, and the rest are examples of $f = ma$, and it is being examples of $f = ma$ that makes them similar, like each other. Without having been exposed to them or some equivalents as examples of $f = ma$, students could not learn to see either the similarities between them or what it was to be a force or a mass; they could not, that is, acquire the concepts of force and mass or the meaning of the terms that name them.[19] Similarly, the three examples of the fable—marten/grouse, fox/marten, and wolf/fox—are concrete illustrations of what, lacking a better term, I shall call the "power situation," the situation in which terms like 'strong', 'weak', 'predator', and 'prey' function. It is their illustrating the same aspect of the situation that simultaneously makes them similar to each other and makes the situation into the one the fables convey. Without exposure to these or similar situations, a candidate for socialization into the culture that exhibits them could not acquire the social kinds called 'the strong', 'the weak', 'predators', or 'prey'.

Though other resources are available for acquiring social concepts like these, fables and the maxims that accompany them have the particular merit of simplicity, which is presumably why they have played so large a role in the socialization of children. Nancy speaks of them as "thin," a term she also applies to, say, models like the frictionless plane and the point pendulum. The latter are, if you will, physicists' fables (Newton's second law being the maxim juxtaposed to them), and it is their characteristic thinness which makes them so especially useful for socializing potential members of the profession. That is why they figure so prominently in science textbooks.

Except for my insistence on positioning them within the learning or socialization process, these points are all explicit in Nancy's paper, and so is another which she has given me new words to describe. Once the new terms (or the revised versions of old ones) have been acquired, there is no ontological priority between their referents and the referents of the antecedently available terms deployed in the acquisition process. The concrete (pendulum or marten) is neither more nor less real than the abstract (force or prey). There are, of course, both logical and psychological priorities between the members of these pairs. One cannot acquire the Newtonian concepts of force and mass without prior access to such concepts as space, time, motion, and material body. Nor can one acquire the concepts of predator and prey without prior access to such concepts as kinds of creature, death, and killing. But there are not, as Nancy puts it, relations of either fact or meaning reduction between the members of these pairs (between force and mass, on the one hand, and space, time, etc., on the other; or between predator and prey, on

19. The point is more precisely made in my "Possible Worlds in the History of Sciences." 'Force' can be acquired without 'mass' by exposure to examples of Hooke's law. 'Mass' can then be added to the conceptual vocabulary by presenting illustrations either of Newton's second law or of his law of gravity.

the one hand, and death, killing, etc., on the other). In the absence of such relations, there is no basis for singling out one or the other juxtaposed set as the more real. To insist on this point is not to limit the concept of reality but rather to say what reality is.

It is in our response to that shared analysis that Nancy's route and mine diverge, but in a way I am finding especially instructive. Both Nancy and I are pushed to a reluctant pluralism. But she would achieve hers by permitting restrictions on the universality of true scientific generalizations, suggesting, for example, that the truth of Newton's second law does not depend upon its applying to all of its potentially concrete models. The law's scope is, for her, uncertain: in one part of its domain it may be true, while in another some other law may obtain. For me, however, that form of pluralism is barred. 'Force', 'mass', and their like are kind terms, the names of kind concepts. Their scope is limited only by the no-overlap principle and is thus part of their meaning, part of what enables their referents to be picked out and their models to be recognized. To discover that the scope of a kind concept is limited by something extrinsic, something other than its meaning, is to discover that it never had any proper applications at all.

Nancy introduces scope restrictions to account for the occasional failure of the search for workable models of laws she considers true. I would instead resolve such failures by introducing a few new kinds that displace some of those in use before. That change constitutes a change in lexical structure, one that brings with it a correspondingly changed form of professional practice and a different professional world within which to conduct it. Her pluralism of domains is for me a pluralism of professional worlds, a pluralism of practices. Within the world of each practice, true laws must be universal, but some of the laws governing one of these worlds cannot even be stated in the conceptual vocabulary deployed in, and partially constitutive of, another. The same no-overlap principle that necessitates the universality of true laws bars the practitioners resident in one world from importing certain of the laws that govern another. The point is not that laws true in one world may be false in another but that they may be ineffable, unavailable for conceptual or observational scrutiny. It is effability, not truth, that my view relativizes to worlds and practices. That formulation is compatible with transworld travel: a twentieth-century physicist can enter the world of, say, eighteenth-century physics or twentieth-century chemistry. But that physicist could not practice in either of these other worlds without abandoning the one from which he or she came. That makes transworld travel difficult to the point of subversion and explains why, as Jed Buch-wald emphasizes, practitioners of a science almost never undertake it.

One more step will return me to Ernan's paper and take me to the last of the problems to be considered in this one. The developmental episodes that introduce new kinds and displace old are, of course, the ones that in *Structure* I called "revolutions." At the time I thought of them as episodes in the development of a single science or scientific specialty, episodes that I somewhat misleadingly likened to gestalt switches and described as involving meaning change. Clearly, I still think of them as transforming episodes in the development of individual sciences, but I now see them as playing also a second, closely related, and equally fundamental role: they are often, perhaps always, associated with an increase in the number of scientific specialties required for the continued acquisition of scientific knowledge. The point is empirical and the evidence, once faced, is overwhelming: the development of human culture, including that of the sciences, has been characterized since the beginning of history by a vast and still accelerating proliferation of specialties. That pattern is apparently prerequisite to the continuing de-velopment of scientific knowledge. The transition to a new lexical struc-ture, to a revised set of kinds, permits the resolution of problems with which the previous structure was unable to deal. But the domain of the new structure is regularly narrower than that of the old, sometimes a great deal narrower. What lies outside of it becomes the domain of another scientific specialty, a specialty in which an evolving form of the old kinds remains in use. Proliferation of structures, practices, and worlds is what preserves the breadth of scientific knowledge; intense practice at the horizons of individual worlds is what increases its depth.

This is the pattern that led me, at the end of my remarks on Ian Hacking's paper, to speak of specialization as speciation, and the parallel to biological evolution goes further. What permits the closer and closer match between a specialized practice and its world is much the same as what permits the closer and closer adaptation of a species to its bio-logical niche. Like a practice and its world, a species and its niche are interdefined; neither component of either pair can be known without the other. And in both cases, also, that interdefinition appears to require isolation: the increasing inability of the residents of different niches to crossbreed, on the one hand, and the increasing difficulty of communica-tion between the practitioners of different specialties, on the other.

That pattern of development by proliferation raises the problem to which, in a more standard formulation, most of Ernan's paper is devoted: what is the process by which proliferation and lexical change take place, and to what extent can it be said to be governed by rational considerations? On those questions, more than on any of those I've discussed above, my views remain very close to those developed in *Structure*, though I can now articulate them more fully. Indeed, Ernan has already articulated most of them for me. There are only two points in his presentation of my position from which I've any inclination to dissent. The first is the use he makes of the distinction between shallow and deep revolutions: though revolutions do differ in size and difficulty, the epistemic problems they present are for me identical. The second is Ernan's understanding of the intent with which I refer to Hume's problem of induction: I share his intuition that the developmental approach to science will dissolve (not solve) Hume's problem; the object of my occasional references to it was simply to disclaim responsibility for a solution.

In other areas what I need to do is explain what Ernan sees as equivocations and inconsistencies in my position. That will require my presupposing, at least for the sake of the argument, that you have already set aside the notion of a fully external world toward which science moves closer and closer, a world independent, that is, of the practices of the scientific specialties that explore it. Once you have come that far, if only in imagination, an obvious question arises: what, if not a match with external reality, is the objective of scientific research? Though I think it requires additional thought and development, the answer supplied in *Structure* still seems to me the right one: whether or not individual practitioners are aware of it, they are trained to and rewarded for solving intricate puzzles—be they instrumental, theoretical, logical, or mathematical—at the interface between their phenomenal world and their community's beliefs about it. That is what they are trained to do and what, to the extent they retain control of their time, they spend most of their professional lives doing. Its great fascination—which to outsiders often seems an obsession—is more than sufficient to make it an end in itself. For those engaged in it, no other goal is needed, though individuals often have a number of them.

If that is the case, however, the rationality of the standard list of criteria for evaluating scientific belief is obvious. Accuracy, precision, scope, simplicity, fruitfulness, consistency, and so on, simply *are* the criteria which puzzle solvers must weigh in deciding whether or not a given puzzle about the match between phenomena and belief has been solved. Except that they need not all be satisfied at once, they are the "defining" characteristics of the solved puzzle. It is for maximizing the precision with which, and the range within which, they apply that scientists are rewarded. To select a law or theory which exemplified them less fully than an existing competitor would be self-defeating, and self-defeating action is the surest index of irrationality.[20] Deployed by trained practitioners, these criteria, whose rejection would be irrational, are the basis for the evaluation of work done during periods of lexical stability, and they are basic also to the response mechanisms that, at times of stress, produce speciation and lexical change. As the developmental process continues, the examples from which practitioners learn to recognize accuracy, scope, simplicity, and so on, change both within and between fields. But the criteria that these examples illustrate are themselves necessarily permanent, for abandoning them would be abandoning science together with the knowledge which scientific development brings.

The pursuit of puzzle solving constantly involves practitioners with questions of politics and power, both within and between the puzzle-solving practices, as well as between them and the surrounding nonscientific culture. But in the evolution of human practices, such interests have governed from the start. What further development has brought with it is not their subordination but the specialization of the functions to which they are put. Puzzle solving is one of the families of practices that has arisen during that evolution, and what it produces is knowledge of nature. Those who proclaim that no interest-driven practice can properly be identified as the rational pursuit of knowledge make a profound and consequential mistake.

20. These points are elaborated in two papers of mine that Ernan cites: "Objectivity, Value Judgment, and Theory Choice," in my *Essential Tension* (Chicago: University of Chicago Press, 1977), pp. 320–29, and "Rationality and Theory Choice," *The Journal of Philosophy* 80 (1983): 563–70, reprinted in this volume as essay 9. The themes developed in the second of these papers are still another product of my interactions with C. G. Hempel, the one to which, in the first section of this paper, I promised to return.

A Discussion with Thomas S. Kuhn

A Discussion with Thomas S. Kuhn

ARISTIDES BALTAS
KOSTAS GAVROGLU
VASSILIKI KINDI

"A Discussion with Thomas S. Kuhn" is an edited transcript of a tape-recorded three-day discussion—essentially an extended interview—between Kuhn and Aristides Baltas, Kostas Gavroglu, and Vassiliki Kindi. The discussion took place in Athens on October 19–21, 1995. The occasion was the awarding of an honorary doctorate to Kuhn by the Department of Philosophy and History of Science of the University of Athens, and a symposium in his honor at the University. The symposiasts included, in addition to the above discussants, Costas B. Krimbas and Pantelis Nicolacopoulos. The proceedings of the symposium, as well as this discussion, were published in a special issue of Neusis: Journal for the History and Philosophy of Science and Technology *(1997). The interview has been lightly edited again for this volume.*

K. GAVROGLU: Well, let's start from your school days, especially the kinds of courses that intrigued you, the kinds of courses you hated, the kinds of teachers you met . . .

T. KUHN: I started my education—or started going to school, which is another matter—in New York, in Manhattan. And I was there for some years in progressive school, from kindergarten and then on through the fifth grade. Progressive school encouraged a sort of independent thinking. On the other hand, it didn't do much to teach subject matter. I remember at a time when I was probably already in second grade, my parents were getting very discouraged because I didn't seem to be able to read; my father held letters up for me and then I got on to it pretty fast. Then, as I went on to sixth grade the family moved out to the country, about forty-fifty miles out of New York to Croton-on-Hudson, and I went to a small progressive school there called the Hessian Hills School. It no longer exists. But it was particularly good in terms of teaching me to think for myself. It was a very left-oriented school; the woman who was its principal founder was called Elizabeth Moos. She was the mother-in-law of a man called William Remington—you may remember him, he was somebody who was ultimately put away for having been a Communist courier, this was something that came out during the McCarthy period. So, there were various radical left teachers all over, except that we were all encouraged to be pacifists. There was no Marxist training or anything of the sort; we were told by our parents that this was a radical school but we didn't quite see it that way ourselves. There I had one teacher who influenced me. I had several teachers who influenced me, but I had a math teacher called Leon Sciaky.[1] Everybody loved him dearly, he was very good at teaching mathematics. What I had with him was mostly elementary algebra but I had always been . . . not very bad but only mediocre in arithmetic, I made too many mistakes, I'd add things up and they wouldn't come out the same way two days in a row. I could do my multiplication tables but I've never really gotten past nine times nine. But suddenly turning to more abstract, with variables, I came alive for mathematics and that was in his hands. And I loved it, I was quite good at it, and that was a rather special experience. I was also rather good I guess in others. . . . There were no grades at the school; when it turned out that I was doing particularly well I was quite surprised—I didn't know that. I was there through sixth, seventh, eighth, and ninth grades—four years. I had a good social studies teacher. Here the radicalness of the school came out a little bit —we read as a group substantial parts of Beard's *Economic Interpretation of the Constitution of the United States,* and talked about it.

---

1. Leon Sciaky was born in Salonica; he wrote a memoir, *Farewell to Salonica: Portrait of an Era* (New York: Current Books, 1946).

In my class there were six or seven people; it was that sort of very
much hands-on education; hands-on in order to have hands-off education. And I thought that made a major contribution to my independence of mind.

A. BALTAS: Could you tell us some more things about the notion of
progressive school: is it a particular kind of school?

T. KUHN: No. Progressive education was a movement which—to the
best of my knowledge—really sprung from some proposals of John
Dewey's. It emphasized subject matter less than it emphasized independence of mind, confidence in ability to use one's mind. So it was standard to say it doesn't teach spelling—there was very little drill. We
started having French lessons; after three years I still couldn't remember any French—that sort of thing. But I was getting to be bright.
I'm not sure that the Hessian Hills School—which I really think was
an important formative influence—a small-scale education with very
little in the way of set subject matter, a good deal of work by oneself,
or what one thought one was doing by oneself. One thing I would
say is that when I got to MIT, I found that by the time they were
ready to graduate, students, many of them, had never written a ten-
or twelve-page paper, the sort of paper I would try to assign. I wrote
at least one twenty-five-page paper while I was in sixth grade, or seventh grade. So, there was more of that. And that flavor of work, that
flavor of encouragement was I think very important to things that
happened later. Now, that school went only through ninth grade. In
fact, it often went only through eighth grade; but for our group it
went on through ninth grade and after that I went away to boarding
school. My parents were worried that I would find the transition difficult, so they sent me to a small school.

K. GAVROGLU: The fact that there were quite a few left people in the
school, or the overall climate was left, was that something at that
period which was looked down upon, or was it, among a certain group
of people, something good?

T. KUHN: I'm sure there were circles in which it was looked down upon;
I was more radical than my parents but they did not look down on
it. On the other hand, it's worth getting me to say just a little bit about
that. This was an age and an age group when people were beginning
to join something that was called the American Student Union. A
prerequisite for being a member of the American Student Union was
willingness to sign the Oxford Oath. It's an oath that you would not
fight, even for your country. And I remember talking to my father

about this, because I didn't really feel that I was happy about saying that I would not fight for my country. I wanted to be a member of the Student Union but I wasn't sure I could do that. And I remember his saying to me, "I have signed a lot of oaths and then later violated them. But I don't think I've ever signed an oath thinking I was going to violate it." And I took that very seriously and I did not join the Student Union. On the other hand, there was a meeting of students from numerous progressive schools at my school, and I don't remember what we were talking about, what the official subject was, but I wound up being reported in the Peekskill, New York paper. I don't know that it was by name, but it was I who had stood up and said, "Who profits from our national possessions? Not you, not I, just the capitalists. Let the Philippines go." So, that gives you some flavor.

A. BALTAS: Approximately when did these incidents occur?

T. KUHN: Well, I left Hessian Hills in ninth grade in 1937 and this would have been one of the two years preceding that, and that gives you the general time scale. Then, I went to spend one year at a school in Pennsylvania called Solebury. That was all right. I didn't have special difficulties there, but the idea was that I should try that for a year and I could stay if I really liked it, but otherwise I would be moved to a still more advanced sort of preparatory school. I liked it all right, but I wasn't enthusiastic, I was missing the sort of interactions I'd had in Hessian Hills. So the next year I was moved to a school which I liked rather less, which really was a preparatory school—mostly a Yale preparatory school. It was a school called Taft, in Watertown, Connecticut. There I think there's nothing special to be said, about either of those. I mean I had good teachers and less good teachers, but nothing special except a very good English teacher in eleventh grade at Taft School. I remember we read Robert Browning a lot. That was important to me. Otherwise, the science teaching there was lousy. I remember a physics course which was taught by somebody who knew some chemistry but not much, and didn't know any physics, or not much. And I suddenly found myself suggesting—I can't have suggested that heat was the mean kinetic energy of the molecules but I sort of made up some kinetic theory and I brought it to the teacher, twice I think, and the second time he said, "Look, just wait till you're ready!" It was clear that was a sort of encouragement that I was not going to get because I don't think he knew the answer. And it was relatively elementary physics.

So, those schools gave me more formal training, more language—

although I was never any good, I've never been any good really at foreign languages. I can read French, I can read German, if I'm dropped into one of those countries I can stammer along for a while, but my command of foreign languages is not good, and never has been, which makes it somewhat ironic that much of my thought these days goes to language.

After those two schools in which I did well, I mean, let that be a matter of record, my grades were good, I went on to Harvard.

K. GAVROGLU: What was your father's profession?

T. KUHN: My father had been trained, and this is of some importance for me, as a hydraulic engineer. He'd gone to Harvard. Briefly, there was a joint program in which he took a five-year bachelor's and master's from Harvard and MIT, and he did that program. It was set up under a will and then the will was thrown out by the courts or something of the sort, so the program was divided again. He finished this five-year program, I think in 1916, and then the *Lusitania* went down and he was off to the wars. And he wound up in the Army Corps of Engineers and I think had the happiest, most productive period of his life. After the war—his father had died while he was away—he came back, sort of hung around Cincinnati to help his mother, and did some civil engineering which was not very interesting. Then, he got married. What he used to say, and I don't trust everything he said, was that he felt he couldn't compete with the younger people right out in hydraulic engineering, so he used his talents, which were very considerable, elsewhere. I was born in Cincinnati. That was his home—he brought my mother there, she was a New York girl. When I was six months old I was moved to New York with my parents. He went into what later became industrial engineering. He worked for a while for a bank, as somebody who investigated investment possibilities and gave advice both to the bank and for their clients. He was very active during the days of the national reconstruction administration, did stuff on the tobacco industry, he gave testimony before Congress and other things of this sort. But he was never, I think, the sort of success he had expected to be and under other circumstances might have been. And I think those around him thought of him—I think, but I've had only one person say this to me—as somebody who had not fulfilled his promise, who was very bright, who could have done much more, and that there was a real waste of talent involved. I think that's right. He would never have said that, but I think he felt it. I admired him greatly. I thought for many years that except for James Conant, he was the brightest person I had ever known. He wasn't much of an intellectual, but he had a very very sharp mind. He used to catch me out all the time and that didn't do me a lot of good either.

K. GAVROGLU: Did he take an active interest in your education apart from trying to send you to good schools and keeping an eye on how you were doing? But did he actually get into the details of your education?

T. KUHN: No. He was the person who said, "Well, what can you say in French?" and I said "l'éléphant"! He was the one who was afraid I wasn't learning to read and held up letters or words—I don't remember—for a bit. But he was not actively involved except for points of that sort.

K. GAVROGLU: What about your mother in that matter?

T. KUHN: My mother wasn't actively involved either. But in a funny way, although she was not nearly as bright as my father, she was more intellectual. And sometimes in a somewhat flighty way—but she read more books. She did some professional editing. I was brought up to believe—everybody said—I took after my father; and this was fine, I admired him greatly, I was afraid of him and all the rest of it. I was also told that my younger brother was just like my mother. Later I realized this was exactly backwards. My brother was much more like my father, and I was much more like my mother. Now, part of realizing this had to do with the following. I, as you know, and I'll say more how I got there, I was a physics major. I was insistent upon being a *theoretical* physicist. But I loved working with my hands and I later built ham radios. I was never a ham-radio operator but I used to be in the old tube and battery days; I also did shopwork. And I could never finally figure out why it was that I had decided I had to be a theoretical physicist rather than an experimental physicist. And I finally realized that it was because theoretical physics was more nearly an intellectual activity and I was following my mother at this point, not my father. It came to me as a considerable shock to realize this and in the long run of course I have no regrets about how this worked out. But it was a puzzle to me, and it was quite explicitly a puzzle, as to why is it that I had this determination to do this.

A. BALTAS: Do you have only one younger brother?

T. KUHN: I've only the one younger brother. Well, after I finished at the Taft School, I went on from there to Harvard, which had been my father's university. At this point my life changed quite markedly, because what I have not said up to this point but should say now is that throughout school I had really had almost no friends. I was isolated. I've been isolated since; but I was quite unhappy about it. I somehow wasn't a member of the group and I wanted terribly to be a member of the group. Harvard was big enough, and sufficiently intellectually oriented, and had a large variety of groups. You didn't have to be a member in order to be a part of one, and you could be a part of several. And I began to feel as though I was much more nearly part of something than I had been before, and I began to have happier social relations. It didn't go as far as it might have gone, but it went far enough to give me a very different sense of myself. Now, at Harvard—I had a lovely conversation with my father that you will be interested in, in the summer before I went to Harvard. I had really been good at high school mathematics. And I was a year ahead of myself, I'd done a year of calculus, with one other person at my preparatory school in my last year, I had had a year of physics that had been badly taught. I didn't hold this against physics but it had prevented my getting . . . so I talked with my father. There was no question but that I was going to major in science or math, or physics or math. Should I be a physicist or should I be a mathematician? And this will also tell you something that you know perfectly well, about how the situation has changed in the years since. This would have been in the summer of 1940. He said to me, "Look, if you prefer one of these, significantly, by all means do that. But if you are really torn between the two, I think you should probably do physics. Because, if you do mathematics, unless you wind up in one of the very good mathematics departments, all that's left is teaching high school or being an actuary for an insurance company. Whereas if you do physics, it's not very different but there are a few places like the General Electric Lab and the Naval Research Laboratory where you can still do research even though you are not in one of the research-oriented universities." So, I did physics.

K. GAVROGLU: And at that time you actually applied for a particular department. It was not that you were applying to the Science Faculty.

T. KUHN: No, at the end of your freshman year you announced your major. And that was a major in a department. So, [then] I enrolled in physics. Look, I had a strange experience in this respect in my first year and it's one that I think probably also had a formative influence. I think part of my emphasis on solving problems, puzzles, may come from, or have been affected or prepared by this. I'd always been very good. I'd been a straight-A student in school, that sort of thing. Somehow, I was having trouble in my physics course. It was a rapid physics course, it was a two-year physics course for the future majors. And I think this was in my freshman year, I didn't do very well on the tests; in the middle of my first semester I had a C average. And I went to talk to the professor and said "Can anybody ever be a physicist with this?" and he encouraged me to give it a try, he didn't say "no" or anything like that, he didn't say "of course" either. But he told me what I should do is prepare myself better for the exams, and I started really learning how to do problems. We called them problems, I call them puzzles. And I got an A-minus in midyear and As from there on. But boy, it was perplexing not to be getting my A. I mean that had also a considerable influence somewhere along the line. So, here I was, in my first year I declared myself as a physics major. Of course the next year . . .

K. GAVROGLU: Before going to the next year, did you contemplate applying to another place, or was Harvard more or less given because of your father being already there?

T. KUHN: I think the answer is that Harvard was the place I wanted to go, various uncles and so forth had been there, what I knew of it I liked, I'd visited and liked it. I'm sure I applied to at least one or two other places to be sure that I would get in somewhere, and I remember I was away visiting when acceptance came in from Harvard and I was terribly pleased.

Years later, a study was published in which my class was a baseline. I had been something like one of 1,016 people admitted out of a total number of eligible applicants of 1,024. I mean, it got to be terribly competitive later. I thought it was terribly competitive then, but it really wasn't. That was a myth. Now, my freshman year was '40–'41. In the fall of my sophomore year was Pearl Harbor; and we all sort of prepared ourselves to go to war or whatever. And in my case, since I was a physics major, the Physics Department turned itself largely into training people in electronics, and I had much more concentration in physics, but somewhat oddly distributed physics. There were some courses that every normal physicist would have taken that I didn't get till graduate school and not very fully then. And I turned myself to doing that as rapidly as possible. I got out in three years instead of four by going to summer school two summers. The result was that on the whole I got less of other subjects beyond science than I might have and that showed up in two ways in particular. In my freshman year I took another course that deeply influenced me, which

I should say something about. I decided I wanted to take a philosophy course. I didn't know what philosophy was, but I had a strange uncle who was a Spinozist. He was a private Spinozist. He collected, he wrote about Spinoza. The family did not for the most part like him, but I sort of liked him. And I wanted to find out about philosophy. There were some philosophy courses regularly open to freshmen and there was another course called History of Philosophy, which of course wasn't history, but it was detailed studies of, in the fall, Aristotle and Plato and in the spring, I think, Descartes, Spinoza, Hume, and Kant. And they gave me a hard time about getting into it, but I sort of insisted and I again had a hard time with this course, at the beginning; and I think that was partly not my fault. I had a very odd, although very well known section man, not the person who lectured. The person who lectured was a Greek, Raphael Demos, who's written about the scientists of the Greek Enlightenment. In any case, I had a section man, Isenberg; and he kept knocking me down, I mean, I didn't do very well on the quizzes, but I kept asking questions. One of them in particular that got to me terribly—he was teaching Plato and Plato's idea of the good. And now I'm not even sure that I will remember exactly which Platonic doctrine I am quarreling about, but it was the notion which got illustrated by the person who was a candidate from medical school and wanted to be a doctor. But the night before the exam he went out and stayed up too late and didn't do well on his exams and if he had thought about that and evaluated, he would not have done that and would have gotten into medical school. And I said, "Look, suppose he would have improved his *chances* of getting into medical school surely, but suppose they still weren't going to be very good statistically and suppose he really wanted to go out to that movie that night, why would it be irrational (although that was not the relevant word) for him to do that?" And the instructor treated this as a terribly strange question. He didn't understand it. I asked the same thing next week and he sort of led the class in laughing at me. It was an absurd episode. I mean, I know you are physicists, you see what the question was, but it was not a stupid question. And I guess in the middle of the year I got only a B− in that course. Then he left to go somewhere else and the course was taken over for the spring by another man. I said [just now that] I hadn't been doing very well: that was the course in which I really learned something more about how to study. I went back and made detailed notes and really pinned myself to it and I started doing better in the quizzes. And [then]

I asked to be in the honors section [of that course] and I was allowed to. I was quite fascinated by this stuff although I didn't understand it terribly well. But that was Descartes, Spinoza, Hume, and Kant. Spinoza didn't hit me very hard, Descartes and Hume were both in the current, I could understand them easily; Kant was a revelation. And I remember everybody gave a presentation in the section meeting, and I gave a presentation on Kant and the notion of preconditions for knowledge. Things that had to be the case because you wouldn't be able to know things otherwise. It [my presentation] was thought very well of, but it just knocked me over, that notion, and you can see why that's an important story.

K. GAVROGLU: Could you elaborate on it a little?

T. KUHN: Oh, it's an important story because I go round explaining my own position saying I am a Kantian with moveable categories. It's got what is no longer quite a Kantian a priori, but that experience surely prepared me for the Kantian synthetic a priori. And I do talk about the synthetic a priori. Well, I had thought I would take more philosophy while I was an undergraduate. I thought also that I'd take more literature as an undergraduate. I had not much liked this English literature survey that I took in my freshman year. I thought the professor was not treating us with very much respect—he was making jokes, not at us but at the things we were supposed to be studying. But I had a very good American literature course in my second year, and I would have gone on with that; I mean, I liked the study of literature and I wanted to do more philosophy. But here I was; we were at war, I was already well into my sophomore year, I was going to be there only for one year after that, so I didn't go on with either of those things. Then I graduated a year ahead of time and that was not uncommon at the time. One other thing that I did then that I should mention. I went out for the paper, the Harvard *Crimson*. In the beginning of my sophomore year I had a roommate who had already gone out in his freshman year and made the news staff; I went out for the editorial board, and I got to be the head of the editorial board in my last year. And that was the year we went into the war, so there I was writing editorials about our presence in the war and what Harvard should do and so forth. I even then had the problem that I have had ever since, of finding it very hard to write. So that it would always take me forever to write an editorial, and I never got that journalist's ability to sit down and turn something out. It was more than a little of a disadvantage.

K. GAVROGLU: Was it an elected position?

T. KUHN: Yes, it was an elected position and there was a competition for people who were interested. The senior board picked me as the candidate. I was part of the slate. But there was then a good deal of opposition to the slate from some of the members; and I remember sitting downstairs with the other person who had been nominated by some of the members, while the discussion went on upstairs, wondering who was going to get it—this was not unheard of, but it was not the usual thing either. And it tells you something about the extent to which I had only been partially socialized by my transition to Harvard and yes, I did get the job. But it was a strange experience.

A. BALTAS: You mentioned writing and you are considered one of the very few people in the business who have a sense of writing. And because you mentioned a difficulty in writing, this fits well with the picture I have of you because writing is something difficult. It's a good thing you did not succumb to the journalistic style. The question is your relation to writing.

T. KUHN: This is one my mother had a lot to do with. She did do some editing and she did read. As a kid, and still, I don't write letters to people, I write the business letters I have to write. I am a terrible correspondent, and I used to get in some difficulty about it. My mother once said to me, "You can say anything you like, but be very careful what you write down." My mother said a lot of things to me, not all of them wise, but all of them stuck. My mother was an extraordinarily tactless woman. She couldn't *not* say what was on her mind, and what was on her mind was not always very well thought out. I remember when I first started going with a woman; I had not had the normal number of dates by the time I was out of graduate school, and there was a woman I saw more than occasionally. My mother, who had not met the woman, saw the two of us on a New York street and just said to me a few days later, "I saw you and G . . . , and she's not the right person for you."Agh!

K. GAVROGLU: Before we get to the graduation. You are going through the undergraduate years when there was a war in Europe. And American society, at certain levels, is a torn society, there are very high emotions as to how America will deal with the situation. How do you relate to this situation, and how is the situation among the people at the university? Is this an issue among the people at the university? Obviously as time nears, it's an issue, Pearl Harbor is an issue, but what about before, what about your own position?

T. KUHN: Look, I'm surprised to find out how little I remember. I will tell you something about it. I indicated that I was quite radical through ninth grade at the school. We used to march in May Day parades, solidarity with labor. After I went away to school, that rather dropped away. I maintained my liberal convictions, but I was no longer an activist at all and I've never been an activist since. It embarrasses me sometimes [that I haven't]. So far as my own attitudes and my family's were concerned, we were very glad when Roosevelt made the arrangements to help the British; we rather thought that America should join the war; but that wasn't a happy feeling. Remember this is a Jewish family—not very Jewish, but I mean we were all certified Jews. Nonpracticing Jews. My mother's parents had been practitioners, not Orthodox practitioners. My father's parents had not, the Cincinnati branch of the family. So, it was not a big issue but it undoubtedly put us more ready to go after Hitler than we might otherwise have been. And the fact of the matter is, I'm sure I was around people who felt very differently, but I don't remember that at all. I feel as though all the people I was around felt more or less the same way. Then of course it became irrelevant after Pearl Harbor.

K. GAVROGLU: Were there no voices even among the students not to get into the war after Pearl Harbor?

T. KUHN: If there were I sure don't remember them, and I don't think there really were.

K. GAVROGLU: Could you tell us some things about courses at Harvard? Which courses did you like more, and some of the teachers at Harvard that had at the time one of the best physics departments, not the best I guess. Columbia maybe was better.

T. KUHN: Or Chicago. Harvard was not known for a particularly good physics department.

K. GAVROGLU: Not even at that time?

T. KUHN: No. I think Harvard physics began to get good only after the war. John van Vleck was there then, but I didn't study with van Vleck, I didn't know van Vleck until later. Wendell Furry was the person who I had for that freshman course I spoke to you about. I liked him, he was a good teacher. Street took the second year of that course. I never got to know him well, but he was a known physicist for spectroscopy.

K. GAVROGLU: You had nothing to do with Slater, he was down at MIT.

T. KUHN: Yes, I had nothing to do with him. He became a name to me and I met him during the Quantum Physics Project, but no, I had

nothing to do with him. I had more to do, before
who were teaching electronics. [Leon] Chaffee, an
[Ronald W. P. King] who was very good on ant
of these made a deep dent. Chaffee was an incret
King was a very good teacher. In mathematics I w
but I didn't quite dare go into the second-year calcu.
took the first-year calculus course and found it so eas
go to class, I used to do my problems and send them in wi
else in the course. I don't mean that I never went to class, ... i very
rarely went to class after the first few weeks, and I did fine. The second
year I moved a little bit ahead of my level in the second-year course.
I mean, I sort of skipped directly into the second half of the second-
year course. It was when I went into the third-year course that suddenly things were terribly hard for me. It was taught by George Birkhoff, a famous mathematician and one of the worst teachers you can
imagine. And we were doing multiple integrals and partial differentiation, and I couldn't quite see what was going on. I did all right, but
never really felt I was in control of the material. I had a friend who
was very good, and when we were nearly in the usual level the next
year I said to him, "How are you doing?" and he said good. I told
him I had a lot of trouble understanding the course, and he said, "How
could you have, it was just what we've all done before, only with
more variables." And I said, "Oh," and something clicked and it all
fell into place. And although I still don't always get multiple integrals
right, yes, he was right. That's what it was and Birkhoff kept me from
seeing it.

K. GAVROGLU: You didn't say much about theoretical courses.

T. KUHN: You've got to remember, I was there only three years. The
first and second year my main physics courses were this two-year
sequence, which is a tough sequence, and a good sequence. I can't
remember what else I may have taken, I don't remember exactly what
I took in physics in my third year. What I did do was to take a lot
of electronics and get physics credit for it. I took some electromagnetic
theory, I'm sure; I took a course in electricity and magnetism, which
was all right, I wasn't terribly excited by it. It was not yet really Maxwell: Page and Adams, is that a book, do you remember that?[2] It was
that level. I can't remember—I took a course in thermodynamics with

2. L. Page and N. I. Adams, Jr., *Principles of Electricity: An Intermediate Text in Electricity and Magnetism* (New York: Van Nostrand, 1931).

w.] Bridgman when I was in graduate school; I may have
~~~n a previous undergraduate course in thermodynamics, in which
case that would have been with Philipp Frank, but I'm not sure
whether I did or not. I've always rather liked thermodynamics; that
sense of a subject which is largely mathematical but which gives you
important physical consequences is a strange, tasteful experience.

K. GAVROGLU: What about relativity courses?

T. KUHN: I took a relativity course in graduate school. You have to
remember I had very little physics as an undergraduate. I didn't have
an optics course which I would ordinarily have had. I'm not sure I had
a thermodynamics course; I probably had an intermediate mechanics
course.

A. BALTAS: What about history courses, except the one you mentioned
of philosophy, humanities in general?

T. KUHN: I took one history course. History was not a thing I thought
of myself as being very fond of, it was a course in summer school in
British nineteenth-century history. Why I took that I don't know, the
teacher was well liked, and I liked him myself, but the subject matter
wasn't going to do anything for me. In summer school also I took a
course in political science with Max Lerner, who was also, in some
sense, an older friend of the family's. But I took very little, as you
see, that wasn't either electricity or electronics and I can't now tell
you just what it was. On the other hand, I was very busy with the
paper [the *Crimson*] and my friends were people who were on the
whole in literature or one thing or another. Mostly they were not
physicists, mathematicians, engineers, although there were some of
those. So, to my considerable surprise I was in my sophomore year
elected a member of something called the Signet Society, which was
not really one of the Harvard clubs, but it was sort of an intellectual
discussion society, which had lunches and so forth. And then in my
senior year I was president of it. And I was probably the only physicist
who was ever president of the Signet. Although I was terribly over-
concentrated, though I didn't take very many literature courses (I
guess I had only two years, one of English literature which I didn't
like much, and one of American literature with Matthiessen and Mur-
dock, two very famous Harvard professors whom I admired greatly)
I was known for this combination. And that is something to hold onto,
because it's going to play a role later.

And then I graduated and I went to work for something called the
Radio Research Laboratory. Radio Research Laboratory was physi-

cally at Harvard, in the north wing of the biology building including two extra wooden-structured floors built on top of the north wing. I was in the theoretical group and my boss was van Vleck. We were doing radar countermeasures. King was now there designing special antennas including one rotating antenna that flew on a plane and was supposed to be able to triangulate radar sites. I was largely cooking standard formulas (whose derivations I couldn't have begun to understand, or at least I didn't think I could, and I wasn't given time to find out) on doing radar profiles as a function of distance—there is a standard formula about the square root of the height of the two antennas and so forth with various allowances for propagation conditions and so on. I would produce graphs showing when you would pick up such and such a plane, [and] do maps. I think I did one set, or did some that had to do with radar coverage of Kamchatka. How close could the Japanese get or we get to the Japanese. I'm not even sure now which it was.

A. BALTAS: Was this a job, I mean were you hired there?

T. KUHN: Yes, I was hired there because of my degree, my training and because anybody with that sort of training was needed for the war. And it got me a deferment: I was not drafted under these circumstances. That's not why I was doing it, but I never had any regrets. I mean, it isn't that I wanted to be drafted and that they wouldn't let me be. I started that in the summer or fall of '43 and I was there about a year. They had an advanced base laboratory at Great Malvern, in England, and after about a year, I think, I asked to be sent over there; I had never been abroad, and I went there, and I must say it was my first plane flight. I got on a plane at LaGuardia, we landed once somewhere in Nova Scotia or Iceland, I think, and then came down in Scotland, in Glasgow. I had never been in a plane, and there we were! I kept reciting from Saint-Exupery's "Night Flight"—it was thrilling! And so then I was at Malvern for a while, and then sort of farmed out to a technical intelligence unit at the United States Strategic Air Force Headquarters which was in Bushy Park outside of London— I lived in London. And that was hard. I was having trouble, sort of fitting in and getting interested in what I was being asked to do. But it was all right and there was some fun involved.

Then, I went from there into uniform as a civilian, and went to France, or I went to the Continent—again for the first time I'd ever been there. I was in uniform so that if I were captured I would not be held as a spy. I went to look at radar sites and to bring back further information about them. That was one of the most fantastic experiences of my life. Because I got on this plane and landed at the base of the Cherbourg Peninsula. I was supposed to be going to the submarine pens at Rennes, where there was supposed to be a major German radar installation. Now, this was just during the breakthrough, with [General] Patton rushing across France, and nobody knew quite where the army was. But I was supposed to join a group that was already there. And I was in a command car with a captain who'd been in this group, he knew them, he was going to come there with me and then he was going on to—I forget where. We got there, the group wasn't there anymore. Nobody knew where they were, but the betting was pretty good that they had headed out for Paris and that we could find them there. Nobody was quite sure what was going on in Paris. But we started out and it then turned out with the driver and this captain and I, I was the only one who had ever studied any French. And I hadn't done that for a long time and I kept trying to remember my French. And I would say, "soldier-soldat," no that can't be right, that's German. So, we drove and we drove and I guess we stayed overnight somewhere on the way and then the next morning early we were up and driving again. And another one of those sights that I will absolutely never forget, we were driving across a plain and suddenly coming up over the horizon there was something like this—We were watching it, it was getting higher . . . Chartres! It was those two odd towers of Chartres Cathedral. We drove right in the town and never got out [of the car], kept on around, but wow, was that exciting! And as we came out of Chartres, we began to pass a convoy that was on the road and we got around the convoy, got out ahead of it, started into Paris, somehow we got to the Petit Palais which turned out to be the place where this group I was supposed to join was set up. They were found. And there we were. We'd been there for about an hour, when this convoy started to come down the Champs Elysées. It was de Gaulle entering Paris! And suddenly there was rifle fire from somebody on the roof of the building across the street. Somebody fired, some member of the milice who was shot down. And there was still fighting out at Le Bourget, the other side of Paris. So, it was an exciting time.

Oh, I'm telling you stories of my life, this is quite irrelevant to anything that happened to me later, except that it may tell you something further about me which will not surprise you.

I was working with, teamed up with, an RAF radar man; he was called Chris Palmer. He and I were told to go up the Eiffel Tower and see what sort of installations there were up there. So, we started out, we got to the Eiffel Tower and we talked to somebody there who told us, "L'ascenseur ne marche pas." We said, "Gee, we'll have to climb it" and they had wire wrapped around the stairs at the bottom level and we went up over the ladder—over the wire, and on to the stairs, and we climbed up to the third level—and here came the elevator. It stopped and invited us aboard. And God damn it, we got in. I've always kicked myself I've never climbed the Eiffel Tower!

K. GAVROGLU: Then you go back to England?

T. KUHN: I was there [in Paris] for a few weeks, and they were exciting weeks. I went back to England for a while and then I came back again [to France]. The transformation of the French, in the interim! I was stumbling over my French that first time, and people kept telling me how well I spoke French; and there was dancing in the streets, and so on and so forth . . . I got back there, not more than about six weeks later, I think, they wouldn't talk to you! Just a total change. And at that point I was assigned to the Ninth Bomb Division in Rheims as an advisor on radar countermeasures. There was a . . . what did you call it . . . industrial engineering group?—that's not quite what it was called. These people who applied mathematics and science to strategic and other such problems in a rather unsystematic way. At that point it was often called the main advantage of having such things that they could talk back to the general, as nobody within the army structure really could. So, I did that and then later still, as we went on into Germany, I was sent on again looking at radar installations, trying to talk to people in Germany and find out what had been going on there. Of course I didn't find out very much; but I saw the flattened city of Hamburg, I'll never forget it. I also saw Saint Lô the day we arrived in France and I'll never forget it. None of this has a lot to do with what happened later, except that as all of this was going on I increasingly realized that I was not all that interested in radar work. This gave me a somewhat bad taste of what it was going to be like to be a physicist. It was of course totally misleading. A number of my classmates, in somewhat similar positions, instead of going on to radar and radar countermeasures, wound up at Los Alamos. And I don't think it's out of the question that if I had gone to Los Alamos I might still be in physics. I doubt it, I mean I think there were too many other factors involved, but certainly an increasing distaste— which is already too strong a word—but an increasing number of doubts as to whether this was for me began to pile up. I will find it very hard to weigh the various factors that went into this, that went into my decisions, but that was certainly a part of it.

A. BALTAS: Let's talk about your starting doubts about doing physics. Were the doubts related to the war?

T. KUHN: I had been a "physicist." I put that now in quotation marks, because in some sense I wasn't trained to be a physicist in view of what had happened, but it was leading into this, and I was finding it fairly dull, the work was not interesting. I still believed in science and I remember there was somebody with whom I used to talk about the need for the redesign of science instruction, and one thing or another. But I was by no means sure, I was beginning to get doubts as to whether a career in physics was what I really wanted—particularly in theoretical physics. And I guess it may well be at that time, though it may be later, that this question, why did I insist on being a theoretician? began to arise. But those doubts weren't all that big, or anything of the sort, but they were there. I mean, I had been frustrated by not having gotten back to do some philosophy. So, in 1945, shortly after VE Day, I returned to the States, and there I was back at Harvard again, because the war wasn't over; there was some question at that time, whether we were going to be shipped on out to Japan. But it wasn't so long before that was over too. Meanwhile, in that intervening time (I mean, I guess I came back in late spring–early summer of '45, I think) I went back into the lab, and when the fall came, the fighting was still going on in Japan, though it looked as though it would not be going on forever, and I got permission—because work was slowing down at the lab—to sign up for a physics course, or for two physics courses, while I was still employed by the laboratory. One of the courses that I then took was group theory with van Vleck. And I found that somewhat confusing. I had a first course in quantum theory while I was an undergraduate—I have to have had. Group theory is interesting stuff, and although I never felt that I was controlling it, that sense of mathematics giving physical results was very appealing. Van Vleck was not a terribly good teacher.

Then the war was over in Europe [and I signed up to do graduate physics at Harvard]. I can't remember quite when I made this deal. Because I had taken so much physics as an undergraduate, because I was back at Harvard where I hadn't planned to go to graduate school—but it would have been silly not to, given the continuity— it would have cost me at least a year to pick myself up and go somewhere else. I petitioned the department to let me use half of my first year outside of physics, to explore other possibilities. And it was philosophy in particular that I had in mind, and I took a couple of philosophy courses. Well, so here I was first year in graduate school, I had permission to take half of my courses in philosophy. Well, I did that for a semester, probably the fall semester of '45, and unfortunately it was the case that most of the good people who would have interested me, in philosophy, were somewhere else. They hadn't gotten back yet. I took two courses, and I realized that there was just a lot of philosophy I hadn't been taught, and didn't understand, and was not finding it very palatable to pick up this way. I didn't know quite why people were doing the things that they were doing. And I fairly rapidly decided, yes, I was interested in philosophy, but my God, I was a graduate, I had been through war in some sense or other, I couldn't go back and sit still for that undergraduate chicken-shit and go on from there. So, I decided I'm going to take my degree in physics. But it also was clear, and became increasingly clear, that I was not being very much fulfilled by my graduate physics teaching. And that was not as distinct from the undergraduate. Partly, I think, although I continued to do well enough, I was no whiz kid any longer, and it wasn't clear that I was good enough to do it . . . I mean, to really shine. I surely could have been a professional physicist . . . In retrospect I think I was wrong. In retrospect, in view of what I've learned putting myself through stuff as a historian of science and learning more about what the career was like, I think I would have been a damn good physicist; I do not think I would have been Julian Schwinger or you know that first level, but I think I could have done quite respectable work. Whether I'd have liked it much I don't know. But certainly questions about my ability had something to do with a growing disenchantment. A sense of not being focused—there was this whole thing about having been president of the Signet, and liking literature and philosophy—and now, if I was going to go through graduate school successfully and on, I really had to focus my attention in one spot and give my full energies to it; and I found that hard to do. So, my grades remained entirely respectable but I think I began to get some Bs, that sort of thing, and I was very much of two minds; partly it was simply I didn't know what I would do if I didn't do physics. I was looking and thinking about other things, none of which turned me on all that much. I used to talk to my father about this—science journalism or something else of the sort. And then, of course, I had this extraordinary experience which I've talked about, of being asked by Conant to assist in his course. Who the hell wouldn't have taken the chance to work with Conant for a semester?

K. GAVROGLU: Before we get into Conant, did you have any people from physics encouraging you for your searches, as it were, in philosophy, or any other place outside physics?

T. KUHN: No. They gave me permission and they understood, so they knew I had something less than the ultimate in dedication; but I came back after one semester and so forth.

A. BALTAS: I'd like to ask you a rather strange question. Did you have when you started, you decided to do physics, in graduate school or afterwards, was there any kind of utopian dream, in the sense of "I discover the secrets of nature," "I do this for something big independently if I manage to do it or not"? Or was it just related to conditions, to work?

T. KUHN: No, I think initially—you know I would have been very glad to get the Nobel Prize—I certainly wanted fame in some sense. I don't remember it that way, but it has to be the case.

K. GAVROGLU: You studied solid-state physics with van Vleck, which obviously was not one of the trendiest things to do. So, were you interested in the subject itself or in working with van Vleck?

T. KUHN: It was neither. By the time I decided on a thesis topic, I was quite certain that I was not going to take a career in physics, and I didn't want to prolong my time in graduate school. Otherwise I would have shot for a chance to work with Julian Schwinger, but there were a lot of things I didn't know, and would have had to study, if I were to do it that way. I wanted the degree—it would have been stupid to have gone that far and not to get the walking papers that would result. But I didn't want to do the amount of extra training. Now, in practice that degree with van Vleck took a long time, I spent a lot of time punching buttons on a calculating machine. But it was that that motivated the decision.

K. GAVROGLU: And that decision also led you to van Vleck . . .

T. KUHN: Yes, I mean, I liked van Vleck, but he's not the person I would have worked with if I had not wanted to finish this off. I think it would have been Schwinger or I would have tried to be with Schwinger.

K. GAVROGLU: Schwinger at the time was what? A young, bright professor doing things you considered fundamental?

T. KUHN: I first saw Schwinger while I was still in the Radio Research Laboratory. We used to go down to MIT to the Radar Laboratory

for lectures sometimes. He gave a lecture on integral calculus of variations and wave guide computations. I didn't really get it all and I
wasn't much interested in doing wave-guide calculations, but there
was a degree of elegance in the presentation, and a degree of control
over deeply technical material, that was just fascinating to watch. And
then I think I had electromagnetic theory with him and maybe audited
quantum theory courses or something like that afterwards. And he
was a phenomenon, no question.

K. GAVROGLU: Okay, Conant.

T. KUHN: I was asked by Conant.

K. GAVROGLU: How did Conant find you?

T. KUHN: Remember, I had been an undergraduate editorial chairman
of the *Crimson* paper at a time when we were entering the war. So,
I had gotten to know not Conant, who was away all the time, but the
Dean of the Faculty. And when the general education report came
out, which Conant had set up to have done for him, I was asked by
the Dean of the Faculty to write a précis of the report for the alumni
bulletin; and also I was one of several people writing comments on
the report; I was the student among those writing comments on the
report. So, I was known for this range of interests. Who fed my name
to Conant I'm not sure—there are various people that it could have
been. But I had a reputation as the physicist who was president of the
Signet Society, there were various things of that sort in my record. I
was one of the two people Conant then asked to assist him. First time
he gave this course out of that little book called *On Understanding
Science*, which had been the Terry Lectures at Yale. I accepted with
alacrity; and I've never quite forgotten that first time I met him. Here
I was, not finished with my physics thesis and being immune to this
sort of material—I have by then read the page proofs of *Understanding
Science*—being asked to go out and do a case study on history of
mechanics for this course? Wow! And that was also Conant; he would
do that sort of thing. So, that was the first time—I had sat in on some
lectures of Sarton's as an undergraduate and found them turgid and
dull. And I was not in my bones a historian; and I *was* interested in
philosophy. But I had no real interest in history, and this Aristotle
experience[3] was terribly important. Conant, in case histories of his
own and in his teaching, never I think saw to the extent that I did
the need to say what people had believed *before*. He would always

3. See essay 1, "What Are Scientific Revolutions?" in this volume.

start in more or less with the beginning of the work. There would be something about it, but there was very little preparation for getting to the person. I always felt you had to do more; and that meant you had to do a stage set, within another conceptual framework, in order to get at these things. And that was what this did for me. But the main thing is, it didn't really get me *interested* in history of science; and there are those who feel, and feel with some justice, that I never really did get to be a historian. I think in the end I did get to be a historian, but of a rather special narrow sort. I used to think—forgive me—that with the possible exception of Koyré, and maybe not with the exception of Koyré, I could read texts, get inside the heads of the people who wrote them, better than anybody else in the world. I loved doing that. I took real pride and satisfaction in doing it. So, being a historian of *that* sort was something I was quite willing to be and got a lot of kicks out of being, and did my best to teach other people to do. I'll come back to that. But my objectives in this, throughout, were to make philosophy out of it. I mean, I was perfectly willing to do the history, I needed to prepare myself more. I wasn't going to go back and try to be a philosopher, learn to do philosophy; and if I had, I'd have never been able to write that book! But my ambitions were always philosophical. And I thought of *Structure,* when I got to it finally, as being a book for philosophers. And, boy, did I get fooled for quite a long time!

K. GAVROGLU: So you started preparing for the course.

T. KUHN: So I started preparing for the course, that got me to the reading of Aristotle, and I taught with Conant in the course for one semester. At the end of that time I knew what I wanted to do. I wanted to teach myself enough history of science to establish myself there in order to do the philosophy. I went to Conant—I had developed a relatively warm relationship with him, to the extent that people had warm relationships with him. He was a quite reserved person, rather cold—not so much cold, he was very reserved. I asked him whether he was willing to sponsor me for the Society of Fellows. Officially that's a question you didn't ask; but I felt able to, and he did, and I got in. I had to postpone my entry a little while I finished my thesis. And in some sense or other, after that I never looked back.

K. GAVROGLU: Before that . . . Of course the atomic bomb had already been dropped in Japan. What were your feelings and of the people that you were directly connected with, people on the verge of Los Alamos? Did you have any relation to Los Alamos?

T. KUHN: I basically had no relation to Los Alamos although I did have some relations to a few people who had some relation. These were high-level, but not necessarily very old, government consultants who flew around and touched base on a lot of these things. And one of them told me about the Los Alamos project. And indeed the context in which he did that was when the V2s began to come down in England, the fear had been that they were carrying atomic warheads. Of course, they weren't, the Germans weren't ready, they were not going to be ready in anything like that time, but that was the fear I didn't know about—that people had. And a man—I think his name was David Griggs, he was a knowledgeable figure—told me about the atomic bomb. So I remember I was in a train going to Washington, to go I guess to the Naval Research Laboratory for some tests, or something of the sort, and on the platform of the Pennsylvania Station in New York I looked out and saw this headline on the paper. I knew what it had to be, and it was the atomic bomb. I guess I would say, yes, I knew that there were people who felt that we should not simply have dropped it, that we should have demonstrated it, but the general feeling was: Look, we had to get out of this. And I was sympathetic with those who felt that maybe we should have taken another technique. But I didn't know enough about it to really feel any great convictions on that score, or any great sense that it would have worked; and probably it wouldn't have worked. So I'm not one of those who has been terribly upset by the behavior of the government. I don't know that I knew anybody who was deeply, deeply upset, although there were plenty of people who admired that group and agreed with them and wished they had been able to do more than they had. And I guess I would have associated myself with them, but it wasn't a big issue for me, I mean I could suppose that the time had come to get this over with.

A. BALTAS: I'd like to come back to what you said before. What did philosophy mean at that time for you?

T. KUHN: I will tell you a story. I had a classmate, also on the *Crimson* at a time when I guess I was in graduate school, I think he was also taking a graduate degree. He got married and to my surprise he asked me to be one of the ushers at his wedding—I'd never been an usher at a wedding, but I was. I met his bride, of whom I got very fond, but I met this woman—G . . . , the one my mother told me wasn't for me—she was a bridesmaid and it was that that established the relation between us. After a while she gave a cocktail party in New York for me, to meet some of her friends. And I went, and I got to talking to a very beautiful—not so much beautiful but very striking, buxom, well-turned-out woman. I don't know what the conversation was, but suddenly as happens occasionally all the voices in the room dropped and I was heard saying (including by me), "I just want to know what Truth is!" So, that's what it meant to me. And this may well be before I was associated with Conant. I can't date it quite that accurately, it certainly can't be long after. I may already have been in the Society of Fellows, but I think perhaps not.

A. BALTAS: It is very well connected to the Aristotle incident. They connect very well together.

T. KUHN: Yes, and it could have happened either before or after. My Aristotle experience certainly made it problematic, and I'm not sure quite what the problem had been earlier, if this was before that. So I really can't give it to you in terms that will be developmental. But from an early stage, that tells you something; I don't mean that that was the single goal, but that's what it meant to me to be studying philosophy or to have philosophical ambitions, that was the type of thing it meant to me.

K. GAVROGLU: It is not uncommon for a lot of people to start to search for truth through physics first and then going to philosophy as a next stage.

T. KUHN: But remember, when I said that, I wasn't saying that I want to know what is true; I was saying I want to know what it is to *be* true. And that's not something that one gets to through physics.

K. GAVROGLU: No, no. You are right.

T. KUHN: We stopped [the previous tape] just as I had decided that I was now going to move into history of science with an eye to doing something philosophical with it and I asked President Conant to recommend me for the Society of Fellows. He did recommend me, I was elected to the Society of Fellows, that is a three-year term, of which I didn't really, as it turned out, get all of the three years. I had to take some time at the beginning to finish my thesis and to publish some articles out of it, at least to finish the thesis. But in November of 1948, I guess, I began to work in the Society of Fellows. It was terribly important to be there, because it relieved me of other responsibilities and what I was trying to do was to train myself to be a historian of science. And it was partly just reading, and I didn't mostly read history of science. It was in those years—I mean, I don't remember what got me to it, I guess in reading Merton's thesis[4]—somehow or other, I think it was there that I discovered Piaget. And I read a good deal, beginning with his *Mouvement et vitesse*.[5] And I kept thinking, my, these children develop ideas just the way scientists do, except— and this was something I felt Piaget did not himself sufficiently understand, and I'm not sure that I realized it early—they are being taught, they are being socialized, this is not spontaneous learning, but learning what it is that is already in place. And that was important.

A. BALTAS: Can you tell us a few words about the Society itself?

T. KUHN: The Society itself in those days, and it's still more or less the case, I'm not sure how much it has changed, was a group of twenty-four fellows, usually, eight elected each year. And a group of senior fellows who did the election. The whole group dined together every Monday evening. And the dinners were quite good dinners, so that there was a certain element of ceremony as well as of sociability. The fellows also lunched together I think twice a week and that was less ceremonial but it brought them together and the amount of interaction varied quite a lot. I don't even quite remember the people who were part of my group, but I don't think there was anybody that I talked to in the Society who was terribly important to my development, although the talk was good and I got some sense of support and so forth. A senior fellow at that time was Van Quine. And this was just—I don't remember the dates—this was just about the time that his analytic-synthetic paper was coming out.[6] And that as I said the other day had a considerable impact on me because I was wrestling already with the problem of meaning, and at least to discover that I didn't have to be looking for necessary and sufficient conditions was extremely important. Quine has been important to me for that piece, and for the problems that *Word and Object*[7] presented me in trying to figure out why I was so sure it was wrong (outside the fact that

4. R. K. Merton, "Science, Technology, and Society in Seventeenth Century England," *Osiris* 4 (1938): 360–632; reprinted with a new introduction by the author (New York: Harper & Row, 1970).

5. J. Piaget, *Les notions de mouvement et de vitesse chez l'enfant* (Paris: Presses Universitaires de France, 1946).

6. W. V. O. Quine, "Two Dogmas of Empiricism" (1951). Reprinted in *From a Logical Point of View: 9 Logico-Philosophical Essays* (Cambridge, MA: Harvard University Press, 1953).

7. W. V. O. Quine, *Word and Object* (Cambridge, MA: Technology Press of the Massachusetts Institute of Technology, 1960).

there isn't much of an argument), where he was going off the rails. We can come back to that later. I've only really been able quite recently to formulate it, in a way that I find satisfactory. But, for those three years in the Society I was beginning to read my way into the field and establish myself; and also doing something else, which I think I *should* put on the record. I've said something yesterday about . . . until I got to Harvard not having had many friends; I was clearly a neurotic, insecure young man. It was also the case that somehow or other my parents, my mother I think in particular, worried about this: I was not having dates and that sort of thing. My relations with women were almost nonexistent. But that was in some part because my environment was a male environment. The result was that I was persuaded, without a lot of difficulty, to go into psychoanalysis. I'd had some experience as a child with child psychiatry which I did not think very much of and don't carry fond memories of. The analysis in the Harvard years was with a man I, in retrospect, hate, because I think he behaved extremely irresponsibly with me. He used to fall asleep and then when I would catch him snoring he would act as though I had no business being at all angry or upset about it. On the other hand, I'd previously read Freud's *Psychopathology of Everyday Life*. I do not for a moment like the theoretical categories that he introduces, or feel that for me, at least, they have any force. The *technique* of understanding people and enabling them to understand themselves better—I'm not sure that it produces real therapy of any sort—but it sure as hell is interesting. And I think myself, I'd have great trouble documenting this, but I think myself that a lot of what I started doing as a historian, or the level of my ability to do it—"to climb into other people's heads," is a phrase I used then and now—came out of my experience in psycho-analysis. So in that sense I think I owe it a tremendous debt. I think it's too bad that it is getting the very bad reputation that it's getting these days, although I think it richly earned it; but I think what gets forgotten is that there is a craft, hands-on aspect to it, that I know no other route to, and that is intellectually of vast interest.

The psychoanalysis must have been mostly before I got into the Society of Fellows, because it terminated when two things happened: I got married and my psychoanalyst moved out of town. At that point I finished my thesis which was typed by my then wife. That was a marriage that went on for just about thirty years, which produced three lovely children, whom I find immensely rewarding.

I think I produced nothing while I was in the Society; I did a good deal of reading. And also of course, as I said, in the first year, I missed the beginning because of the need to finish my thesis. The second year I had unencumbered. And then, in the third year, Conant decided to stop giving the course and he invited Leonard Nash, a chemist and famous teacher, and me to take it over. Leonard Nash I had not known before. It was good for me, I could not have refused and at that point, knowing that I was going to have very little time the next year, my wife and I went off to Europe. It was not uncommon for members of the Society to spend a last year in Europe to do their research. We went on a two-month trip to meet colleagues abroad; I wasn't even ready for them, because I hadn't got far enough into the history of science. But we were in England for a bit and in France for a bit. I don't think we went further in Europe than that.

K. GAVROGLU: Let me ask you something. You told us that what you really felt was challenging was philosophy and then you got into the Society of Fellows and you were up to your neck with the history of science. Obviously that had something to do with the course that you were teaching, of course, but was it only that?

T. KUHN: I had made that attempt to investigate going into philosophy immediately after the war when I first came back and got into graduate school and I decided I wasn't going to go back to fulfill undergraduate philosophy. And in certain respects I'm extremely glad I didn't, because I would have been taught things that would have given me a cast of mind which would have, in many ways, helped me as a philosopher, but they'd have made me into a different sort of philosopher. So I had decided, when I applied to the Society, to do history of science. My notion was, and my application indicated, that there was important philosophy to come out of it; but I needed first to learn more history, to do more history, and to establish myself professionally as a historian before I let the cat out of the bag.

K. GAVROGLU: And what were your relations with the Department of History of Science at Harvard, which was a well-established department?

T. KUHN: No, it wasn't. Look, at this time there was no Department of History of Science at Harvard.

K. GAVROGLU: What about Sarton and his group?

T. KUHN: There really wasn't a group; I mean Sarton drove people away who wanted to study with him. He would tell them, "Sure, but you'll have to learn Arabic, and Latin, and Greek" and so forth, and very few people were going to do that.

K. GAVROGLU: Why didn't you associate yourself directly with Sarton then, since you wanted to do history of science?

T. KUHN: Look. My notion was that there was a sort of history of science to do that Sarton wasn't doing. I mean, I would not have said then the sorts of things I would say now about him, and I recognize that in some very important sense he was a great man, but he certainly was a Whig historian and he certainly saw science as the greatest human achievement and the model for everything else. And it wasn't that I thought that it was *not* a great human achievement, but I saw it as one among several. I could have learned a lot of data from Sarton but I wouldn't have learned any of the sorts of things I wanted to explore. Anybody who took a degree in history of science at that time, he went to talk to Sarton and got it that way, there was no program; and that was not a way for me to do it. Look, when I fairly shortly joined the History of Science Society, there were perhaps then fewer than half a dozen people in the United States—I've written about this somewhere—who were employed in order to teach history of science. There were a number of other people who taught it within one or another of the science departments. But what they taught often was not quite history—in my terms, at least, not quite history; it was textbook history. I have sometimes said that some of the greatest problems that I've had in my career are with scientists who think they are interested in history.

A. BALTAS: There's a phrase in the *Structure*, "If history is viewed as a repository for more than anecdote or chronology . . ." Could you comment on it?

T. KUHN: Yes—of course anecdote and chronology were done by people who were not scientists as well as by scientists. But the stuff I did was, potentially at least, subversive of things that were for very good reasons part of the ideology of scientists. I mean, I'm saying things at this point that I've only gradually learned as I've tried to understand my situation over the years. And by and large my relations with scientists (until the Planck book came out[8]) have with a few exceptions been very cordial; and from a number of them, including a number of physicists, *Structure* got a very good hearing. It was of course not widely read by scientists. I used to say that if you go through college in science and mathematics you may very well get your bachelor's

---

8. T. S. Kuhn, *Black-Body Theory and the Quantum Discontinuity 1894–1912* (1978; reprint, Chicago: University of Chicago Press, 1987).

degree without having been exposed to the *Structure of Scientific Revolutions*. If you go through college in *any* other field you will read it at least once. That was not altogether what I had wanted.

K. GAVROGLU: You said reading Merton's thesis was a relatively important experience.

T. KUHN: I got the reference to Piaget there, and that was important. There are just a few things like this . . . It was I think in Reichenbach's *Experience and Prediction*[9] that I found a reference to a book called *Entstehung und Entwicklung einer wissenschaftlichen Tatsache*.[10] I said, my God, if somebody wrote a book with that title—I have to read it! These are not things that are supposed to have . . . they may have an *Entstehung* but they are not supposed to have an *Entwicklung*. I don't think I *learned* much from reading that book, I might have learned more if the Polish German hadn't been so very difficult. But I certainly got a lot of important reinforcement. There was somebody who was, in a number of respects, thinking about things the way I was, thinking about the historical material the way I was. I never felt at all comfortable and I still don't with [Fleck's] "thought collective." It was clear it was a group, since it was collective, but [Fleck's] model [for it] was the mind and the individual. I just was bothered by it, I could not make use of it. I could not put myself into it and found it somewhat repugnant. That helped me keep it somewhat at arm's length, but it was very important that I read that book because it made me feel, all right, I'm not the only one who's seeing things this way.

K. GAVROGLU: Did you have any rapport with any of the other historians of science either by correspondence or at least intellectually? I include Europeans as well as Americans.

T. KUHN: While I was still in the Society of Fellows I didn't know any others. I'd met Sarton, I knew Bernard Cohen; Bernard has done a lot of good for the history of science but he is not someone who thinks about development at all in the way I do. We've not seen eye to eye. In any case, in the third year I started out with Nash offering a general education course: *Science for the Nonscientist*. It was a strange experience and certain things that happened to me that year I think have made a lot of difference [to me] since. Lots of people had come to it when Conant was giving it; they wanted to hear the president of the

---

9. H. Reichenbach, *Experience and Prediction* (Chicago: University of Chicago Press, 1938).
10. L. Fleck, *Entstehung und Entwicklung einer wissenschaftlichen Tatsache* (1939), reprinted as *Genesis and Development of a Scientific Fact*, ed. T. J. Trenn and R. K. Merton; trans. F. Bradley and T. J. Trenn; foreword by T. S. Kuhn (Chicago: University of Chicago Press, 1979).

university in action. I do not myself think, but it would be hard to be sure, that they got an awful lot out of it. I mean they had an experience which was of some importance, because they listened to things that the president—a very very bright man—wanted to say to them. But I don't think they had deep intellectual involvement with it. Nash and I wanted to increase the intellectual involvement. But what happened was the enrollment at the course, of course, immediately plummeted, though not to a terribly low level, and we realized suddenly that we were not getting across to the students—they were not really seeing or understanding what we were trying to do—or rather to most of them. There was a sort of top cream of students in that course, that got excited, in ways I love to get students excited, and they remember it and they talk about it still. But most of the people were sitting there like a lump. That was when teaching began to be difficult for me. I had taught in Conant's course when I first gave it, I had given this case history. I had given lectures on some other things also, it had always been easy: I wrote a few notes and went in the class and taught. And I didn't do that badly. Now I started spending much too much time preparing, getting very nervous in advance, and I've never altogether gotten over that. I mean I've never regained the original freedom to just go in with rough notes—knowing I knew the stuff—and start talking. Which has cost me some things; I think probably including some facility and sitting down and writing easily, although writing was always different from talking for me, as I said.

Just before that last year as I said, my wife and I had gone to England, and there I met people, the group particularly at University College, which was then one of the two places in the world where there was a history of science program, the University of Wisconsin really being the other. And we went to France. At some point I had already met Koyré—I guess he'd been some in the United States. My French wasn't good, the French were not all that hospitable, but [Koyré] gave me a letter to Bachelard, and said I should definitely see Bachelard. I delivered the note, was invited to come over, climbed the stairs. The only thing of his I'd read was that *Esquisse d'une Probleme Physique*,[11] I think it's called. But I'd heard he did brilliant work on American literature, and on Blake and other things of the sort; I as-

---

11. *La Philosophie du non: Essai d'une philosophie du nouvel esprit scientifique* (Paris: Presses Universitaires de France, 1940).

sumed he would greet me and be willing to talk in English. A large burly man in his undershirt came to the door, invited me in; I said, "My French is bad, may we talk English?" No, he made me talk French. Well, this all didn't last very long. It is perhaps a pity, because although I think I have read a bit more of the relevant material since, and have real reservations about it, nevertheless he was a figure who was seeing at least some of the thing. He was trying to put it in too much of a constraining . . . He had categories, and methodological categories, and moved the thing up an escalator too systematically for me. But there were things to be discovered there that I did not discover, or did not discover in that way. The English connections established then really were somewhat with the University College group. Mary Hesse, and Alistair Crombie, I met Mackie, I met Heathcote, I met Armytage. But I had more to do with Mary Hesse obviously, somewhat more with Alistair Crombie, and in France really with nobody, except Koyré of course who was not in France during this trip. I came back to the U.S. then, late in the summer. That was the year that America went to war in Korea. And all of the planes were called up for military purposes. We had a hell of a time getting back. But I had to get back and start teaching. So that was that summer.

K. GAVROGLU: Concerning Koyré and Mary Hesse in particular, do you remember some of the things you discussed?

T. KUHN: Look, I realize I've left out something extremely important. When Conant asked me to do this case, which is my first work in history of science in a sense, I started by reading Aristotle to find out what the beliefs had been *before*. And very shortly after that, and it was at Bernard Cohen's suggestion, I picked up Koyré's *Etudes Galiléennes*,[12] I loved them. I mean, this was showing me a way to do things, but I just hadn't imagined it was there. In a sense it wasn't quite so strange as it might have been, because I had read and admired a good deal Lovejoy's *Great Chain of Being*.[13] But that you could do that with *science* had not quite occurred to me, and this is what Koyré was in some sense showing me. And that was important. I liked Mary Hesse and we talked some. The thing I remember best in my interactions with Mary Hesse—and this is of course much too early in this story to tell this—after *Structure* came out she wrote a very nice review

---

12. A. Koyré, *Etudes Galiléennes* (Paris: Hermann, 1939–1940).
13. A. O. Lovejoy, *The Great Chain of Being* (Cambridge, MA: Harvard University Press, 1936).

of it in *Isis,* a very favorable review. When I next saw her we were in England, and I remember walking with her and going into the Whipple Museum—it's another one of those imprinted images. She turned to me and she said, "Tom, the one problem is now you've got to say in what sense science is empirical"—or what difference observation makes. And I practically fell over; of course she was right but I wasn't seeing it that way. Another story out of sequence I don't want to forget: shortly before Alexandre Koyré died—which is now a good many years later, he died shortly after *Structure* came out— I had a last letter from him. We had not really corresponded much, but he wrote me—he was sick and known to be probably dying at that time. He said, "I've been reading your book," and I don't know what adjective he used, but it was a thoroughly agreeable one. He said, and again I had not seen this coming—when I thought about it, I thought he was right—he said, "you have brought the internal and external histories of science, which in the past have been very far apart, together." Now, I hadn't thought of that at all as what I was doing. I saw what he meant, and coming from him it was particularly agreeable because he had been so anti-external history; his gifts were as an analyst of ideas. And that made an impression, or at least it pleased me tremendously.

A. BALTAS: Could you tell us how you met him?

T. KUHN: I didn't see him much in person. I met him through the *Etudes Galiléennes;* then he was in the United States and I think he was visiting Harvard and I think Bernard probably introduced me to him. I saw him from time to time, but never at all closely, never on a continuing basis of interaction. So it was not the personal interactions that made a difference.

After this trip to Europe, I came back, I started to teach the Conant course, which I've already talked about, and one of the things that happened, although it was not the first thing that happened, was that Karl Popper gave the James Lectures, I think they were, at Harvard. I had had reason to think I was going to like these and I was clearly interested in it. I was introduced to Popper at a fairly early stage and we saw a little bit of each other. Popper was constantly talking about how the later theories *embrace* the earlier theories, and I thought that was not just going to work out quite that way. It was too positivist for me. But Popper did me a tremendous favor. This is another example of getting the books that meant something to me from people I would not have expected. He sent me to Emile Meyerson's *Identity and Reality*.[14] I didn't like the philosophy at all. But, boy, did I like the sorts of things he saw in historical material. He went into those briefly and I mean he didn't do it as a historian but he was getting it right in ways that were different from the ways that history of science was being written. Somebody else I discovered during the trip to France—I had not been aware of her before, and [by then] she was no longer around—whose work I thought extremely well of and it was of some importance to me was Hélène Metzger. Another person whose work was of some importance to me—[although] I did not read [her work] that much, and I never met her—was a medievalist working in Rome out of the Vatican, Anneliese Maier. It's hard to say which of these were important, but these were works in history that I admired. A sort of history, and an approach to history that I admired and which I encountered fairly early.

K. GAVROGLU: My mind is hooked on something that you said before. It's not the time now, but it might as well be put . . . You said that Koyré sent you a letter saying in one form or another that *Structure* merges, as it were, the internalist and externalist approaches. Did you say that you were not aware of that? I find it difficult to accept that you had not realized it.

T. KUHN: I hadn't thought of it as doing that. I mean, I saw what he meant . . . I thought of it as pretty straight internalist. It constantly surprises people in England that I'm an internalist. They cannot get their heads around it. Now, there is something I've left out which should go in here: In that Conant course that I worked in the first time—and I think we did the same thing later—Conant put in a significant social dimension. And it came from him, and I liked it, although I never got myself at all involved with it. He had published a little article on Cambridge versus Oxford during the Restoration, and why science in England developed the way it did. In any case, we read a significant selection from Hessen, wè read Merton, that was where I first was introduced to the Merton thesis, we read G. N. Clark's *Science and Society in Seventeenth-Century England,* and that we probably read all of, and maybe one other thing, I'm not sure. I've read some Zilsel, and by and large thought well of it. And I may have met him at that time, but I'm not sure. But there were things of this sort that were also going on with me. If you look at the introduc-

tion to the Copernican Revolution book, you'll see that I sort of apolo-
gize for the lack of much of any external things and point out that if
I were going to do that I'd say more about the importance of the
calendar, and other things of this sort.

Just to clear that up: although I've never really done external work,
and although I am deeply conscious of and have talked a bit about
the differences of the research techniques, the sources and so forth,
it's a very different state of mind. I have done a bit of methodological
writing about relations of internal and external, particularly in that
article, "History of Science," in the *Encyclopedia of Social Science,* and
somewhere else. I've always been conscious of it, I've always wanted
to see the two things put together, and I think they still almost never
have been. And I think there are important difficulties . . . The thing
that seems to me to be, among the things I've read, the best example
of putting them together is also a very special one. And that's the
Great [Devonian] Controversy book,[15] which I think is splendid. But
there it just was begging to be done in both ways at once, and the
science was such that you could do it that way; and I don't know how
to manage that problem.

Okay, look. Let me jump back or ahead. I had talked about teaching
the course with Leonard Nash when we took it over from Conant.
After that I became an instructor for one year and then an assistant
professor for several more years at Harvard. My primary assignment
continued to be the general education course, but I began to give a
little history of science elsewhere. I developed a course of my own,
a sort of advanced undergraduate course which really was formative [for me], and which is still one of my favorite courses, though I
haven't given it for years. I forget whether it was called . . . *The
Development of Mechanics from Aristotle to Newton.* But I started
out by getting people to read Aristotelian texts and talk about what
motion was like and what the so-called laws of motion were and why
that was not the thing to call them, and did a certain amount of
medieval material and then wound up with Galileo and a little bit of
Newton. That was a course I liked. I started that—initially I think I
gave it a couple of times at Harvard—I gave it at Berkeley, and so
on. I had an undergraduate tutorial group, which is a way of teaching
students in small groups, majors in small groups—there hadn't been

15. M. J. S. Rudwick, *The Great Devonian Controversy:* The Shaping of Scientific Knowledge
among Gentlemanly Specialists (Chicago: University of Chicago Press, 1985).

any of that in history of science previously. I can't think just what else I did.

A. BALTAS: How long did you stay in Harvard after your thesis till you left?

T. KUHN: My thesis was in my first year in the Society of Fellows, which was 1949. I think '47 was the year in which I'd met Conant. I was there from then until '57—'56 or '57. I think '57 was the year I went to Berkeley.[16]

A. BALTAS: The reason for the change?

T. KUHN: Oh, the reason for the change is Harvard didn't want me. And it's in many ways a very good thing they did not. I didn't like that and I was one of those people who was at least in real danger of breaking up because Harvard didn't want them there. That was something that happened to people who'd spent too much time around Harvard.

Something else that happened while I was still in the Society—it was not uncommon for people in the Society of Fellows to be asked to give the Lowell Lectures. There have been some famous ones including, I guess, *Science in the Modern World*, by Whitehead. And there were other very well known ones. The series is still given, but by then it was pretty pro forma and the audience was no longer the intellectual elite of Boston and so forth. And I undertook to give those and I think I gave them in the year after I'd gotten back from Europe. I guess under the title of "The Quest for Physical Theory." I had a dreadful time preparing it and I nearly cracked up. But I got through them. What I was trying to do was to write the *Structure of Scientific Revolutions* in three lectures, and there were various other attempts as time went on. And there are copies of those in the archives now. They are not very good, but they sure give indications of what I was trying to do. One thing that happened in the course of doing those: I gave a lecture which was intended to trace the role of atomism in the development of science. I was persuaded in one respect or another that it had been a transforming influence in the seventeenth century. I still think that. I think in many ways the nature of the transformation has not yet been fully appreciated, although there are things I've learned since. For the record, the part of it that I think has not been fully appreciated is the extent to which atomism helped say, like other sources as well, that you can learn things about nature not simply by

16. In fact, it was 1956.

looking at things as they happen, but by what Bacon called "twisting the lion's tail." And that is terribly important to the development of an experimental tradition and it attaches very comfortably to atomism and not a bit comfortably to any sort of essentialism. It's something that . . . I've taught it, I've said just a word about it in print in one other piece, but I think it's critically important. And I think it's one of those things that's been missed. I always was going to go back sometime and write an article which was really going to be on Bacon and Descartes, about the emergence of epistemology for the first time as a subject in the seventeenth century, which has also got to do with the fact that atoms did not speak for themselves.

v. KINDI: And was that an influence of philosophy on science or perhaps of science on philosophy?

T. KUHN: Look, one of the things I now insist on, that is not at all well handled in *Structure of Scientific Revolutions*, is that you must not use later titles for fields. And it's not only the ideas that change, it's the structure of the disciplines that are working on them. So, you can't separate this sort of philosophy from science yet in the seventeenth century. That separation begins to come after Descartes, but it isn't there in the early Descartes, it's only partly there in Leibniz . . . it's not there in Bacon. The British empiricists begin to force it . . . Locke in particular. That was something I wanted to write a book about. Well, other people have written a book since, and I got too busy with other things—that book is never going to appear from my hand. But that was one of the things that emerged. Now, in the course of this, thinking about atomism . . . if you believe that the atomism of the seventeenth century was like Epicurean and Democritean atomism in some important ways, but what it was *not* like was those ancient and medieval atomisms which took the atoms to be indivisible but which had built into them Aristotelian qualities, or something like Aristotelian qualities, so that atoms are fire, air, earth, and water—[instead], this was a matter and motion atomism. It suddenly occurred to me that if you believed *that*, you would believe that you can make anything out of anything—it's a natural basis for transmutation. I told this idea to Leonard Nash and he said, "I don't know, it's very plausible, but the way you should find it out is of course look at Boyle." So, bright and early on a Monday morning, I was standing outside Widener waiting to get in. And I dashed into Widener and I went to the shelves that had editions of Boyle, and I pulled out one of the *Collected Works*, and found the *Skeptical Chymist* and started to read. Very early there is a remark in which one of the interlocutors says to the major figure who represents Boyle, "It sounds very much to me as though you don't believe in the elements," or something like this. And Boyle says, "That's a very good question. I'm glad you asked me." And then he proceeds to say, "I mean by element those things out of which all things are made, and into which they can be divided." Now, that is taken to be, and it isn't quite, the definition of an element. And Boyle is given credit for the first definition of an element, but what he's doing at this point . . . he says, "I mean by an element, *as I take it all chemists do*"—[but that phrase is] replaced by dot dot dot, when that definition is quoted! And he says, "I'm going to give you reasons for believing that there are no such things." And that was almost my first article.[17] It is, I think, a very good article—it's totally unreadable because I thought I had to persuade a very learned group of historians of chemistry out there. And what I gradually discovered was that nobody knew nearly as much about this problem as I did. And I shouldn't have encumbered it to that extent with supporting evidence and with lots of quotations. In the course of that I also discovered or saw a strangeness in Newton's thirty-first query, which has to do with *aqua regia*, something that dissolves silver and not gold, and something else that dissolves gold and not silver. I thought there was a misprint here, and I still think there was. And this is anomaly showing up. The anomaly about Boyle's is the first. That actually was published first,[18] it's a short piece—those were my first two articles. And during this time at Harvard, when we—Nash and I—took over the course, I started it with lectures on the Copernican revolution. The book really, though it's got more detail, was modeled very precisely [on those lectures]; it's an extended case history. And it illustrates this thing that I'm deeply convinced about. Sometimes you have to go way back in order to find the starting point, to write something that indicates how powerful these prior beliefs were and why they ran into trouble. And I couldn't have started earlier than prehistory; I had to go practically back that far. So, I put in—it's still extraordinary—it was during those years that I was approached by Charles Morris. He was an author of the *Encyclopedia of Unified Science,* and he wrote a very influential book whose name I can't now remember, that grew out of his mono-

17. T. S. Kuhn, "Robert Boyle and Structural Chemistry in the Seventeenth Century," *Isis* 43 (1952): 12–36.
18. T. S. Kuhn, "Newton's 31st Query and the Degradation of Gold," *Isis* 42 (1951): 296–98.

graph and the encyclopedia; and he asked me whether I would take over a volume in the encyclopedia. That volume had originally been assigned to, I think, he's an Italian who wound up in Argentina—Aldo Mieli probably. If you go back and look at the list, history of science was not one of the projected volumes at the very beginning, but it was listed for a long time before anything appeared, under various different authors. They had gone to Bernard [Cohen] who'd suggested that I do it. And I, thinking that I would use it to produce the first version, a short version of *The Structure of Scientific Revolutions*, said yes, and I put in a proposal for a Guggenheim Fellowship, toward the end of my time at Harvard. My project—I was already writing *The Copernican Revolution*—was to finish that, and to write the monograph for the encyclopedia. Well, I didn't finish *The Copernican Revolution* and the monograph for the encyclopedia appeared only fifteen years later. No, not fifteen years later—it was fifteen years between the time these ideas *started* and the time I was finally able to write *Structure*. So that's sort of where I was in those years; I had published my first book just at the end of this time.

v. KINDI: *The Copernican Revolution* was published in . . .

T. KUHN: '57, I think.

A. BALTAS: Why did you choose the Copernican revolution?

T. KUHN: Oh, I was already writing it—I had been giving it as lectures. I needed a book, I had this material, I could do a book, and I didn't think it was a stupid book to do. I mean, it was not what I mainly wanted to be doing, but it was something worth getting done. But that's why I chose it, because I had been giving lectures on the subject.

v. KINDI: And the application to the Guggenheim, was it earlier?

T. KUHN: I got a Guggenheim Fellowship probably in '55–'56 [actually '54–'55]. But something about my unreality with respect to my project and what I could do . . . I mean, I've never been any good in saying how long it will take me to do things, for ten years I've been saying, I think it will take me about another two years to finish this book I'm still working on, and I still think it will take me another two years. But that's what I was under way with.

v. KINDI: And you had already some ideas about writing *Structure? The Structure of Scientific Revolutions?*

T. KUHN: Oh, look, I had wanted to write *The Structure of Scientific Revolutions* ever since the Aristotle experience. That's why I had gotten into history of science—I didn't know quite what it was going to look like, but I knew the noncumulativeness; and I knew something about what I took revolutions to be. I mean, I think in retrospect I was wrong, in the ways I talked about the other night; but that was what I really wanted to be doing. And thank God, it took me a long time, because I managed to get myself established in other ways meanwhile, and the ideas—I didn't let go of them *too* prematurely. I did let go of them somewhat prematurely, but . . . thank God!

v. KINDI: Some of your ideas are similar to the ones developed by Hanson in his book *Patterns of Scientific Discovery,* especially chapter 1 on "Observation."

T. KUHN: Yes. It's the logic of discovery I still don't myself believe in, although I think you can talk about not the *logic,* but the *circumstances,* in ways that illuminate discovery.

v. KINDI: What about *seeing?*

T. KUHN: It was the gestalt switch aspect of it, it was the conceptual framework aspect of it.

v. KINDI: Which you had gotten from where?

T. KUHN: I had gotten that from the Aristotle experience. I had it on other occasions, too. When I taught Galileo, I used to teach it in a way in which key things in it were the relatively anomalous things. I thought I understood why . . . You know, there is an argument in Galileo that the freely falling body which starts at the top of the tower moves in a semicircle at a constant rate, and winds up at the center of the earth. That was for me very important. I thought I'd figured out why he was saying that. There is also an argument which people take to be better, which is about why bodies would never be rejected from the earth regardless of how fast it was spinning. Now, that is a mistake and I think I know what the mistake is; if you know how medieval "latitude of forms," so called, analyzed these problems of motion, you can spot it. It's a standard mistake in the early history because people didn't used to have both the notion of accelerated motion and . . . this depends upon medieval . . . So, it was those questions of frameworks which illuminated anomalies, which were right at the center of what I was doing.

A. BALTAS: We are at the point where you are leaving Harvard and you are going to Berkeley. Let's take a lead there, I mean what did you do there, whom did you meet, whom did you interact with . . .

T. KUHN: An important story. I had a friend, who was a tutor as I was at that time in Kirkland House, at Harvard, who was a friend of a man called Steven Pepper, who was chairman of the philosophy department, at Berkeley. He knew I was leaving and looking for a job, and told Steven Pepper about me; Steven Pepper came to call. The philosophers at Berkeley wanted to hire a historian of science. They didn't know that they didn't want one, they didn't know that this was not a philosophical discipline—I jumped at the change, because I wanted to do philosophy. And I got offered that job, and they asked me at the last minute, would I like to be in history too, and I said, of course. I hadn't proposed that, I didn't know that it was possible, but it was clearly a better, in certain ways a better locus. I went out there jointly in the history and in the philosophy departments. It then turned out that Berkeley could not list a course in both departments at once; I had to divide my courses. It could have "see also the philosophy courses" or vice versa, but there was no way to give a history course a number in philosophy. I thought I knew computers could do that and I kept being assured they couldn't, and it wouldn't work, and I was sore as hell. In any case. I did do things that way. And I taught, I guess, two courses in history and two in philosophy. Two of the courses were survey courses. I'd never given a survey course before in history of science, I'd never *had* a survey course in history of science. So that every lecture I gave was a research project and it was very good for me. After a while I couldn't get that much [more] out of the survey course, but I learned a lot of history of science, I learned how to look at books that didn't feel the way I did, and nevertheless I figured out what must have gone on. I mean, that's the way I learned to do history of biology for the survey course. And I learned some of the problems of trying to organize the development of science. And one of the things that really I finally did, that has shown up also in some of my written work, and that I think is very important: the standard division of history of science—ancient-medieval as one, and then modern science starting in the seventeenth century—it just doesn't work. There is a group of sciences which starts in antiquity and comes to a first major culmination in the sixteenth-seventeenth century, and that's mechanics, parts of optics, and astronomy. And then there are a whole lot of fields which scarcely exist in antiquity and have not yet got much identity, and they are the experimental fields. So I used to go through Newton in the fall, and then drop back in the spring to the beginning of the seventeenth century and pick up Bacon and Boyle and the experimental movements. That's a hell of a lot better way to organize a one-year survey than the standard way—and what it really is, is the origin of the article of mine on *Mathematical versus Experimental Traditions in the Development of*

Physical Science.[19] That's a very schematic article, but it's there that I pushed this business about "don't name fields by their subject matters, [but] look to see what the fields were," which I had not done in *Structure*—it's a bad aspect of *Structure*. Then, in the philosophy department I gave this course that I have mentioned before, on Aristotle to Newton. Also it was there that I gave a graduate seminar each year. You couldn't really give graduate seminars in this field at Berkeley. I mean you got enough students, but very few of them had the preparation. So, one had to pick an area and then let people work at all sorts of levels and report it. There were some useful things that came out of this, but it really wasn't until I got to Princeton that there was a group you could count on, you could name a subject that you were going to work on and get people who would work on it . . .

Early at the time at Berkeley I was invited—in fact, the year I went out there I'd been invited to come to the Behavioral Sciences Center [at Stanford]. And I couldn't, because I'd just accepted a new job at Berkeley, that would have been the inaugural year. But after I'd been at Berkeley for a year or two, I was invited again. I took leave and I went there [to the Center] and that's the year I devoted to preparing *Structure*. And I had an impossibly difficult time. I had one very formative paper that I should have mentioned, that I had done earlier. Earlier at Berkeley I was asked to do a command performance appearance in something the social sciences were doing on "the role of measurement in xyz." That was where I met your prime minister, Andy [Andreas Papandreou]. He gave one on economics, I gave one on physical science. The paper that ultimately emerges is *The Function of Measurement in Physical Science,*[20] and that really was extremely important. Just that little phrase very early on about an extended mopping-up operation—I don't even remember quite how it gets introduced, but that's where the notion of normal science enters my thinking. It isn't that I had thought everything was revolutionary—revolution in perpetuity is a contradiction in terms. But somehow or other I saw normal science as puzzle solving, although it wasn't already all there; it was something that came out at that point, and which helped me be ready, I thought, to write *Structure,* which was my enterprise for the follow-

19. T. S. Kuhn, "Mathematical versus Experimental Traditions in the Development of Physical Science," *Journal of Interdisciplinary History* 7 (1976): 1–31; reprinted in *The Essential Tension,* 31–65.
20. T. S. Kuhn, "The Function of Measurement in Modern Physical Science," *Isis* 52 (1961): 161–93; reprinted in *The Essential Tension,* 178–224.

ing year. Well, what happened was, I wrote a chapter on revolutions, slowly but not with excessive difficulties, and talking about gestalt [switches] . . . Then I tried to write a chapter on normal science. And I kept finding that I had to—since I was taking a relatively classical, received view approach to what a scientific theory was—I had to attribute all sorts of agreement about this, that, and the other thing, which would have appeared in the axiomatization either as axioms or as definitions. And I was enough of a historian to know that that agreement did not exist among the people who were [concerned]. And that was the crucial point at which the idea of the paradigm as model entered. Once that was in place, and that was quite late in the year, the book sort of wrote itself. I wrestled for the whole year and I got sort of two chapters and one article, or something, done that year. But I went back and I wrote the whole of that monograph very quickly while also teaching in the next twelve–sixteen months at Berkeley. That was the key to it. Now, a question I don't know the answer to—this is a point at which my work is often linked to Polanyi's. Polanyi came to the Center that year and gave a lecture on *tacit knowledge*. I liked the lecture all right, and it's possible that it helped get me to the idea of paradigm, although I'm not sure. There is no great reason why it should have, because tacit knowledge was also propositional knowledge in some sense or other. It's there you will recognize the remark I made about your paper, Aristides, that we need to find something . . .

A. BALTAS: . . . something that's not propositional . . .

T. KUHN: Yes. But I couldn't have said that. So I just don't know. It's perfectly possible; we did read some Polanyi in the Conant course. Conant introduced him to the course, and I liked it quite a lot—I don't remember just what it was, except that I kept feeling terrible at those points where he sort of spoke as though extrasensory perception was the source of what the scientists did. I didn't believe that. That [flavor] gets into the tacit knowledge thing also. I don't know. But Polanyi was certainly an influence. I don't think a great big one, but it was very helpful to me to have him out there. In that connection, another story—two books that came out while I was trying to write *Structure*. One of them was Polanyi's *Personal Knowledge*[21] and another one was Toulmin's *Foresight and Understanding*. Particularly with *Personal Knowledge*, I looked at it and said, I *must* not read this book now. I would have to go back to first principles and start over again,

21. M. Polanyi, *Personal Knowledge* (London: Routledge & Kegan Paul, 1958).

and I wasn't going to do that. I also said that with *Foresight and Understanding*, which I think I could have dealt with more. Later, when I did try to read *Personal Knowledge*, I discovered I didn't like it. I never got through that early bit about statistics, which seems to me just way off, quite wrong. I did later read Toulmin's *Foresight and Understanding*,[22] and I understand why Toulmin might have been sore at me for stealing his ideas, but I don't think I did. Let me be perfectly clear: I'm not at all sure he felt that, he has never said that. Toulmin was one of the people I had met during this trip to England at the end of my time in the Society of Fellows—I got along with him fine, he showed me around Oxford one day, but we hadn't gotten at all close. But since the time he came to the States, he and I have not gotten along very well.

A. BALTAS: What about colleagues at Berkeley . . .

T. KUHN: Very good. I would say only one—I mean there were sympathetic people, on the whole not in the philosophy department. The person who was *extraordinarily* important was Stanley Cavell. My interactions with him taught me a lot, encouraged me a lot, gave me certain ways of thinking about my problems, that were of a lot of importance.

V. KINDI: Did you meet him there?

T. KUHN: He had also been in the Society of Fellows. I met him just before he and we left for Berkeley. The Society of Fellows had a softball game every spring, and he was just back from Europe. We were just going off, and I met him there. But I didn't get to know him at all, until we got to Berkeley. And that was a very close and meaningful relationship at that time. We are both in Cambridge and I don't see him anymore, and I regret that.

V. KINDI: Was Feyerabend there?

T. KUHN: Feyerabend was there. He came late in my time. My recollections are not as precise as I would like them to be. I think I remember a talk with Feyerabend. He was sitting behind his desk and I was standing at the door of his office, which was very close to mine. Now, I'm not sure this is right, I mean this is the sort of thing I could easily have constructed. I said something to him about my views, including the word *incommensurability*, and he said, "Oh, you are using that word too." And he showed me some of the things he was doing, and *Structure* came out in the same year as his big article in Minne-

22. S. Toulmin, *Foresight and Understanding* (Bloomington: Indiana University Press, 1961).

sota Studies. We were talking about something which was in some sense the same thing. I messed it up more than he did; I think it's now *all* language and I associate it with change of values. Look, values are acquired together with language, so it isn't that bad a mistake, but it surely made it harder for people to see—or me to see . . . I didn't know enough about meaning, so I was leaning hard on gestalt switches; I think I talked about meaning change in *Structure*, but I looked to find the passages recently, and I was surprised at how few of them there are.

V. KINDI: How did you come up with the terms *paradigm* and *incommensurability?*

T. KUHN: Look, *incommensurability* is easy.

V. KINDI: You mean mathematics?

T. KUHN: I don't remember to whom I told this story recently, but I think it's since I've been here. When I was being a bright high school mathematician and beginning to learn calculus, somebody gave me— or maybe I asked for it because I'd heard about it—there was a sort of big two-volume calculus book by, I can't remember whom. And then I never really read it, but I read the early parts of it. And early on it gives the proof of the irrationality of the square root of 2. And I thought it was beautiful. That was terribly exciting and I learned what *incommensurability* was then and there. So, it was all ready for me, I mean, it was a metaphor but it got very nicely at what I was after. So, that's where I got it. *Paradigm* was a perfectly good word, until I messed it up. I mean, it was the right word at the point where I said, you don't have to have agreement about the axioms. If people agree that this is the right application of the axioms whatever they are, that this is a model application, then they can disagree about the axioms; just as with logic, without its making any difference, they can disagree about the axioms, they can switch axioms and definitions quite freely back and forth, and sometimes do. Here in physics, if you switch axioms and definitions you change to some extent the nature of the field. But the notion that you could have a scientific tradition in which people agreed that this problem had been solved, although they could still disagree vehemently about whether there were atoms or not, or something of that sort. Paradigms had been traditionally models, particularly grammatical models of the right way to do things.

A. BALTAS: This is your first relation to the word—I mean that's why you took it over.

T. KUHN: That's right.

v. KINDI: You were not aware of perhaps Lichtenberg's use of *paradigm*, or Wittgenstein's use of the term . . .

T. KUHN: I certainly was not aware of either of them. Lichtenberg has been called to my attention, and I'm a little surprised that I haven't had my nose dragged through Wittgenstein's use of it. But no, I was not. Now, a very bad thing happened immediately after that. The first time it's introduced into a published work of mine, was in a paper called *The Essential Tension*,[23] which I read at a conference. And there I use it right. But I had been seeking to describe what scientists—the way a tradition worked in terms of *consensus*. And what consensus was about. Consensus was about models, but it was about a hell of a lot of other things [as well] that weren't models. And I proceeded to use the term for the whole lot, for all of the things, which made it very easy to miss what I thought of as my point entirely, and to simply make it the whole bloody tradition, which is the main way it has been used since.

v. KINDI: What about Masterman's twenty-one uses?[24]

T. KUHN: All right, I'll tell you a story. This story comes from a little later day. There was an International Colloquium in the Philosophy of Science held at Bedford College, London. The Proceedings appeared in the volume titled *Criticism and the Growth of Knowledge*. At that meeting I read a paper, Popper was in the Chair and Watkins commented on the paper and there was to have been further discussion, on the original plan for what was going to happen. One of the people who had been invited to participate in this further discussion was Margaret Masterman—whom I'd never met, but of whom I'd heard, and what I'd heard about her was not altogether good, and it was largely that she was a madwoman. She got up at the back of the room in the discussion, strode toward the podium, turned to face the audience, put her hands in her pockets and proceeded to say, "In my sciences, in the social sciences" (she was running something called the Cambridge Language Lab), "everybody is talking about paradigms. That's the

23. T. S. Kuhn, "The Essential Tension: Tradition and Innovation in Scientific Research," in *The Third (1959) University of Utah Research Conference on the Identification of Creative Scientific Talent*, ed. Calvin W. Taylor (Salt Lake City: University of Utah Press, 1959), pp. 162–74; reprinted in *The Essential Tension: Selected Studies in Scientific Tradition and Change* (Chicago: University of Chicago Press, 1977), pp. 225–39.
24. M. Masterman, "The Nature of a Paradigm," in *Criticism and the Growth of Knowledge: Proceedings of the International Colloquium in the Philosophy of Science, London 1965*, vol. 4, ed. I. Lakatos and A. Musgrave (Cambridge: Cambridge University Press, 1970), pp. 59–89.

word." And she said, "I was recently in hospital and I went through the book and I think I found twenty-one," twenty-three, whatever, "different uses of it." And, you know, they are there. But she went on to say, and this is the thing that people don't know, although it's more or less in her article, "And I think I know what a paradigm is." And she proceeded to list four or five characteristics of a paradigm. And I sat there, I said, my God, if I had talked for an hour and a half I might have gotten these all in, or I might not have. But she's got it right! And the thing I particularly remember, and I can't make it work quite but it's very deeply to the point: a paradigm is what you use when the theory isn't there. And she and I interacted then, during the rest of my stay, quite a lot.

A. BALTAS: This conference is in London, in '65, I think, so your book is out three years already. It's published when you reached London . . . What is the initial reception?

V. KINDI: It was published already in the Encyclopedia, right?

T. KUHN: Yes, that appeared in '62. Look, I'll tell you a story about that also. I told you that I had written very quickly, after I got back from the Stanford Center, the manuscript. I hoped it was important. I had wanted to do this, I was not altogether satisfied, but I was pretty excited about it. I didn't know how it was going to go over. I began to have vast reservations about putting it in the *Encyclopedia of Unified Science,* because the encyclopedia had been an exciting thing fifteen years before, but its reputation had declined considerably, and it was no longer in the forefront. But I had a commitment. I went and talked to a friend of mine at the [University of] California Press as to what was right to do under these circumstances. And he said, look, the assistant director of the [University of] Chicago Press is a lovely man called Curley Bowen. Write to him telling your problems and see what he says. So I wrote a long letter to Curley Bowen and I had by then just got a dittoed copy of the manuscript, which was going to need some revisions but I think not very many. And I described the problem—what I was seeing as the problem—and I also said, "would you by any chance . . . it's twice as long as any of the other monographs and I don't know how to cut it. But if you would bring it out full-length more or less, independent of the encyclopedia, I'll cut it somehow or other for the encyclopedia." This went into the mail I think late Sunday afternoon, or Monday morning. Wednesday, as I was leaving the house in California, the telephone rang and it was Bowen. He said, "Don't worry about a thing, we'll . . ." Wow, what an experi-

ence to have with a publisher! My relations with Chicago, although
he left fairly early, have been very good ever since. He said, "We'll
publish it, you don't have to cut it." And they brought it out, they
also brought out a hardcover version initially, from which they omitted
the encyclopedia stuff. After that it's sort of taken care of itself.

I think that with the exception of one story—that I think I do want
to put on the record—that we've about gotten to the end of the time
at Berkeley. But a strange, and to me quite destructive thing happened
at this point. I had been invited to take a job at Johns Hopkins. The
job at Johns Hopkins would have promoted me to full professor, given
me a significantly higher salary, and the opportunity to appoint three
or four more people; it was a major offer. And I went east, I told the
chairmen of both my departments [philosophy and history] that I was
going to go and look at it. I said, I don't think there is anything for
you to worry about, and I will tell you if there is, but I simply want
to put it on record that I'm going off to do this. And I did go, and
in fact I found it extremely attractive. I got back, and I told the chairman that I didn't know how it was going to come out, and that I
proposed to argue with myself about it. But in practice I was finding
it extremely attractive. Then I was asked what it would take to make
me stay [at Berkeley], and I said, look, if under these circumstances
you don't give me a full professorship, I would at least want to know
why. The title doesn't mean that much to me, I'm sure you won't
match the finances, but I don't need to ask you to, though I could
use a raise. But, I said, the thing I *must* have is an expansion; I got
permission to appoint one other person—that was all I could get, one
junior appointment. As I sat around thinking about this, I said to myself, look, maybe five years from now I'd have gone back to Hopkins,
but I've only been here two or three years. It's a very rich institution,
I mean rich in terms of very good people. I decided I can't leave this
at this point. I told the chairman of the history department, and I went
to the chairman of the philosophy department and told him, and he
said, "Don't make up your mind too fast. Hold off." And in fact I
had already written Hopkins that I wasn't coming. I don't think I told
him that. So, I went on with what I was doing, my teaching, and some
weeks and weeks later coming out of a class I had a call asking me
to come down to the chancellor's office. The acting chancellor, who
was my colleague in philosophy, Ed Strong, who had done some historical work himself, wanted to talk with me. I got down there and
he said, "The recommendation for your promotion has now gone all

the way through, it's favorable, and I have it on my desk. There is just one thing. The senior philosophers voted unanimously for your promotion—in history." And I said, "Suppose I don't accept that." He said, "You get it anyway, but . . ." And then I said, "You mean, but why would you want to stay where you are not wanted?" He sort of nodded. I was extraordinarily angry, as you can guess, and very deeply hurt, I mean that's a hurt that has never altogether gone away. The fact that I'd been asked by philosophers, and was in a philosophy department . . . I knew it wasn't altogether congenial to them, but I sure as hell wanted to be there, and it was my philosophy students who were working with me, not on philosophy but on history, were nevertheless my more important students. And I said, "I'll have to think about it." And Strong said, "Oh, but I have to get it to the Regents on Friday if it's to go through before the next meeting." I said, "You will have my decision before Friday." And I went up and had a terrible fight with the chairman of the philosophy department. And I finally said, "All right, what else can I do, I'll accept it." I've been very sorry since. I mean, I felt what I should have done is to say, look, I will accept it after sitting down and discussing the situation with the senior members of the philosophy department. If they still want to do it that way, I will accept it. But I won't accept it on these terms. I think if I'd said that, they wouldn't have faced me. In any case, I think I should not morally have simply let myself be treated that way. But that hurt a lot. I stayed for another year or so, and it wasn't because of that that I left Berkeley. But there were certain things that had happened at Berkeley that had diminished my pleasure in being there, though they hadn't taken it all away. I got an offer to go to Princeton and at Princeton I was going to have a senior colleague who'd set up the program. The two of us were going to work together, there were going to be other people. And that was just a much more manageable situation. I got that offer while I was in Denmark. I said, I can't answer until I've gotten back, but I will work on it as hard as I can when I get back, and I'll come visit Princeton. So when we got back, which was probably the fall of '63, my wife and I went to Princeton, visited, and I decided I was going to take this offer—which I did.

K. GAVROGLU: Why were you in Denmark?

T. KUHN: Shortly after I finished the manuscript of *Structure*, I got asked by a committee of members of the American Physical Society to direct an archival project on the history of quantum theory. One of the people who asked me was the man with whom I'd done my Ph.D. thesis, van Vleck. And I did. If I hadn't finished *Structure* at that point, I wouldn't have accepted. But the thing I had been dying to get done— my big commitment to myself—I'd known what I wanted to do next, and it was just this book about science and philosophy in the seventeenth century. But I thought, that one I can put aside. So I did. And I accepted the job; and the rest of that, basically you know. There is nothing particular to tell about it, except one thing: that project has probably had some real influence. We brought back a lot of microfilm of archives and we got manuscripts and letters deposited in various places. And we catalogued them. And that was probably the more important part of it. Interviewing was frustrating as hell! Some of the interviews are really very good. But the physicists, including the ones sponsoring the project, really wanted to get the development of ideas, and that's of course what I wanted also. I knew from experience as a historian that scientific autobiographies are invariably inaccurate, they tell the wrong story. But it's usually the case that if you sit down with the published papers and whatever else there may be, and then ask, why did he tell that story instead of this story . . . you get very important clues to a reconstruction. What I hadn't anticipated was the number of times people would say, "I don't know, I can't remember; how, why would you expect me to remember that?" In that sense, for that sort of thing, we got much less than I had hoped. The other side of this is that what you can get scientists to talk about quite freely and richly is what it was like to be at Munich, and so forth, and who the important teachers were, and what your first experience was when you went from there to Göttingen or vice versa, or whatever. And that you could get some talk about. If you started back at the places I used to try to start at and say—how did you get into science in the first place, did your parents approve—too often you'd get, "That's not physics." So, that's the quantum physics project; and I came back from it to look at Princeton, and the next year we went to Princeton.

v. KINDI: Would you like to talk about the students who worked with you?

T. KUHN: I'd never had a lot of graduate students. Presumably that's partly because there haven't *been* so many graduate students, and partly because I tend to scare them away. I criticize. My first two graduate students—although one officially didn't take his degree with me—the first one was John Heilbron, and [the second was] Paul Forman, who was the one who took finally his degree with Hunter

Dupree after I left, though he'd got into it through my courses. John was all but finished with a physics thesis, but he'd been ill and while he was ill he had read Thorndike's *History of Magic and Experimental Science* and decided he wanted to be a historian of science.

I may be mixing up two stories when I tell you that, because my first experience at Berkeley sitting in my office, not yet having taught a course, was a graduate student in philosophy—oh, no, I'm not mixing it up—who came down, wanted to find out about the course, and he said to me, "What do you think of Dampier?" Dampier is a one-volume history of all of the sciences,[25] and I said something like, "I've never quite been able to read all the way through it, I think it's supremely dull." And he said, "Oh, I think it's wonderful!" But you see why I thought I might have confused these two stories.

So, that was the beginning of my production of students. It must be said: there had been other students—I think none who emerged with the authority and reputation that John has . . . I am still a devotee of Paul's *Weimar Culture and the Quantum Theory*.[26] One knew it couldn't be right altogether, but I think he's given much too much away as he sort of backed away from the criticisms. And I remember when I first read that. I was at Princeton, I went and put a note on the bulletin board of the department office, saying, "I have just read the most exciting piece that I've read since I discovered Alexander Koyré!" So, that's my first two students. There have been some others at Princeton and I'd like to say something about them when we get there. But as both John and Paul indicate, I have exposed my students to the sort of history of science I do, they are in principle capable of doing it and they've each showed that in early work. But they've both turned entirely away from it. So, in that sense, I haven't produced any children. With one exception, and that exception is in fact Jed Buchwald—who was not a graduate student of mine, but an undergraduate student of mine. I turned him on to history of science, and that was the point at which he decided to do it. But it's always been a source of some chagrin that these things that I like to do, and have taught people to do, don't get pushed along. But there are a lot of

25. W. C. Dampier and W. M. Dampier, Cambridge Readings in the Literature of Science: Being Extracts from the Writings of Men of Science to Illustrate the Development of Scientific Thought (Cambridge: Cambridge University Press, 1924).
26. P. Forman, "Weimar Culture, Causality, and Quantum Theory, 1918–1927: Adaptation by German Physicists and Mathematicians to a Hostile Intellectual Environment," *Historical Studies in the Physical Sciences* 3 (1971): 1–115.

reasons for that. One of them, of course, is that people have to get away from their doctor-father; and another is that the field has been moving way away from the sort of history of science I do . . . did. But I still don't altogether like it!

V. KINDI: Did you have students doing philosophy of science?

K. KUHN: No. I have never had a philosophy graduate student. At Princeton I wouldn't have had, at MIT I might have, but I'm so far away from the main sources of jobs in the philosophical tradition that are being fostered by my colleagues. I had one or two people start to do a degree with me, but I've sort of driven them away. One of them finally looked at me in one discussion and said, and he was very good about it, "You really think this is off the wall, don't you?" and I said, yes, I do. He picked himself up and went to find another thesis director and did it on a different thing. The other one we finally asked to accept a master's, and go do something else. Those are the two people who were in philosophy departments, and that was at MIT. I have given seminars for philosophers, occasionally at Princeton and regularly [at MIT?], and I've had some very good interactions there. That's something we can come back to.

One thing I realize I left out before, that should be filled in, and that is the question as to where I got the picture that I was rebelling against in *The Structure of Scientific Revolutions*. And that's itself a strange and not altogether good story. Not altogether good in the sense that I realize in retrospect that I was reasonably irresponsible. I had been, as I'd said, vastly interested, caught a real interest in philosophy in my freshman year, and then had no opportunity to pursue it—initially, at least. It then turned out that after I graduated and went to the Radio Research Laboratory, and indeed continuing most of the time while I was in Europe—I was no longer having school assignments and papers to write—I had what was basically a nine-to-five job; and I suddenly had time to read. And I started reading what I took to be philosophy of science—it seemed the natural place to be reading. And I read things like Bertrand Russell's *Knowledge of the External World*,[27] and quite a number of others of the quasi-popular, quasi-philosophical works; I read some von Mises; I certainly read Bridgman's *Logic of Modern Physics*[28]; I read some Philipp Frank; I read a little bit of Carnap, but not the Carnap that people later point

27. B. Russell, *Our Knowledge of the External World*, 2d ed. (London: Allen & Unwin, 1926).
28. P. W. Bridgman, *The Logic of Modern Physics* (New York: Macmillan, 1927).

to as the stuff that has real parallels to me. You know this article that recently appeared.[29] It's a very good article. I have confessed to a good deal of embarrassment about the fact that I didn't know it [the Carnap]. On the other hand, it is also the case that if I'd known about it, if I'd been into that literature at that level, I probably would never have written *Structure*. And the view that emerges in *Structure* is not the same as the Carnap view, but it's interesting that coming from what were partially different . . . Carnap staying within the tradition had been driven to this—I had rebelled already and come to it from another direction, and in any case we were still different. But that was the state of affairs in my mind at the time that I had this experience of being asked to work in the Conant course. And it was against that sort of everyday image of logical positivism—I didn't even think of it as logical empiricism for a while—it was that that I was reacting to when I saw my first examples of history. We have gotten me to Princeton . . .

A. BALTAS: Yes, you have *Structure* out, you have started the project about the quantum mechanics sources, you are moving to Princeton . . .

T. KUHN: Yes. Well, I really have finished the Archives, I mean I still had to do with some of the putting together of the catalogue after I got to Princeton, and that took a good deal of my time during the first year at Princeton.

A. BALTAS: Perhaps a good way to continue is to ask you the following question: You have *Structure* out in '62. The way we have perceived things, which may be wrong, is that the big boom, as it were, the big explosion of the reception of *Structure*, arrives after '65, more or less— when you had this London thing. I mean *Criticism and the Growth of Knowledge* gets published in the '70s or thereabout, or there are rumors already circulating about your debate with Popper, or things like that. This is a kind of image we have, which may be completely wrong.

T. KUHN: I can't tell you you are wrong, I'm a little surprised at it; I would not have told the story that way. But it's very possible that the evidence will show me wrong. I would say myself . . . It built up a year at a time more or less under its own momentum, and I would not have thought that there was any particular burst in connection with '65. On the other hand, what may well have happened in '65,

29. G. Irzik and T. Grunberg, "Carnap and Kuhn: Arch Enemies or Close Allies?" *British Journal for the Philosophy of Science* 46 (1995): 285–307.

or as a result, is that the philosophers began to pay more attention. I mean, a lot of the early audience was social scientists. But not just social scientists. I mean, the book was well reviewed by Shapere in the journal that's published from Cornell.[30] Well reviewed, except for some strong reservations which I thought were largely incorrect. People picked on 'paradigm' from the beginning, and I don't think they were wrong to do that. It made it harder for me to recall people to what I was really after; and if I had seen what I had done myself, I could have done that better. I have impressions but I'm not at all sure that they're accurate, and some of them come from certain feelings of disappointment, or whatever. The early reactions—the book got good reviews.

A. BALTAS: In what kind of journals, philosophy mostly or . . .

T. KUHN: I'd almost have to go back and look at the file in which I have reviews. Pretty widely; probably not mostly in philosophy journals. But it wasn't only the Shapere review. Mary Hesse wrote a review in *Isis*, that I do remember . . . I gradually realized that a lot of the response was coming from social scientists, and I was quite unprepared for that; I thought of the book as directed to philosophers. And I think not a lot of them read it, I think it was picked up much more widely than that; it was no particular force for some time in philosophy, although the philosophers surely knew it. But I remember—I guess it was Peter Hempel's telling me that he had been to a meeting, I think it was in Israel, at which groups had said, "That book should be burned!" and "All this talk about irrationality! . . ." Irrationality in particular, irrationality and relativism—the thing that bothered me about the Shapere review is all the talk about relativism. I saw why he said that, but I thought, if he'd thought a little bit more seriously about what relativism was and what I was saying, he would not have said anything like that. If it *was* relativism, it was an interesting sort of relativism that needed to be thought out before the tag was applied. In practice, I would say it's not a relativistic book. And although I would have had trouble initially, I tried in the end of *Structure* to say in what sense I thought there is progress. I largely squeezed out the answer to that, talked about the accumulation of puzzles, and I think I would now argue very strongly that the Darwinian metaphor at the end of the book is right, and should have been taken more seriously than it was; and *nobody* took it seriously. People passed it right by.

30. *Philosophical Review* 73 (1964): 383–94.

This question of stopping to see us, i.e., ceasing to see us, as getting *closer to* something, but see us instead as moving *away from* where we were—that was beyond anything I'd really quite grasped until the point at which I had to really wrestle with that problem. But saying that was important to me and it led to things that have happened since. And I think it might have been picked up and recognized more. And what went with all of this, Vasso, I saw in one of your papers,[31] which talks about how just the things that made me unpopular in the sixties make me popular in the eighties. And that's I think a very revealing and very apt remark, but it's wrong in one respect: the sixties were the years of the student rebellions. And I was told at one point that "Kuhn and Marcuse are the heroes at San Francisco State [University]." Here was the man who had written two books about revolutions . . . Students used to come to me saying things like "thank you for telling us about paradigms—now that we know what they are we can get along without them." All seen as examples of oppression. That wasn't my point at all. I remember being invited to attend and talk to a seminar at Princeton organized by undergraduates during the times of trouble. And I kept saying, "But I didn't say that! But I didn't say that! But I didn't say that!" And finally, a student of mine, or a student in the program who had sort of helped get me into this, and had come along to listen, said to the students, "You have to realize that in terms of what you are thinking of, this is a profoundly conservative book." And it is; I mean, in the sense that I was trying to explain how it could be that the most rigid of all disciplines, and in certain circumstances the most authoritarian, could also be the most creative of novelty. And to cut my way through that aporia, I had to set it up; but of course to set it up as an aporia ran into all sorts of resistance. So, it's hard to say how I felt. I thought I was being—I want to say badly treated—badly misunderstood. And I didn't like what most people were getting from the book. On the other hand, I did not for a moment think that that's all that was going on. There were people who picked it up and really did seem to be moving it ahead and getting on with it, probably initially more in the ways that at the beginning some of the sociologists picked it up. I got very good initial responses from scientists.

A. BALTAS: Physicists, biologists . . .

31. V. Kindi, "Kuhn's *The Structure of Scientific Revolutions* Revisited," *Journal for General Philosophy of Science* 26 (1995): 75–92.

T. KUHN: Yes, both. Various people reported to me that this book was the first thing in philosophy that they'd read, that really felt to be about what they were doing. And I valued that, and there were other things . . . I mean, clearly, I wanted it to be an important book; clearly it was being an important book—I didn't like most of the ways in which it was being an important book, but on the other hand, I recognized that if I had to do it all over again, I could, if I had the opportunity, eliminate some of the misunderstandings. But if I couldn't do that, I'd do it all over again the way it was. I mean, I had disappointments, I didn't have regrets.

A. BALTAS: Are there any incidents in your discussions with philosophers that were significant, in both ways as regards your own perception of what you had been doing, as well as regarding the overall reception of the book? Some incidents in conferences, or people who talked to you about it and then gave you a new light . . .

T. KUHN: Initially not a lot. I got invited to talk at a couple of places, and I was glad to, but I wasn't very well received. I was not really getting through to philosophers, although some of them were very interested. When I got to Princeton, I began to work a good deal with Peter [Hempel]. This was the first philosopher, I guess of any sort, but certainly the first philosopher in the logical empiricist tradition who began to respond, and to respond seriously to what I was doing. And his position along the way has not become mine and there is no Wittgenstein in it. But it shifted markedly in ways that I think are important. And when I used to try to compare the two traditions, I used to point to the time—which I'm not sure is my responsibility, but I think it may well be—at which Hempel, instead of talking about theoretical and observational terms, started to talk about *antecedently available terms*. And that in itself is already putting things in a sort of historical developmental perspective. I don't think he saw it quite that way, but it was a very important step.

V. KINDI: What about the other philosophers of science of the period—Feyerabend or Lakatos—you were all received together, in a sense.

A. BALTAS: The historicist turn, as it was dubbed.

T. KUHN: It's hard to talk about that. Certainly there were some philosophers who took it up and more than a few. And they began to talk about the historical philosophy of science. From my point of view, I was glad to see that, but it struck me very forcefully that all of them entirely dropped the problem of *meaning* when they made that turn, and that they therefore dropped incommensurability, and that they therefore were back having eliminated [what for me was] the philosophical problematic. With Feyerabend I had strange experiences. He was at Berkeley, I gave him this draft manuscript of the book that I'd sent out to Chicago. I think he liked it in one sense, but he was terribly upset by this whole business of dogma, rigidity, which of course is exactly counter to what he believed himself. And I couldn't get him to talk about anything except that. And I tried, and I tried: we would have lunch together, or something—he'd always come back to it. I got more and more frustrated and I finally just stopped trying. So, he and I never really did have good talk about these problems. The quasi-sociological elements of my approach were overwhelmed by his desires for society in the ideal. And we really never made contact.

v. KINDI: What about the ones that come after, like Laudan or van Fraassen? It seems that the field has stopped dealing with the phenomenon of science as a whole, with the kind of questions you raised, and has now gotten back to the standard problems of philosophy of science, induction, confirmation, Bayesianism . . .

t. KUHN: I'm surprised that you include van Fraassen.

v. KINDI: I don't mean that he belongs to the historicist tradition. I include him because he dealt with issues like the theory/observation dichotomy.

t. KUHN: But that was long before me.

v. KINDI: Didn't you contribute to its being undermined?

t. KUHN: It was being undermined already. He in a sense, I think, was trying to get it back again, to show that it was still a viable notion. And I wasn't the primary underminer of it. The theory/observation distinction was in trouble already before me. Putnam was undoubtedly a more important person than I for the philosophers in undermining the distinction. There are several very important Putnam papers.

a. BALTAS: There is a certain incommensurability going at this moment, because in the following sense I think it's rather different in the States. It's better to clarify it. I think that in Greece for sure, and I think also in places like Italy, or France, I don't know much about the other places, the perception of this period is the following: you have logical positivism with its own problems, etc., some criticism within the tradition, and then *you* come along and change the paradigm, so to speak. And by changing that, people who already had done parallel things, let's say, criticizing the main logical empiricist framework, join forces, as it were, with you, not in the real sense of writing papers together, but you are all perceived . . .

V. KINDI: You opened up the field . . .

T. KUHN: I'm sure that's right, but I nevertheless was quite surprised to have van Fraassen and Laudan put into the same . . . Laudan is somebody who said he was doing historical philosophy of science. He says things about me that are absolutely not the case. People point it out to him and he goes right on saying them. He tries to hold on to the traditional view of scientific progress, closer and closer to the truth, absolutely dropping the problems that [I had] pointed out. From my point of view, that's very bad stuff!

Somebody who has done both philosophy and history, and who has encouraged me and whom I'm very fond of, is Ernan McMullin. He's really been a supporter over the years. He doesn't like some things I would try to persuade him he ought to like, but that's been helpful. What I have found is that now—and this I delight in—as historians of science turn further and further from scientific substance, a number of important *philosophers* of science have gotten more and more involved with doing some history. And they do it more nearly the way I would like to see it done. And that's been a very agreeable route to watch.

V. KINDI: Who are the philosophers you refer to now?

T. KUHN: Well, John Earman has done some of it. Clark Glymour has done some of it, but I don't think that comes from me (though Earman probably does). John was one of the products of the Princeton program in philosophy and history of science from the philosophy end. I had him in a seminar the first year I was at Princeton, and I talked with him some after that, and he went on. So that had a role. Ron Giere is another person who began to do some of this. There was an increasing influence of a different approach to historical examples among the philosophers of science gradually. I first discovered how much the change had been when I was suddenly elected president of the Philosophy of Science Association, about five or six years ago. And I had not even been a member of the Association, except for one year after which I dropped out, or something of the sort. The scene among philosophers of science was just very very different. And I had clearly had something important—though not I alone—to do with that. It becomes important to remember Russ Hanson, and to some or lesser extent Polanyi, Toulmin. I think Russ Hanson was probably more important than either of those. Feyerabend and so forth. There were a whole bunch of people moving in this direction. I don't think that the people who were doing history, by and large, saw everything in it that I was seeing in it. They were not coming back asking "What does this do to the notion of truth, what does it do to the notion of progress," or if they did, they were finding it too easy to find answers that seemed to me superficial. It isn't that I knew the answers, but I didn't think that theirs were ones that were going to withstand the scrutiny that they needed. I was worrying about it, I mean I was back to writing history for a change. But I wanted nothing more than to get back and straighten out these problems—and I really didn't know how to do it—and I kept saying, it's like going around on a stage set, opening doors, to see which ones have just a painted canvas behind them and which one will lead into another room. Well, I gradually found some that led into another room, or part way into another room: the causal theory of reference. Kripke made a very big difference,[32] because I was absolutely persuaded that it was a breakthrough with respect to proper names—but then it didn't work for the other things, common nouns. Putnam's stuff also helped—but I simply couldn't reconcile myself to saying, "If heat is molecular motion, then it always was molecular motion." That just wasn't where the action was. But I got some very important tools out of that, and one of them was to go back and think about the Copernican revolution, and suddenly realize, look, you can trace the individual planets, Mars, heavenly bodies through the Copernican revolution—what you can't trace through it is 'planets'. Planets are just a different collection before and afterward. There was a sort of localized break that fitted very closely. And now it turns out that some people, to an extent that surprises me and others, simply say, "in the Ptolemaic system the planets go around the earth and in the Copernican system they go around the sun." But that's an incoherent statement! And it is! It's too easy to get around it, because you then start doing: there is a finite number of planets, and they have proper names, and you do it that way. But it isn't wrong that the statement is incoherent. It's highly suggestive of this sort of thing, and I think one needs to talk about it. It's always been clear to me also, or clear to me for some time, that the two people I was sure were taking the problems I was looking at seriously were me and Hilary. When Hilary started talking about internal realism. I thought, hell, *now* he's talking my language. Well, he sort of stopped talking my language. But at this point these problems were getting to be important in philosophy in ways that they had not been before. Nobody

32. S. Kripke, *Naming and Necessity* (Cambridge, MA: Harvard University Press, 1980).

could reasonably show anything but respect for Putnam; but they could joke about him a bit, because he went so far and then took so much back, and wrote the same thing so many times changing it each time. For the Putnam who had written a paper about incommensurability called *How Not to Talk About Meaning*,[33] in which the whole thing was the rope—you could change one strand or change another strand but it was still the same rope, and therefore there wasn't the sort of problem that Feyerabend and I were talking about—that was a big step . . . internal realism and the things that went with that. And causal theory seems to me—thinking about causal theory had been very important to me. I don't think it works for common nouns. But it's terribly interesting to see why it seems to work. And that becomes clearer in this thing I'm working on now—the senses in which it almost works. It doesn't work across periods of revolution or whatever, but it'works very well between those. And by the time you reconstruct what has occurred after a revolution, you get it back to seeming to have worked again. In this paper I mentioned the other day of mine, called *Possible Worlds and History of Science*,[34] I talk about what's wrong with Putnam's "water is and always was H_2O." And this has gradually been sneaking into philosophical discussion. And I have felt rather good about that, and I think I am more read in philosophy courses than I used to be, and more talked about, and I have more influence. But I have to say, as I've already been saying, I have never been in the sort of lovely situation that you people have put me in here. And it's extra agreeable by virtue of this background.

A. BALTAS: What about the *Black-Body* [book]? I mean you have become a big success with the *Scientific Revolutions*, some might expect that you go on from there, explain yourself better than the *Postscript*[35] and things like that; and a book comes out which, apparently at least, does not look like one much expected . . . for example, in application let's say, in quotation marks.

T. KUHN: I have said repeatedly, and I will say again: you cannot do

33. H. Putnam, "How Not to Talk about Meaning," in *Boston Studies in the Philosophy of Science*, vol. 2, *In honor of Philipp Frank*, ed. R. Cohen and M. Wartofsky (New York: Humanities Press, 1965); reprinted in *Mind, Language and Reality*, Philosophical Papers, vol. 2 (Cambridge: Cambridge University Press, 1975).

34. T. S. Kuhn, "Possible Worlds in History of Science," in *Possible Worlds in Humanities, Arts and Sciences: Proceedings of Nobel Symposium 65*, ed. Sture Allén, Research in Text Theory, vol. 14 (Berlin: Walter de Gruyter, 1989), pp. 9–32; reprinted in this volume as essay 6.

35. T. S. Kuhn, "Postscript," in *The Structure of Scientific Revolutions*, 2d ed., rev. (Chicago: University of Chicago Press, 1970), pp. 174–210.

history *trying* to document, or to explore, or to apply a point of view that is as schematic . . . Clearly, I do history differently, because of things that I think I have learned, that lie behind *The Structure of Scientific Revolutions*, and that were perhaps further developed. I like doing history, and I've gone back and forth—you can't do the two things at the same time. The philosophy has always been more important, and if I had seen a way to go back and work on the philosophical problems straightforwardly at the time I wrote the *Black-Body* book, I probably would have. Look, I'll tell you just how far this goes. Before the book came out, I'd agreed to talk with a group of people about this book, and when I got to this small group that was supposed to sit around the table, it turned out the room was full, or more or less full. So, I had to give an impromptu lecture, and I did. When it was done, somebody held up their hand, and said, "It's all very interesting, but tell me, did you find incommensurability?" I thought, "Jesus! I don't know, I haven't even thought about that." Now, yes, I mean I *had* found it, and I later recognized what it was, recognized it particularly when I began to get reviews from people, like Martin Klein: it had to do with the energy element *hv*. I mean, there is an aspect of that that's talked about in the book. I talk about Planck's letter to Lorenz in 1910 or 1911 in which Planck says it's the switch from resonator to oscillator. He says, "you will see I have stopped calling them resonators, they are oscillators"; and my view is that that is a very significant switch. Resonators respond to a stimulation, oscillators just go back and forth. And others, I mean . . . the energy "element" of Planck's was not to be understood the way the energy "quantum" of Planck's is to be understood, and so forth. So, it's in there, but I wasn't looking for it particularly. And the reason for telling you this story about the question, was simply to tell you I hadn't thought about it! It was a perfectly good question; I later realized how to answer it, but it just floored me at the time, and I sort of stammered around.

V. KINDI: Because you don't apply the philosophical theory to do history.

T. KUHN: No. If you have a theory you want to confirm, you *can* go and do history so it confirms it, and so forth; it's just not the thing to do.

A. BALTAS: Because there has been a lot of discussion regarding the relation between history and philosophy of science, what kind of advice would you give, what is your position in the sense of giving some advice to younger people who want to do either or both.

v. KINDI: You said something at the lecture, like getting into their minds . . .

T. KUHN: Yes, and that's what I think the intellectual historian has to do. It's exactly what the philosophers systematically resist doing. But the way they tell the story—history of philosophy telling the story of Descartes, what he got right and what he got wrong, and what could have been done to bring the two together.

I wrote a paper on the relations of history and philosophy of science[36] in which I insist that, although I'm the chairman of a program in "history and philosophy of science," there is no such field. And I tried to talk a little bit about my experience of having philosophers and historians and scientists in the same classroom. The philosophers and the scientists are much closer to one another, because they all come in being concerned about what's right and wrong—not about what happened—and therefore tending to look at a text and simply pick out the true and the false from a modern point of view, from what they already know. The historian, at least doing it my way, insists on saying: that's a respectable human being, [so] how could he ever have thought anything of the sort? A remark of Vasso's in her work that I particularly like is, just this. . . . I mean, yes, people treated me as though I were a fool! I want to say, how the hell could anybody ever have thought that I would believe anything like that! That really was fairly destructive, and I fairly early simply stopped reading the things about me, from philosophers in particular. Because I got too angry. I knew I couldn't answer, but I got too angry trying to read them and I would throw them across the room, and I wouldn't finish them, and I would miss anything that might have been helpful in it, by virtue of this rage. It was too painful.

About history and philosophy . . . So, I said these are very different fields. I speak of it as different ideologies, as different goals, and correspondingly different methods, different senses of what it is obligatory to be responsible for. Both of them will say "yes, but that's trivial, but that doesn't matter." But the historians and the philosophers feel licensed, and able, to say that at very different places. On the other hand, it is my sense that there can be quite a lot of cross-fertilization, if you can get some interaction between the two, which becomes harder rather than easier in one sense, because the historians have

36. T. S. Kuhn, "The Halt and the Blind: Philosophy and History of Science," *British Journal for the Philosophy of Science* 31 (1980): 181–92.

stopped dealing with technical issues. But it is the case, I think, that there is quite a lot to be done in interaction, and at least I produce myself as an example, because I'm never a philosopher and a historian at the same time, but the two do interact. And that's the ideal arrangement, from my point of view.

K. GAVROGLU: After the appearance of *Structure* . . . and not totally independent of it, history of science branched into quite a few well-articulated approaches. What has come to be known as the Strong Program has been the most controversial. Although you have, not very systematically, expressed your thoughts on the Strong Program, I think it may be interesting to tell us your views as regards the scholarship of the Strong Program.

T. KUHN: Let me tell you two stories. One is when I read that lecture on the relations between the history and philosophy of science, a philosopher came up to me afterwards and said, "But we have such good *scholarship!* We have such good *scholars* in the history of philosophy!" Yes, but they're not doing history. I mean, I didn't say that then. But I say this because you used the term, what do I think of the scholarship. The scholarship is often damn good! You and I have talked about *Leviathan and the Air Pump*,[37] in which I think the scholarship *is* very good, and I think it's quite a fascinating book. It upsets all hell out of me that they [the authors] can't understand what everybody now learns in high school, or even elementary school about the theory of the barometer . . . They talk basically about the "emptiness" of the dialogues between Hobbes and Boyle, and they get it all very badly wrong. I said to you when we talked about this before, they talk about Boyle's switching back and forth between talking about "pressure" and talking about "the spring of the air." It's not a consistent way of talking, but there's a very important reason why he hasn't yet quite gotten out the way to talk about it; it's not things that don't matter. He's using a hydrostatic model. Hydrostatic models deal with an incompressible fluid. Air is not an incompressible fluid. So, what you get in one case from straight pressing down, you can get in the other also by compression; and you can get these two together. But the fact that he is going back and forth between the two . . . they are not incompatible, they are different ways of talking about the same thing,

37. S. Shapin and S. Schaffer, *Leviathan and the Air Pump: Hobbes, Boyle, and the Experimental Life* (Princeton: Princeton University Press, 1985).

but they had better get more deeply integrated than they are at the point that Boyle's talking about it; there is an incompleteness there. And again, they talk about the impossibility of proving that the barometer explanations which talk about subtle fluid coming in and filling up the top—you can't indeed show those to be wrong—these are the antiperistalsis explanations. Of course you can't show them to be wrong, but what [Shapin and Schaffer] totally miss is the vastly greater explanatory power that comes, including that straightforwardly with the Puy-de-Dôme experiment and many others. So that there is every rational reason to switch from one of these ways of doing it to the other, whether you think you've shown that there is a void in nature or not. And that sort of thing bugs me. And as I also said to you, Kostas, what really bothers me most about it is that history of science students themselves now don't care. I talked to Norton [Wise] about this, and he had sort of gotten me to read *Leviathan and the Air Pump*. And I think it's in many ways an extraordinarily interesting and good book. So it's not the scholarship that's bothering me. Norton, who is a physicist himself, thought about it, he felt I was right, so he told his class about them. He told me that nobody in the class could see that it was of any importance. At that point I'm bothered. I mean, that to me is a bother. Now, the whole thing is turning around in some ways again, and I don't know where it comes out. It isn't that I think it's all wrong. I said to you that the term "negotiation" seems to me just right, except that, when I say "letting nature in," it's clear that that's an aspect of it to which the term "negotiation" applies only metaphorically, whereas it's fairly literal in the other cases. But you are not talking about anything worth calling science if you leave out the role of [nature]. Some of these people simply claim that it doesn't have any, that nobody has shown that it makes any difference. Now, I don't think they are saying that anymore, but I don't think it's coming back to the point where they've really got room for it . . . I haven't read Pickering's latest book, *The Mangle of Practice*.[38]

v. KINDI: What about the Stegmüller group?

T. KUHN: Look. I don't know, it's moot how much I had to do with Sneed. I met their work through Stegmüller. Stegmüller sent me a copy of the journal—I guess I talked to Aristides about this, and I

38. A. Pickering, *The Mangle of Practice: Time, Agency, and Science* (Chicago University of Chicago Press, 1995).

think I'd be glad to have it on tape. He sent me a copy of *Theorien-strukturen und Theoriendynamik*,[39] with a very nice inscription, and a card in which he described himself as a Carnapian who was perhaps becoming a proto-Kuhnian, or something of the sort. I started looking at the book and I realized, I've got to learn to read this, but it was all in German, and it was all in set theory, and I didn't know set theory, and I didn't know what a function was, represented as in set theory. I still don't really know model theory and I didn't have the German vocabulary for them either. But I realized I had to read the book; it took me two or three years—a year and a half maybe. I used to carry it on planes, and so forth, and nibble at a little more of it. I thought it was immensely exciting! The thing that I could not support about it was the reduction thesis, which was basically a one-language thesis again. I thought what he said about paradigms came closer to what I had had in mind than anything I had seen by a philosopher! Or anybody else for that matter—it was paradigms as examples, and this was a point of view that said, you don't have a structure unless you include in it at least a few examples. Now, those were terribly exciting things, and furthermore some of the things I have been doing since have fed off of their discussion in certain ways. I mean, the stuff about learning force, mass, and so forth in that paper really I probably would never have written without having been exposed some years before to the Sneed-Stegmüller stuff. And I think it's absolutely first class. Certainly it's had an impact on me, and that will show again, and I will talk about it a little bit in this book that I'm doing. I said to Sneed, I don't think you got to this through me, and he said, "don't be too sure, I had read you!"

I tried to get philosophers more interested in that stuff. And on the whole, for a long time, I couldn't succeed at all. Now everybody is talking about the semantic view of theories—but on the whole leaving Sneed and Stegmüller out. And I think now I see the reason; I'd been looking at Fred Suppes's book, and I think I see what that's about. They don't want to get back to anything that looks like . . .

A. BALTAS: Models?

T. KUHN: Well, it's not that. I mean, structures are formal. They see the Ramseyfication, the use of Ramsey sentences, as reintroducing—

---

39. W. Stegmüller, *Probleme und Resultate der Wissenschaftstheorie und analytischen Philosophie*, vol. 2, *Theorie und Erfahrung*, part 2, *Theorienstrukturen und Theoriendynamik* (Berlin: Springer-Verlag, 1973); reprinted as *The Structure and Dynamics of Theories*, trans. W. Wohlhueter (New York: Springer-Verlag, 1976).

and this is why Sneed does it—something like the theory/observation distinction. And they think that's got no place anymore. But unless you have something like that—it's not just theoretical/observational, I don't believe in that either—but what it sure is, is antecedent vocabulary, or shared vocabulary . . . If you take a dynamic view, you've got to have something that talks about revision of terminology and introduction of new terminology as part of the introduction of a new theory, of a new structure. And I don't think you can do it without that, and that's why I would still point back to the Sneed-Stegmüller version as the one that best fits with what goes on. It adapts itself to a historical developmental approach.

K. GAVROGLU: Even though you had at least two students in the history of science, you had none in the philosophy of science.

T. KUHN: I've never directed a philosophy graduate student.

K. GAVROGLU: We can go to the last question.

T. KUHN: Yes. Look, there is only one of my students—he didn't take a degree with me—and that's Jed Buchwald, who does the sort of history of analytic ideas that I do, and I love doing. I was the person who turned him on to history of science as an undergraduate; he took his degree at Harvard. And things have gone on since. But all of my other students have, in one way or another, gone more toward . . . no, this isn't quite true . . . but mostly they have turned to things that are much more science and society oriented, social environment of science, institutions, and so forth. Which is a natural development given where the field has gone, and given the need of any graduate student to get his independence from Papa. But I'd have been glad if Jed weren't the only one who was carrying this on.

A. BALTAS: I have two questions—the one is not really a question, but you said you have not finished with Princeton yet, you may want to add something about your colleagues, the atmosphere of the students . . .

T. KUHN: Look. There is only one thing I would really like to add. I liked Princeton a lot. I had good colleagues, I had good students. I didn't get very far trying to talk to the philosophers, and one of the advantages of being at MIT is that the philosophers are not quite so sure they're as good as the ones in Princeton. So, it's easier to get through to them—it's not very easy anyway, because they *are* very good. Things worked for me there very well, and the reason I left and went to MIT was because I was divorced. It wasn't anything about MIT versus Princeton as such. Look, there is one thing that should be mentioned. After I had been there . . . I'm not sure what the date would have been, fairly late, but not the end of my time in Princeton, Princeton announced that it would be willing to let people negotiate reduced workload for reduced salary. My mother had died in the interim, I could afford to do that, and I wanted more time to do my own work. And I did do that; and then I got invited to have a long-term membership, which was not on the faculty, at the Institute for Advanced Studies. So I had an office over there. That led me to interact with people some of whom perhaps I would not have known at all, and some of those interactions had anything from a little to a lot of importance. The one that had a lot of importance, I think, was with Clifford Geertz, the anthropologist. A couple of other people whom I knew and liked and got encouragement from, but I don't think there was much in the way of ideas feedback—one is Quentin Skinner, who is a philosopher, political scientist at Cambridge, England, and the other is a young historian now in the political science department at Chicago called William Sewell. Both of whom do things in ways that are deeply empathetic.

K. GAVROGLU: Why was it MIT and not Harvard?

T. KUHN: Harvard didn't want me. Furthermore, if [Harvard had asked me and] I could have resisted, I would have been very well advised to. When it was known that I was looking, Harvard did not ask me, MIT did. But you know enough about the Harvard department to see why that might have been the case.

A. BALTAS: The other question is a kind of political statement vis-à-vis traditions in philosophy. Because one of the ways you've been perceived—and I think in a private discussion we've more or less agreed upon that—is that your work has not explained perhaps partly why it had this influence. It's a kind of work that crosses boundaries among philosophical traditions. You cannot be branded as a continental metaphysician, but on the other hand, you cannot be branded as somebody who doesn't know his logic and theories of explanation, and things like that. So, given that this split is somehow getting bridged, how do you see yourself in this?

T. KUHN: Oh, I thought you were going to ask me about traditions in philosophy. Look, there's got to be something right about that, I mean you start out by saying, here's a man who was never trained as a philosopher, who's been an amateur increasingly learning things about it by himself, from interactions and so forth—but not a philosopher. A physicist turned historian for philosophical purposes. The philosophy I knew and had been exposed to, and the people in my environment to talk to, were all of them out of the English logical empiricist tradition, in one way or another. This was a tradition which by and large had no use for the continental and particularly the German philosophical tradition. I think, in some sense or other, I can be described as in some part having reinvented that tradition for myself. And clearly it's not the same, and there are all sorts of ways in which it goes in other directions and past and so forth—there's a whole body of work there that I don't even know very well. But when people say, didn't Heidegger say that, or something of the sort, yes, he probably did, and I haven't read it, and if I had I'd like to think this was going to help bridge the gap. And I think that's part of what it's doing. It's a sort of view I've had, but this is not a statement about philosophical traditions in general; you wouldn't have philosophy without traditions.

K. GAVROGLU: Who was the public Kuhn?

T. KUHN: The answer to the question, Kostas, is complicated. I knew someone at Princeton, who congratulated me on avoiding being a guru. It would be too much to say quite that I didn't want it, although in all sorts of ways I didn't want it, it scared the shit out of me. I mean, I am an anxious, neurotic—I don't bite my nails but I don't know why I don't bite my nails . . . So, I tend quite hard to avoid invitations to be on TV; I've had a few, I haven't had very many, but that's partly because the word gets around that I turn them down. I feel somewhat the same way about interviews, although I have given a few interviews, but I try to set up conditions (a) that I'm not being interviewed by anybody who doesn't know my work, including my more recent work, and (b) that I get a look at the script before it's published and I retain some control. And those are not agreeable conditions in the great world in which interviewing gets done. So there haven't been many interviews, which is a break. That's not a comment on this one . . . Look, I'm saying some things that I'm glad to think will be around somewhere.

K. GAVROGLU: But I remember once you had mentioned to me, and then somehow we couldn't continue the discussion, this trial about creationism to which you were requested to go, in Arizona.

T. KUHN: Look, that one I declined for I think an excellent reason. [The people who approached me were resisting the creationists. I was sympathetic, but] I didn't think there was a chance in the world . . .

I mean I was being used by the creationists, for God's sake![40] At least to some extent. And I didn't think there was any way in the world in which somebody who didn't quite believe in Truth, and getting closer and closer to it, and who thought that the essence of the demarcation of science was puzzle solving, was going to be able to make the point. And I thought I would do more harm than good, and that's what I told them.

K. GAVROGLU: What about when you were at the National Science Foundation and the guidelines in research in the history and philosophy of science?

T. KUHN: I was certainly on various committees way back in the old days. On committees, particularly fellowship committees, to go over fellowship applications, but I was on various other committees also. I thought of that as a professional obligation and I did it. But I never tried to play a leadership role. And the few times I did, in some sense or other, get very angry about something that was going on, I was totally ineffective—in part because I was angry.

A. BALTAS: You have not discussed at all your recent work, what you are working on now, but perhaps you can give us an idea of what is the state of the field today.

T. KUHN: Which field?

A. BALTAS: Both. I mean, both history and philosophy of science.

T. KUHN: I'm not close enough to history of science. I mean I really, as I've gone on now in the last ten, fifteen years, really trying now to develop this philosophical position, I have just stopped reading history of science. I have read practically nothing in history of science. Look, the fact of the matter is that the time I stopped reading history of science, not totally but largely, was when I was writing *Structure*. I had to stop reading for that. By the time I got back out of that, the literature had expanded so large. Now, that doesn't mean that I stopped reading it; I went on reading stuff in the fields that I was working on right on through the time at Princeton, and I kept up with the literature some. Nobody could keep up with it in its entirety anymore, and I didn't even try. But now I'm not doing it at all. And I see references to one thing or another, and I think, that sounds very interesting and I didn't know it was there. So, I don't want to comment on the state of the field, except to the sense that I already have com-

40. They quoted Kuhn in support of their anti-science position.

mented by saying, I wish there were more attention to the internalities of science. Beyond that I don't want to go.

A. BALTAS: And philosophy of science?

T. KUHN: [conspiratorial whisper:] I think everybody's waiting for my book!

V. KINDI: We sure need it. Do you have any other interests . . . like listening to music, an interest in painting . . .

K. GAVROGLU: Obsessions. Obsessions except philosophy of science.

T. KUHN: You don't really want to know! I read detective stories.

V. KINDI: Oh, that sounds like Wittgenstein.

T. KUHN: And I like music all right—it took me a long time to discover that I did, in part because I had a musical father and a very musical younger brother, and that was not good for my relation to music. People used to have symphonies on the record player, or I was taken to symphonies; I didn't like that, I mean they bored the hell out of me. When I discovered chamber music, my feelings changed. I don't listen a lot, it's hard for me to sit still, but I do like it. And that we still do, we don't go to concerts much, which is partly for one reason and partly for another. I like the theater, though we don't see much theater. I like, or used to like, to read. But most of what I read now is detective stories. I remember, my kids used to sort of ridicule me, or whatever—not ridicule me, but wish—how could I go on with this sort of thing. And I remember when my daughter who has gone into the academic, there she was reading detective stories, and she said to me, "It's the only thing I can read that doesn't feel like work!" That's it! And Jehane was scornful of detective stories when she married me, and she now reads almost as many of them as I do! I'm a corruptor of the mind!

Publications of Thomas S. Kuhn

An earlier version of this bibliography of the publications of Thomas S. Kuhn was prepared by Paul Hoyningen-Huene and published in his book, Reconstructing Scientific Revolutions: Thomas S. Kuhn's Philosophy of Science *(Chicago: University of Chicago Press, 1993). Stefano Gattei updated and expanded that bibliography for* Thomas S. Kuhn: Dogma contro critica *(Milano: Rafaello Cortina Editore, 2000), which he edited. The editors and the Press thank them both for permitting us to include the bibliography in this volume.*

———◦———

Books and Articles

1945 [Abstract] [on General Education in a Free Society]. *Harvard Alumni Bulletin* 48, no. 1, 22 September 1945, pp. 23–24.

1945 Subjective View [on General Education in a Free Society], *Harvard Alumni Bulletin* 48, no. 1, 22 September 1945, pp. 29–30.

1949 The Cohesive Energy of Monovalent Metals as a Function of Their Atomic Quantum Defects. Ph.D. dissertation, Harvard University, Cambridge, MA.

1950 (with John H. Van Vleck) A Simplified Method of Computing the Cohesive Energies of Monovalent Metals. *Physical Review* 79: 382–88.

1950 An Application of the W. K. B. Method to the Cohesive Energy of Monovalent Metals. *Physical Review* 79: 515–19.

1951 A Convenient General Solution of the Confluent Hypergeometric Equation, Analytic and Numerical Development. *Quarterly of Applied Mathematics* 9: 1–16.

1951 Newton's '31st Query' and the Degradation of Gold. *Isis* 42: 296–98.

1952 Robert Boyle and Structural Chemistry in the Seventeenth Century. *Isis* 43: 12–36.

1952 Reply to Marie Boas: Newton and the Theory of Chemical Solution. *Isis* 43: 123–24.

1952 The Independence of Density and Pore-Size in Newton's Theory of Matter. *Isis* 43: 364–65.

1953 Review of *Ballistics in the Seventeenth Century: A Study in the Relations of Science and War with Reference Principally to England*, by A. Rupert Hall. *Isis* 44: 284–85.

1953 Review of *The Scientific Work of René Descartes (1596–1650)*, by Joseph F. Scott, and of *Descartes and the Modern Mind*, by Albert G. A. Balz. *Isis* 44: 285–87.

1953 Review of *The Scientific Adventure: Essays in the History and Philosophy of Science*, by Herbert Dingle. *Speculum* 28: 879–80.

1954 Review of *Main Currents of Western Thought: Readings in Western European Intellectual History from the Middle Ages to the Present*, edited by Franklin L. Baumer. *Isis* 45: 100.

1954 Review of *Galileo Galilei: Dialogue on the Great World Systems*, revised and annotated by Giorgio de Santillana, and of *Galileo Galilei, Dialogue Concerning the Two Chief World Systems—Ptolemaic and Copernican*, translated by Stillman Drake. *Science* 119: 546–47.

1955 Carnot's Version of "Carnot's Cycle." *American Journal of Physics* 23: 91–95.

1955 La Mer's Version of "Carnot's Cycle." *American Journal of Physics* 23: 387–89.

1955 Review of *New Studies in the Philosophy of Descartes: Descartes as Pioneer and Descartes' Philosophical Writings*, edited by Norman K. Smith, and of *The Method of Descartes: A Study of the Regulae*, by Leslie J. Beck. *Isis* 46: 377–80.

1956 History of Science Society. Minutes of Council Meeting of 15 September 1955. *Isis* 47: 455–57.

1956 History of Science Society. Minutes of Council Meeting of 28 December 1955. *Isis* 47: 457–59.

1956 Report of the Secretary, 1955. *Isis* 47: 459.

1957 *The Copernican Revolution: Planetary Astronomy in the Development of Western Thought*. Foreword by James B. Conant. Cambridge, MA: Harvard University Press, 1957. (Successive editions, 1959, 1966, and 1985.)

1957 Review of *A Documentary History of the Problem of Fall from Kepler to Newton, De Motu Gravium Naturaliter Cadentium in Hypothesi Terrae Motae*, by Alexandre Koyré. *Isis* 48: 91–93.

1958 The Caloric Theory of Adiabatic Compression. *Isis* 49: 132–40.

1958 Newton's Optical Papers. In *Isaac Newton's Papers and Letters On Natural Philosophy, and Related Documents,* edited with a general introduction by I. Bernard Cohen. Cambridge, MA: Harvard University Press, pp. 27–45.

1958 Review of *From the Closed World to the Infinite Universe,* by Alexandre Koyré. *Science* 127: 641.

1958 Review of *Copernicus: The Founder of Modern Astronomy,* by Angus Armitage. *Science* 127: 972.

1959 The Essential Tension: Tradition and Innovation in Scientific Research. In *The Third (1959) University of Utah Research Conference on the Identification of Creative Scientific Talent,* edited by Calvin W. Taylor. Salt Lake City: University of Utah Press, 1959, pp. 162–74. Reprinted in *The Essential Tension,* pp. 225–39.

1959 (with Norman Kaplan) Committee Report on Environmental Conditions Affecting Creativity. *The Third (1959) University of Utah Research Conference on the Identification of Creative Scientific Talent,* edited by Calvin W. Taylor. Salt Lake City: University of Utah Press, pp. 313–16.

1959 Energy Conservation as an Example of Simultaneous Discovery. In *Critical Problems in the History of Science,* edited by Marshall Clagett. Madison: University of Wisconsin Press, pp. 321–56. Reprinted in *The Essential Tension,* pp. 66–104.

1959 Review of *A History of Magic and Experimental Science,* vols. 7 and 8, *The Seventeenth Century,* by Lynn Thorndike. *Manuscripta* 3: 53–57.

1959 Review of *The Tao of Science: An Essay on Western Knowledge and Eastern Wisdom,* by Ralph G. H. Siu. *Journal of Asian Studies* 18: 284–85.

1959 Review of *Sir Christopher Wren,* by John N. Summerson. *Scripta Mathematica* 24: 158–59.

1960 Engineering Precedent for the Work of Sadi Carnot. *Archives internationales d'Histoire des Sciences,* XIII année, nos. 52–53, December 1960, pp. 251–55. Also in *Actes du IXe Congrès International d'Historie des Sciences,* Asociación para la Historia de la Ciencia Española (Barcelona: Hermann & Cie, 1960), I, pp. 530–35.

1961 The Function of Measurement in Modern Physical Science. *Isis* 52: 161–93. Reprinted in *The Essential Tension,* pp. 178–224.

1961 Sadi Carnot and the Cagnard Engine. *Isis* 52: 567–74.

1962 *The Structure of Scientific Revolutions.* International encyclopedia of unified science: Foundations of the unity of science, vol. 2, no. 2. Chicago: University of Chicago Press, 1962.

1962 Comment [on *Intellect and Motive in Scientific Inventors: Implications for Supply,* by Donald W. MacKinnon]. In *The Rate and Direction of Inventive*

Activity: Economic and Social Factors. National Bureau of Economic Research, Special Conference Series 13. Princeton: Princeton University Press, pp. 379–84.

1962 Comment [on *Scientific Discovery and the Rate of Invention,* by Irving H. Siegel]. In *The Rate and Direction of Inventive Activity: Economic and Social Factors.* National Bureau of Economic Research, Special Conference Series 13. Princeton: Princeton University Press, pp. 450–57.

1962 Historical Structure of Scientific Discovery. *Science* 136: 760–64. Reprinted in *The Essential Tension,* pp. 165–77.

1962 Review of *Forces and Fields: The Concept of Action at a Distance in the History of Physics,* by Mary B. Hesse. *American Scientist* 50: 442A–443A.

1963 The Function of Dogma in Scientific Research. In *Scientific Change: Historical Studies in the Intellectual, Social and Technical Conditions for Scientific Discovery and Technical Invention, from Antiquity to the Present,* edited by Alistair C. Crombie. London: Heinemann Educational Books, pp. 347–69.

1963 Discussion [on The Function of Dogma in Scientific Research]. In *Scientific Change: Historical Studies in the Intellectual, Social and Technical Conditions for Scientific Discovery and Technical Invention, from Antiquity to the Present,* edited by Alistair C. Crombie. London: Heinemann Educational Books, pp. 386–95.

1964 A Function for Thought Experiments. In *Mélanges Alexandre Koyré,* vol. 2, *L'aventure de la science.* Paris: Hermann, pp. 307–34. Reprinted in *The Essential Tension,* pp. 240–65.

1966 Review of *Towards an Historiography of Science, History and Theory,* Beiheft 2, by Joseph Agassi. *British Journal for the Philosophy of Science* 17: 256–58.

1967 (with John L. Heilbron, Paul Forman, and Lini Allen) *Sources for History of Quantum Physics: An Inventory and Report.* Memoirs of the American Philosophical Society, 68. Philadelphia: The American Philosophical Society.

1967 The Turn to Recent Science: Review of *The Questioners: Physicists and the Quantum Theory,* by Barbara L. Cline; *Thirty Years that Shook Physics: The Story of Quantum Theory,* by George Gamow; *The Conceptual Development of Quantum Mechanics,* by Max Jammer; *Korrespondenz, Individualität, und Komplementarität: eine Studie zur Geistesgeschichte der Quantentheorie in den Beiträgen Niels Bohrs,* by Klaus M. Meyer-Abich; *Niels Bohr: The Man, His Science, and the World They Changed,* by Ruth E. Moore; and *Sources of Quantum Mechanics,* edited by Bartel L. van der Waerden. *Isis* 58: 409–19.

1967 Review of *The Discovery of Time,* by Stephen E. Toulmin and June Goodfield. *American Historical Review* 72: 925–26.

1967 Review of *Michael Faraday: A Biography*, by Leslie Pearce Williams. *British Journal for the Philosophy of Science* 18: 148–54.

1967 Reply to Leslie Pearce Williams. *British Journal for the Philosophy of Science* 18: 233.

1967 Review of *Niels Bohr: His Life and Work As Seen By His Friends and Colleagues*, edited by Stefan Rozental. *American Scientist* 55: 339A–340A.

1968 The History of Science. In *International Encyclopedia of the Social Sciences*, vol. 14, edited by David L. Sills. New York: The Macmillan Company & The Free Press, pp. 74–83. Reprinted in *The Essential Tension*, pp. 105–26.

1968 Review of *The Old Quantum Theory*, edited by D. ter Haar. *British Journal for the History of Science* 98: 80–81.

1969 (with J. L. Heilbron) The Genesis of the Bohr Atom. *Historical Studies in the Physical Sciences* 1: 211–90.

1969 Contributions [to the discussion of New Trends in History]. *Daedalus* 98: 896–97, 928, 943, 944, 969, 971–72, 973, 975, 976.

1969 Comment [on the Relations of Science and Art]. *Comparative Studies in Society and History* 11: 403–12. Reprinted as Comment on the Relations of Science and Art in *The Essential Tension*, pp. 340–51.

1969d Comment [on *The Principle of Acceleration: A Non-dialectical Theory of Progress*, by Folke Dovring]. *Comparative Studies in Society and History* 11: 426–30.

1970 Logic of Discovery or Psychology of Research? In *Criticism and the Growth of Knowledge: Proceedings of the International Colloquium in the Philosophy of Science, London 1965*, vol. 4, edited by Imre Lakatos and Alan E. Musgrave. Cambridge: Cambridge University Press, pp. 1–23. Reprinted in *The Essential Tension*, pp. 266–92.

1970 Reflections on My Critics. In *Criticism and the Growth of Knowledge: Proceedings of the International Colloquium in the Philosophy of Science, London 1965*, vol. 4, edited by Imre Lakatos and Alan E. Musgrave. Cambridge: Cambridge University Press, pp. 231–78. Reprinted in this volume as essay 6.

1970 *The Structure of Scientific Revolutions*. 2d revised edition. International Encyclopedia of Unified Science: Foundations of the Unity of Science, vol. 2, no. 2. Chicago and London: The University of Chicago Press.

1970 Comment [on *Uneasily Fitful Reflections on Fits of Easy Transmission*, by Richard S. Westfall]. In *The Annus Mirabilis of Sir Isaac Newton 1666–1966*, edited by Robert Palter. Cambridge, MA: MIT Press, pp. 105–8.

1970 Alexandre Koyré & the History of Science: On an Intellectual Revolution. *Encounter* 34: 67–69.

1971 Notes on Lakatos. In *PSA 1970: In Memory of Rudolf Carnap, Proceedings of the 1970 Biennial Meeting, Philosophy of Science Association*, edited by Roger C. Buck and Robert S. Cohen. Boston Studies in the Philosophy of Science, 8. Dordrecht and Boston: D. Reidel, pp. 137–46.

1971 Les notions de causalité dans le développement de la physique. Translated by Gilbert Voyat. In Mario Bunge, Francis Halbwachs, Thomas S. Kuhn, Jean Piaget and Leon Rosenfeld, *Les théories de la causalité*. Bibliothèque Scientifique Internationale, Etudes d'épistémologie génétique, 25. Paris: Presses Universitaires de France, 1971, pp. 7–18. Reprinted in *The Essential Tension*, pp. 21–30.

1971c The Relations between History and History of Science. *Daedalus* 100: 271–304. Reprinted as The Relations between History and the History of Science in *The Essential Tension*, pp. 127–61.

1972 Scientific Growth: Reflections on Ben David's "Scientific Role." *Minerva* 10: 166–78.

1972 Review of *Paul Ehrenfest 1: The Making of a Theoretical Physicist*, by Martin J. Klein. *American Scientist* 60: 98.

1973 Historical Structure of Scientific Discovery. In *Historical Conceptions of Psychology*, edited by Mary Henle, Julian Jaynes and John J. Sullivan. New York: Springer, pp. 3–12.

1973 (editor, with Theodore M. Brown) Index to the Bobbs-Merrill History of Science Reprint Series. Indianapolis, IN: Bobbs-Merrill.

1974 Discussion [on The Structure of Theories and the Analysis of Data, by Patrick Suppes]. In *The Structure of Scientific Theories*, edited by Frederick Suppe. Urbana: University of Illinois Press, pp. 295–97.

1974 Discussion [on History and the Philosopher of Science, by I. Bernard Cohen]. In *The Structure of Scientific Theories*, edited by Frederick Suppe. Urbana: University of Illinois Press, pp. 369–70, 373.

1974 Discussion [on Science as Perception-Communication, by David Bohm, and Professor Bohm's View of the Structure and Development of Theories, by Robert L. Causey]. In *The Structure of Scientific Theories*, edited by Frederick Suppe. Urbana: University of Illinois Press, pp. 409–12.

1974 Discussion [on Hilary Putnam's Scientific Explanation: An Editorial Summary-Abstract, by Frederick Suppe, and Putnam on the Corroboration of Theories, by Bas C. van Fraassen]. In *The Structure of Scientific Theories*, edited by Frederick Suppe. Urbana: University of Illinois Press, pp. 454–55.

1974 Second Thoughts on Paradigms. In *The Structure of Scientific Theories*, edited by Frederick Suppe. Urbana: University of Illinois Press, pp. 459–82. Reprinted in *The Essential Tension*, pp. 293–319.

1974 Discussion [on Second Thoughts on Paradigms]. In *The Structure of Scien-*

tific Theories, edited by Frederick Suppe. Urbana: University of Illinois Press, pp. 500–506, 507–9, 510–11, 512–13, 515–16, 516–17.

1975 Tradition Mathématique et tradition expérimentale dans le développement de la physique. *Annales*, XXX année, no. 5, septembre-octobre 1975, pp. 975–98.

1975 The Quantum Theory of Specific Heats: A Problem in Professional Recognition. In *Proceedings of the XIV International Congress for the History of Science 1974*, vol. 1. Tokyo: Science Council of Japan, pp. 17–82.

1975 Addendum to "The Quantum Theory of Specific Heats." In *Proceedings of the XIV International Congress for the History of Science 1974*, vol. 4. Tokyo: Science Council of Japan, p. 207.

1976 Mathematical vs. Experimental Traditions in the Development of Physical Science. *Journal of Interdisciplinary History* 7: 1–31. Reprinted in *The Essential Tension*, pp. 31–65.

1976 Theory-Change as Structure-Change: Comments on the Sneed Formalism. *Erkenntnis* 10: 179–99. Reprinted in this volume as essay 7.

1976 Review of *The Compton Effect: Turning Point in Physics*, by Roger H. Stuewer. *American Journal of Physics* 44: 1231–32.

1977 *Die Entstehung des Neuen: Studien zur Struktur der Wissenschaftsgeschichte*. Edited by Lorenz Krüger, translated by Hermann Vetter. Frankfurt am Main: Suhrkamp.

1977 *The Essential Tension: Selected Studies in Scientific Tradition and Change*. Chicago: University of Chicago Press.

1977 The Relations between the History and the Philosophy of Science. In *The Essential Tension: Selected Studies in Scientific Tradition and Change*. Chicago: University of Chicago Press, pp. 3–20.

1977 Objectivity, Value Judgment, and Theory Choice. In *The Essential Tension*, pp. 320–39.

1978 *Black-Body Theory and the Quantum Discontinuity 1894–1912*. Oxford: Oxford University Press.

1978 Newton's Optical Papers. In *Isaac Newton's Papers and Letters On Natural Philosophy, and Related Documents*, 2d ed., edited with a general introduction by I. Bernard Cohen. Cambridge, MA: Harvard University Press.

1979 History of Science. In *Current Research in Philosophy of Science*, edited by Peter D. Asquith and Henry E. Kyburg. East Lansing, MI: Philosophy of Science Association, pp. 121–28.

1979 Metaphor in Science. In *Metaphor and Thought*, edited by Andrew Ortony. Cambridge: Cambridge University Press, pp. 409–19. Reprinted in this volume as essay 8.

1979 Foreword to Ludwik Fleck, *Genesis and Development of a Scientific Fact*, edited by Thaddeus J. Trenn and Robert K. Merton, translated by Fred Bradley and Thaddeus J. Trenn. Chicago: University of Chicago Press, pp. vii–xi.

1980 The Halt and the Blind: Philosophy and History of Science. *British Journal for the Philosophy of Science* 31: 181–92.

1980 Einstein's Critique of Planck. In *Some Strangeness in the Proportion: A Centennial Symposium to Celebrate the Achievements of Albert Einstein*, edited by Harry Woolf. Reading, MA: Addison-Wesley, pp. 186–91.

1980 Open Discussion Following Papers by J. Klein and T. S. Kuhn. In *Some Strangeness in the Proportion: A Centennial Symposium to Celebrate the Achievements of Albert Einstein*, edited by Harry Woolf. Reading, MA: Addison-Wesley, p. 194.

1981 What Are Scientific Revolutions? Occasional Paper #18, Center for Cognitive Science, MIT. Reprinted in *The Probabilistic Revolution*, vol. 1, *Ideas in History*, edited by Lorenz Krüger, Lorraine J. Daston and Michael Heidelberger. Cambridge, MA: MIT Press, pp. 7–22; reprinted in this volume as essay 1.

1983 Commensurability, Comparability, Communicability. In *PSA 1982: Proceedings of the 1982 Biennial Meeting of the Philosophy of Science Association*, vol. 2, edited by Peter D. Asquith and Thomas Nickles. East Lansing, MI: Philosophy of Science Association, pp. 669–88. Reprinted in this volume as essay 2.

1983 Response to Commentaries [on Commensurability, Comparability, Communicability]. In *PSA 1982. Proceedings of the 1982 Biennial Meeting of the Philosophy of Science Association*, vol. 2, edited by Peter D. Asquith and Thomas Nickles. East Lansing, MI: Philosophy of Science Association, pp. 712–16.

1983 Reflections on Receiving the John Desmond Bernal Award. *4S Review: Journal of the Society for Social Studies of Science* 1: 26–30.

1983 Rationality and Theory Choice. *Journal of Philosophy* 80: 563–70. Reprinted in this volume as essay 9.

1983 Foreword to Bruce R. Wheaton, *The Tiger and the Shark: Empirical Roots of Wave-Particle Dualism*. Cambridge: Cambridge University Press, pp. ix–xiii.

1984 Revisiting Planck. *Historical Studies in the Physical Sciences* 14: 231–52.

1984 *Black-Body Theory and the Quantum Discontinuity 1894–1912*. Reprinted with an Afterword, "Revisiting Planck" pp. 349–70. Chicago: University of Chicago Press, 1987.

1984 Professionalization Recollected in Tranquillity. *Isis* 75: 29–32.

1985 Specialization and Professionalism within the University [panel discussion with Margaret L. King and Karl J. Weintraub]. *American Council of Learned Societies Newsletter* 36 (nos. 3–4): 23–27.

1986 The Histories of Science: Diverse Worlds for Diverse Audiences. *Academe* 72(4): 29–33.

1986 Rekishi Shosan toshite no Kagaku Chishiki [Scientific Knowledge as Historical Product], translated by Chikara Sasaki and Toshio Hakata. *Shisô* 8(746): 4–18.

1989 Possible Worlds in History of Science. In *Possible Worlds in Humanities, Arts and Sciences: Proceedings of Nobel Symposium 65*, edited by Sture Allén. Research in Text Theory, 14. Berlin: Walter de Gruyter, pp. 9–32. Reprinted in this volume as essay 3.

1989 Speaker's Reply [on Possible Worlds in History of Science]. In *Possible Worlds in Humanities, Arts and Sciences: Proceedings of Nobel Symposium 65*, edited by Sture Allén. Research in Text Theory, 14. Berlin: Walter de Gruyter, pp. 49–51.

1989 Preface to Paul Hoyningen-Huene, *Die Wissenschaftsphilosophie Thomas S. Kuhns: Rekonstruktion und Grundlagenprobleme*. Braunschweig, Wiesbaden: Friedrich Vieweg & Sohn, pp. 1–3.

1990 Dubbing and Redubbing: The Vulnerability of Rigid Designation. In *Scientific Theories*, edited by C. Wade Savage. Minnesota Studies in the Philosophy of Science, 14. Minneapolis: University of Minnesota Press, pp. 298–318.

1991 The Road since Structure. In *PSA 1990: Proceedings of the 1990 Biennial Meeting of the Philosophy of Science Association*, vol. 2, edited by Arthur Fine, Micky Forbes, and Linda Wessels. East Lansing, MI: Philosophy of Science Association, pp. 3–13. Reprinted in this volume as essay 4.

1991 The Natural and the Human Sciences. In *The Interpretive Turn: Philosophy, Science, Culture*, edited by David R. Hiley, James F. Bohman, and Richard Shusterman. Ithaca, NY: Cornell University Press, pp. 17–24. Reprinted in this volume as essay 10.

1992 The Trouble with the Historical Philosophy of Science. Robert and Maurine Rothschild Distinguished Lecture, 19 November 1991, Occasional Publications of the Department of the History of Science. Cambridge, MA: Harvard University, 1992. Reprinted in this volume as essay 5.

1993 Afterwords. In *World Changes: Thomas Kuhn and the Nature of Science*, edited by Paul Horwich. Cambridge, MA: MIT Press, pp. 311–41. Reprinted in this volume as essay 11.

1993 Introduction to Bas C. van Fraassen, From Vicious Circle to Infinite Regress, and Back Again, in *PSA 1992: Proceedings of the 1992 Biennial Meet-*

ing of the Philosophy of Science Association, vol. 2., edited by David Hull, Micky Forbes, and Kathleen Okruhlik. East Lansing, MI: Philosophy of Science Association, pp. 3–5.

1993 Foreword to Paul Hoyningen-Huene, *Reconstructing Scientific Revolutions: Thomas S. Kuhn's Philosophy of Science*, translated by Alexander T. Levine. Chicago: University of Chicago Press, pp. xi–xiii.

1995 Remarks on Receiving the Laurea of the University of Padua. In *L'anno Galileiano*, 7 dicembre 1991—7 dicembre 1992, Atti delle celebrazioni galileiane (1592–1992). Trieste: Edizioni Lint, I, pp. 103–6.

1996 *The Structure of Scientific Revolutions*. 3d ed. Chicago: University of Chicago Press.

1997 Antiphónissi [Reply to Kostas Gavroglu, Honoring Thomas S. Kuhn], translated by Varvara Spiropúlu. *Neusis*, no. 6, Spring-Summer 1997, pp. 13–17.

1997 Paratiríssis ke schólia [Concluding Remarks, at the end of a symposium in honor of Thomas S. Kuhn], translated by Varvara Spiropúlu. *Neusis*, no. 6, Spring-Summer 1997, pp. 63–71.

1999 Remarks on Incommensurability and Translation. In *Incommensurability and Translation: Kuhnian Perspectives on Scientific Communication and Theory Change*, edited by Rema Rossini Favretti, Giorgio Sandri, and Roberto Scazzieri. Cheltenham, U.K. and Northampton, MA: Edward Elgar, pp. 33–37.

Interviews

Paradigmi dell'evoluzione scientifica. In Giovanna Borradori, *Conversazioni americane*, con W. O. Quine, D. Davidson, H. Putnam, R. Nozick, A. C. Danto, R. Rorty, S. Cavell, A. MacIntyre, Th. S. Kuhn. Roma-Bari: Laterza, 1991, pp. 189–206.

Profile: Reluctant Revolutionary. Thomas S. Kuhn unleashed 'paradigm' on the world. Edited by John Horgan. *Scientific American* 264 (May 1991): 14–15.

Paradigms of Scientific Evolution. In Giovanna Borradori, *The American Philosopher: Conversations with Quine, Davidson, Putnam, Nozick, Danto, Rorty, Cavell, MacIntyre, and Kuhn*, translated by Rosanna Crocitto. Chicago: University of Chicago Press, 1994, pp. 153–67.

Un entretien avec Thomas S. Kuhn. Edited and translated by Christian Delacampagne. *Le Monde*, LI année, no. 15561, dimanche 5—lundi 6 février 1995, p. 13.

Thomas Kuhn: Le rivoluzioni prese sul serio. Edited and translated by Armando Massarenti. *Il Sole*-24 Ore, anno CXXXI, no. 324, domenica 3 dicembre 1995, p. 27.

A Physicist Who Became a Historian for Philosophical Purposes: A Discussion between Thomas S. Kuhn and Aristides Baltas, Kostas Gavroglu, and Vassiliki Kindi, *Neusis*, no. 6, Spring-Summer 1997, pp. 145–200. Reprinted in this volume.

Note sull'incommensurabilità. Edited by Mario Quaranta, translated by Stefano Gattei, *Pluriverso*, anno II, n. 4, dicembre 1997, pp. 108–14.

Videorecording

The crisis of the old quantum theory, 1922–25. Science Center, Harvard University, Cambridge, MA, 5 November 1980. 120 minutes.